Uncertainty, Rationality, and Agency

UNCERTAINTY, RATIONALITY, AND AGENCY

Wiebe van der Hoek
Reprinted from *Synthese* 144:2 and 147:2 (2005)
Special Section *Knowledge, Rationality & Action*

 Springer

A C.I.P. Catalogue record for this book is available from the Library of Congress.

ISBN-1-4020-4630-8

Published by Springer,
P.O. Box 17, 3300 AA Dordrecht, The Netherlands.

Cover design by Vincent F. Hendricks

Printed on acid-free paper

Printed in the Netherlands.

Contents

WIEBE VAN DER HOEK

Foreword

This book collects all the papers that appeared in 2005 in *Knowledge, Rationality and Action* (KRA), a journal published as a special section of *Synthese*, which addresses contemporary issues in epistemic logic, belief revision, game and decision theory, rational agency, planning and theories of action. As such, the special section appeals to researchers from Computer Science, Game Theory, Artificial Intelligence, Philosophy, Knowledge Representation, Logic and Agents, addressing issues in artificial systems that have to gather information, reason about it and then make a sensible decision about what to do next.

It will be clear already from the contents pages, that this book indeed reflects the core of KRA: the papers in this volume address degrees of belief or certainty, and rational agency. The latter has several manifestations: often constraints on the agent's belief, behaviour or decision making. Moreover, this book shows that KRA indeed represents a 'loop' in the behaviour of the agent: after having made a decision, the life of the agent does not end, rather, it will do some sensing or collect otherwise the outcome of its decision, to update its beliefs or knowledge accordingly and make up its mind about the next decision task.

In fact, the chapters in this book represent two volumes of KRA: the first appeared as a regular volume, the second contained a selection of papers that were accepted for the Conference on Logic and the Foundations of the Theory of Games and Decisions (LOFT 2004). I will now give a brief overview of the themes in this book and of the chapters in the regular volume, the papers of the LOFT-volume are briefly introduced in Chapter five of this book.

The first two chapters, THE NO PROBABILITIES FOR ACTS-PRINCIPLE and A LOGIC FOR INDUCTIVE PROBABILISTIC REASONING, deal with probabilistic reasoning: one in the context of deliberating about future actions, or planning, and the other in that of making inductive inferences. Chapter three, RATIONALITY AS CONFORMITY and Chapter eleven, A LOGICAL FRAMEWORK FOR CONVENTION, both describe rational agents that reason about the rationality of other agents. In Chapter three the challenge of the agent is to act in conformance with the other, in Chapter eleven the emphasis is on predicting the other agents' decision. Both chapters give an account of the reciprocal reasoning that such a decision problem triggers, using notions like common knowledge, common belief, common sense, common reasoning and common reasons for belief.

Reasons for belief are also the topic of Chapter four, ON THE STRUCTURE OF RATIONAL ACCEPTANCE: COMMENTS ON HAWTHORNE AND BOVENS. The chapter investigates ways to deal with the contradiction that arises from three simple postulates of rational acceptance for an agent's beliefs. Chapter six, A SIMPLE MODAL LOGIC FOR BELIEF REVISION, and Chapter seven, PROLEGOMENA TO DYNAMIC LOGIC FOR BELIEF REVISION, both give a modal logical account of the dynamics of a rational agent's belief. Chapter six introduces a belief operator for initial belief and one for the belief after a revision, and Chapter seven gives an account of update when we have many grades of belief. Degrees of belief are also the topic of Chapter eight, FROM KNOWLEDGE-BASED PROGRAMS TO GRADED BELIEF-BASED PROGRAMS, PART I: ON-LINE REASONING: here, the beliefs can be updated by the agent, but are also used to guide his decision during execution of a program.

Chapter nine, ORDER-INDEPENDENT TRANSFORMATIVE DECISION RULES and ten, A PRAGMATIC SOLUTION FOR THE PARADOX OF FREE CHOICE PERMISSION, take us back to formalisations of rational agents again. In Chapter nine the authors focus on the representation of a decision problem for such an agent: the agent prefers certain representations over others, and uses transformation rules to manipulate them. In chapter ten, the rational agent is a speaker in a conversation, and the author uses some ideas from the area of 'only knowing' to model certain Gricean maxims of conversation in order to formally analyse free choice permission.

Regarding the first four chapters, in the miniature THE NO PROBABILITIES FOR ACTS-PRINCIPLE, Marion Ledwig addresses this NPA principle as put forward by Spohn: *"Any adequate quantitative decision model must not explicitly or implicitly contain any subjective probabilities for acts"*. Ledwig discusses several consequences of the principle which are relevant for decision theory, in particular for Dutch book arguments and game theory: the NPA-principle is at odds with conditionalising on one's actions (as done in diachronic Dutch books) and the assumption that one will choose rationally and therefore predict one's choices (as done in game theory). Finally, she makes clear that the NPA-principle refers not to past actions or actions of other persons, but rather to actions that are performable now and extend into the future.

Manfred Jaeger proposes A LOGIC FOR INDUCTIVE PROBABILISTIC REASONING in the second chapter of this book. In such kind of reasoning, one applies inference patterns that use statistical background information in order to assign subjective probabilities to subjective events. The author sets himself three design principles when proposing a logical language that formalises inductive probabilistic reasoning: expressiveness, completeness and epistemic justifiability. Indeed, the language proposed enables the encoding of complex probabilistic information, and, by putting an elegant semantics based on logarithmic real-closed values to work, a completeness result for the expressive language is obtained. Finally, regarding justifiability, it is the author's aim to model with the inductive entailment relation a well-justified pattern of defeasible probabilistic reasoning, i.e., to use statistical information to refine an already partially formed subjective probability assignment. For this, it is argued, cross-entropy minimisation relative to possible statistical distributions is the adequate formal model.

In RATIONALITY AS CONFORMITY Hykel Hosni and Jeff Paris face the following problem: choose one of a number of options, in such a way that your choice coincides with that of a like-minded, but otherwise inaccessible (in particular non-communicating), agent. In other words, our agent has to 'predict' what an other agent would choose, if that other agent were confronted with the same problem, i.e., to make that choice that coincides with our agent. If a unique option in the space of choices would obviously stand out, that will be the object of choice, and if they are all the same, the best our agents could do is randomise. But what to do in intermediate cases, i.e., where the alternatives are not all alike, but only show *some* structure? In the authors' approach, the agent first singles out a number of outstanding options (called a reason), and then takes a random choice from those. They discuss and mathematically characterise three different reasons: the regulative reason (satisfying weak criteria to choose some naturally outstanding elements: an agent not following them would perform 'unreasonable steps'); the minimum ambiguity reason (a procedural approach based on the notion of indistinguishability of the options) and the smallest uniquely definable reason (take the smallest set of options that is definable in a suitable first-order language). These reasons are then compared and discussed with respect to Game Theory and Rationality.

Gregory Wheeler discusses principles for acceptance of beliefs by a rational agent, in his chapter ON THE STRUCTURE OF RATIONAL ACCEPTANCE: COMMENTS ON HAWTHORNE AND BOVENS. He starts off with observing that the following three principles for rational acceptance together lead to a contradiction: (*i*) it is rational to accept a proposition that is very likely to be true; (*ii*) it is not rational to accept a proposition that you are aware is inconsistent; (*iii*) if it is rational to accept A and also to accept A', that it is rational to accept their conjunction $A \wedge A'$. This is for instance illustrated by the *Lottery Paradox*, in which you rationally accept that each ticket i will not be the winning ticket, but still you don't accept that no ticket will be the winner's. Wheeler's approach is *structural* in the sense that it is deemed necessary to have some connectives in the object language in order to express compound rationally accepted formulas, and to define a notion of logical consequence for such formulas. He then argues that any proposal that solves paradoxes as the one mentioned above, should be structural, in order to bring the conflict between the principles (*i*) and (*iii*) to the fore.

MARION LEDWIG

THE NO PROBABILITIES FOR ACTS-PRINCIPLE[1]

ABSTRACT. One can interpret the No Probabilities for Acts-Principle, namely that any adequate quantitative decision model must in no way contain subjective probabilities for actions in two ways: it can either refer to actions that are performable now and extend into the future or it can refer to actions that are not performable now, but will be in the future. In this paper, I will show that the former is the better interpretation of the principle.

1. INTRODUCTION

Spohn (1977, 1978) claims that his causal decision theory is valuable in part for its explicit formulation of a principle used earlier by Savage (1954, 1972) and Fishburn (1964). This principle, henceforth called the "No Probabilities for Acts"-Principle (or the NPA-Principle) is the following: "*Any adequate quantitative decision model must not explicitly or implicitly contain any subjective probabilities for acts*" (Spohn 1977, 114).[2] Spohn (1978) maintains that the NPA-Principle isn't used in the rational decision theories of Jeffrey (1965) and of Luce and Krantz (1971), and that this lack is the root for the theories' wrong answers in Newcomb's problem, namely taking only one box (cf. Nozick 1969). According to Spohn (1977) this principle is important, because it has implications for the concept of action, Newcomb's problem, the theory of causality, and freedom of will. In a recent paper, Spohn (1999, 44–45) modifies this principle. He postulates that in the case of strategic thinking, that is, in the case of sequential decision making, the decision maker can ascribe subjective probabilities to his future, but not to his present actions without giving a justification for his claim.[3]

I agree with Spohn that the NPA-principle has implications for the concept of action. If the NPA-principle holds, the decision maker has full control over his actions, that is, he assigns a subjective probability of one to the actions he has decided for and a subjective probability of zero to those he has decided against.[4] Furthermore, it

Synthese (2005) 144: 171–180
Knowledge, Rationality & Action 1–10
DOI 10.1007/s11229-004-2010-6

has implications for a theory of causality if one maintains a probabilistic theory of causality as Spohn (1983) himself does.[5] Finally, it
has implications for freedom of will, since an implicit condition for
the application of the NPA-principle is that the decision maker is
free.

In my opinion, the NPA-principle has some additional consequences for Dutch books (cf. Levi 1987) and game theory (cf. Levi
1997, chap. 2). In the case of diachronic Dutch books, the decision
maker must conditionalize on his actions, which violates the NPA-
principle. With regard to game theory, the assumption of common
knowledge of rationality entails that each agent believes he or she will
choose rationally. This means that each agent will be predicting and
therefore also assigning probabilities to his or her own choice counter
to Levi's contention that deliberation crowds out prediction. So if the
NPA-principle holds, game theory has to be built on other assumptions.

I claim that the NPA-principle has some other important consequences:

(1) in opposition to causal decision theories[6] and Kyburg's
(1980) proposal to maximize properly epistemic utility, evidential
decision theories[7] violate the NPA-principle, because the decision
maker conditions his credences by his actions in calculating the
utility of an action. Jeffrey's (1983) ratificationism shows a similar
feature, for the decision maker conditions his credences by his
final decisions to perform his actions. Nozick's (1993) proposal
of combining various decision principles also disagrees with the
NPA-principle by using evidential decision principles. Meek and
Glymour (1994) claim that if the decision maker views his actions
as non-interventions in the system, he conditions his credences by
his actions, so the NPA-principle is violated here, too. Hence if the
NPA-principle is valid, the decision theories which violate it provide wrong solutions to some decision problems and therefore
should be abandoned.

(2) By means of the NPA-principle the decision maker cannot take
his actions as evidence of the states of the world. The decision maker's credence function cannot be modified by the evidence of the
actions, since the NPA-principle demands that the decision maker
shouldn't assign any credences to his actions. Thus Jeffrey's (1965)
logic of decision, which takes actions as evidence of states of the
world, cannot be right if the NPA-principle is valid. Other rational
decision theories also assert that the decision maker cannot take his

actions as evidence of the states of the world. In Jeffrey's ratificationism (1983), for example, the decision maker takes his decisions,
but not his actions as evidence of the states of the world. In Eells'
(1981, 1982, 1985) proposal of the common cause, the decision maker's beliefs and wants and not his actions are evidence of the states
of the world. In Kyburg's (1980, 1988) proposal of maximizing
properly epistemic utility the decision maker doesn't take his free
actions as evidence of the states of the world.

(3) Another consequence of the NPA-principle is to favor Savage's
(1954, 1972) trinitarianism, distinguishing between acts, states, and
consequences, over Jeffrey's (1965) monotheism, where acts, states,
and consequences are all events or propositions, and therefore should
be treated all alike.

Due to the great number of the NPA-principle's implications,
Spohn (1977, 1978) makes his principle more precise, suggests arguments for it (e.g., point (4)), and points out immediate consequences
of it (e.g., point (5)):

(1) The NPA-principle refers to future actions of the decision maker.
(2) Credences for actions do not manifest themselves in the willing-
    ness to bet on these actions.
(3) The NPA-principle requires that actions are things which are
    under the decision maker's full control relative to the decision
    model describing him.
(4) A theoretical reason for the NPA-principle is that credences for
    actions cannot manifest themselves in these actions.
(5) An immediate consequence of the NPA-principle is that uncon-
    ditional credences for events which probabilistically depend on
    actions are forbidden.

Respective objections to these claims are the following (with regard to point (5) no objection came to my mind):

(1) The term future actions is ambiguous; it can either refer to ac-
    tions that are performable now and extend into the future or it
    can refer to actions that are not performable now, but will be in
    the future.
(2) Why could not the decision maker's probability judgments con-
    cerning what the decision maker will do be correlated with the
    decision maker's willingness to bet? There might be some decision
    makers, however, who have an aversion to betting and therefore
    might not be willing to put their money where their mouth is. But

[3]

if one forces them to do so, they surely would bet in accordance with their probability judgments.

(3) We do not have to claim that $P(a_j|a_j) = 1$ is a necessary and sufficient condition for full control in order to claim that options are under the decision maker's full control, for a_j could be a state and not an action. Moreover, one might want to object that Spohn conflates issues about what a person can control with questions about probabilities for actions (Joyce 2002).

(4) Even if credences for actions play no useful role in decision making, Spohn has not shown that they play a harmful role in decision making and should therefore be ommitted (Rabinowicz 2002).

In the following I will explain and criticize in detail only point (1), namely that the NPA-principle refers to future actions of the decision maker. I will begin by presenting Spohn's (1978, 72–73) two examples to provide an intuitive motivation for the NPA-principle: If a friend asks me whether I will be coming to a party tonight and if I answer "yes", then this is not an assertion or a prediction, but an announcement, an acceptance of an invitation, or even a promise. Moreover, if a visitor asks me whether I really believe that I will make a certain move in chess, then I will reply that the question is not whether I believe this, but whether I really want this. That is, in general it can be questioned that, in utterances about one's own future actions, belief dispositions with regard to these actions are manifested. Hence, if I decide to perform a particular action, I also believe I will perform that action.

2. THE NPA-PRINCIPLE REFERS TO FUTURE ACTIONS OF THE DECISION MAKER

The NPA-principle does not refer to past actions and actions of other persons, but only to actions which are open to the decision maker in his decision model, that is, to future actions of the decision maker. Yet "future actions" is ambiguous. It can either refer to actions that are performable now and extend into the future or it can refer to actions that are not performable now, but will be in the future.[8] As I understand Spohn, the NPA-principle refers to actions that are performable now and extend into the future,[9] for Spohn (1977, 115) concedes that decision makers frequently have and utter beliefs about their future actions like the following:

[4]

(1) "I believe it is improbable that I will wear shorts during the next winter."

Moreover, Spohn (1977, 116) points out that "As soon as I have to make up my mind whether to wear my shorts outdoors or not, my utterance is out of place." That is, as soon as I have to deliberate about wearing my shorts outdoors now, I cannot say anymore "I believe it is improbable that I will wear shorts outdoors now." Thus according to Spohn decision makers should not assign subjective probabilities to actions that are performable now, but extend into the future.

Yet Spohn (1977, 115) wants this utterance to be understood in such a way that it does not express a credence for an action, but a credence for a decision situation:

(2) "I believe it is improbable that I will get into a decision situation during the next winter in which it would be best to wear shorts."

Thus Spohn assumes that the embedded sentences "I will wear shorts during the next winter" and "I will get into a decision situation during the next winter in which it would be best to wear shorts" are logically equivalent, which is not true. For while it might be the case that I will not wear shorts during the next winter, it might happen that I get into a decision situation during the next winter in which it would be best to wear shorts. Moreover, identifying an action with a decision situation seems to be problematical, as these are clearly two different things.

However, if we, despite the logical inequivalence, concede this opinion to Spohn for a while, we can observe that something else goes wrong. Observe:

(3) "I believe it is improbable that I will run 100 meters in 7 seconds during the next year."

According to Spohn this utterance should be reformulated, since it does not express a genuine probability for an action:

(4) "I believe it is improbable that I will get into a decision situation during the next year in which it would be best to run 100 meters in 7 seconds."

Yet while (3) might be true, (4) might be false. With regard to (3) I know because of my bodily constitution it would not matter how much I tried, I never would be able to run 100 meters in 7 seconds, so indeed I believe it is improbable that I will run 100 meters in 7 seconds during the next year. At the same time with regard to (4) it could happen that the Olympic Games were to take place next year and luckily I qualified for the Olympic team of my country, so that I was in a decision situation in which it would be best for me to run 100 meters in 7 seconds. Thus my belief that it is improbable that I will get into a decision situation during the next year in which it would be best for me to run 100 meters in 7 seconds would be false. Hence there might be a belief context in which (3) is true, but (4) is false.

One might want to object that this reformulation makes sense under the assumption that the decision maker knows that he is strong-willed and thus knows that he will only do what he thinks is best to do and therefore believes it to be improbable to get into a decision situation during the next year in which it would be best to run 100 meters in 7 seconds. True – yet not all decision makers have that constitution. Hence this objection does not generalize.

What is the relevance of these insights? Not much, as the NPA-principle only refers to actions that are performable now and extend into the future, his reformulation of actions that are not performable now, but will be performable in the future, is of no relevance for the NPA-principle. One can deny that [(1) and (2)] and [(3) and (4)] are synonyms and still accept the NPA-principle. Furthermore, one can even ask why it is so important for Spohn to find alternative interpretations of (1) and (3)? By putting forth alternative interpretations, Spohn seems to defend the view that, even in the case of actions that are not performable now but will be performable in the future, the decision maker should not assign any subjective probabilities to his actions. But we have just seen that this is not so, that is, Spohn allows the decision maker to assign subjective probabilities to his actions that are performable in the future. Yet, strictly put in Spohn's view, even utterances like (1) don't express a genuine probability for an action, only a probability for a decision situation. Thus Levi's (1997, 80) suggestion turns out to be right, namely that these interpretations are meant to express that the decision maker isn't even able to predict his actions that are not performable now, but will be in the future;[10] if utterances like (1) express a probability for a decision situation and not for an action, then the decision maker is not even able to predict his actions that are not performable now, but will be in the future.

3. CONCLUSION

I have clarified that the NPA-principle refers to actions that are performable now and extend into the future.

NOTES

[1] I would like to thank Andreas Blank, Phil Dowe, Alan Hajek, James Joyce, Isaac Levi, Nicholas Rescher, Teddy Seidenfeld, Wolfgang Spohn, Howard Sobel, and especially two anonymous referees from the *BJPS* and two anonymous referees from *KRA* for very helpful comments and discussion. Errors remain my own. I also would like to thank Elias Quinn for correcting and improving my English. A part of this paper was given as a talk in the Fourth In-House Conference in October 2001 during my visit at the Center for Philosophy of Science, University of Pittsburgh, 2001–2002 (cf. also Ledwig 2001).

[2] Trivial conditional subjective probabilities, like $P(a_1|a_1) = 1$ for an action a_1 or $P(a_2|a_1) = 0$ for two disjunctive actions a_1 and a_2, are not considered (Spohn 1977, 1978).

[3] Spohn is not the only one to defend his principle; the weaker thesis that the decision maker should not ascribe subjective probabilities of one or zero to his actions is widely accepted (cf. Ginet 1962; Shackle 1967; Goldman 1970; Jeffrey 1977, 1983; Schick 1979, 1999; Levi 1986). Even the stronger thesis that the decision maker should not ascribe any subjective probabilities to his actions is defended (Levi 1989, 1997; cf. Gilboa 1999). For a discussion of these issues, have a look at Ledwig (forthcoming). As Levi's and Gilboa's arguments for the stronger thesis (with the exception of Levi's betting argument) differ from Spohn's argument, my criticism of the NPA-principle does not hold for these.

[4] One might object that this implication does not hold, if in $P(a_1|a_1) = 1$ a_1 is a state and not an action, because this is simply a consequence of the calculus of probabilities which is true. So this implication only holds, given that one considers only actions as input and does not consider sequential decision problems in which an action may change its status over time from outcome to action to part of the state of the world (cf. Skyrms 1990, 44).

[5] Which causation theory is the adequate one, I want to leave for another paper.

[6] Gibbard and Harper (1978), Skyrms (1980, 1982, 1984), Sobel (1986), Lewis (1981), and Spohn (1978).

[7] Jeffrey (1965, 1988, 1996) and Eells (1981, 1982, 1985).

[8] Spohn does not distinguish between different kinds of future actions.

[9] The extension into the future can be minimal, but needs to be there. Otherwise one could not speak of future actions anymore.

[10] With this we have discovered a further possible implication of the NPA-principle, namely if the NPA-principle holds, the decision maker is not able to predict his own

actions that are performable now and extend into the future. Yet, that deliberation crowds out prediction has already been widely discussed in the literature (Ginet 1962; Jeffrey 1965, 1977, 1983; Shackle 1967; Pears 1968; Goldman 1970, chapter 6; Schick 1979, 1999; Ledwig forthcoming; Levi 1986, Section 4.3, 1989, 1997; cf. Gilboa 1999; Joyce 2002; Rabinowicz 2002). I deal with these authors and their views in Ledwig (forthcoming); moreover, in Ledwig (forthcoming) I defend the thesis that deliberation and prediction are compatible with each other.

REFERENCES

Eells, E.: 1981, 'Causality, Utility and Decision', *Synthese* **48**, 295–327.

Eells, E.: 1982, *Rational Decision and Causality*, Cambridge University Press, Cambridge.

Eells, E.: 1985, 'Causality, Decision and Newcomb's Paradox', in R. Campbell and L. Sowden (eds.), *Paradoxes of Rationality and Cooperation: Prisoner's Dilemma and Newcomb's Problem*, The University of British Columbia Press, Vancouver, pp. 183–213.

Fishburn, P. C.: 1964, *Decision and Value Theory*, Wiley, New York.

Gibbard, A. and W. L. Harper: 1978, 'Counterfactuals and Two Kinds of Expected Utility', in C. A. Hooker, J. J. Leach, and E. F. McClennen (eds.), *Foundations and Applications of Decision Theory*, Vol. 1, Reidel, Dordrecht, pp. 125–162.

Gilboa, I.: 1999, 'Can Free Choice be Known', in C. Bicchieri, R. Jeffrey and B. Skyrms (eds.), *The Logic of Strategy*, Oxford University Press, Oxford, pp. 163–174.

Ginet, C.: 1962, 'Can the Will Be Caused?', *The Philosophical Review* **71**, 49–55.

Goldman, A. I.: 1970, *A Theory of Human Action*, Prentice-Hall Inc., Englewood Cliffs, NJ.

Jeffrey, R. C.: 1965, *The Logic of Decision*, McGraw-Hill, New York.

Jeffrey, R. C.: 1977, 'A Note on the Kinematics of Preference', *Erkenntnis* **11**, 135–141.

Jeffrey, R. C.: 1983, *The Logic of Decision*, 2nd edn., The University of Chicago Press, Chicago London.

Jeffrey, R. C.: 1988, 'How to Probabilize a Newcomb Problem', in J. H. Fetzer (ed.), *Probability and Causality*, Reidel, Dordrecht, pp. 241–251.

Jeffrey, R. C.: 1996, 'Decision Kinematics', in K. J. Arrow, E. Colombatto, M. Perlman and C. Schmidt (eds.), *The Rational Foundations of Economic Behaviour*, Macmillan, Basingstoke, pp. 3–19.

Joyce, J.: 2002, 'Levi on Causal Decision Theory and the Possibility of Predicting One's Own Actions', manuscript.

Kyburg, H. E.: 1980, 'Acts and Conditional Probabilities', *Theory and Decision* **12**, 149–171.

Kyburg, H. E.: 1988, 'Powers', in W. L. Harper and B. Skyrms (eds.), *Causation in Decision, Belief Change, and Statistics*, Vol. 2, Kluwer, Dordrecht, pp. 71–82.

Ledwig, M.: 2001, 'Some Elaborations on Spohn's Principle', *Proceedings of the 4th In-House Conference*, 26th–28th of October, 2001, Center for Philosophy of Science, University of Pittsburgh, on-line at .

Ledwig, M.: (forthcoming), 'Deliberation, Prediction, and Knowledge', *Conceptus*.

Levi, I.: 1986, *Hard Choices: Decision Making under Unresolved Conflict*, Cambridge University Press, Cambridge/ New York/ Port Chester/ Melbourne/ Sydney.

Levi, I.: 1987, 'The Demons of Decision', *Monist* **70**, 193–211.

Levi, I.: 1989, 'Rationality, Prediction and Autonomous Choice', *Canadian Journal of Philosophy* Supplemental. Vol. 19, pp. 339–363, Reprinted in I. Levi (1997), pp. 19–39.

Levi, I.: 1997, *The Covenant of Reason: Rationality and the Commitments of Thought*, Cambridge University Press, Cambridge.

Lewis, D.: 1981, 'Causal Decision Theory', *Australasian Journal of Philosophy* **59**, 5–30.

Luce, R. D. and D. H. Krantz: 1971, 'Conditional Expected Utility', *Econometrica* **39**, 253–271.

Meek, C. and C. Glymour: 1994, 'Conditioning and Intervening' *British Journal for the Philosophy of Science* **45**, 1001–1021.

Nozick, R.: 1969, 'Newcomb's Problem and Two Principles of Choice', in N. Rescher, D. Davidson and C. G. Hempel (eds.), *Essays in Honor of Carl G. Hempel*, Reidel, Dordrecht, pp. 114–146.

Nozick, R.: 1993, *The Nature of Rationality*, Princeton University Press, Princeton.

Pears, D.: 1968, 'Predicting and Deciding', in P. F. Strawson (ed.), *Studies in the Philosophy of Thought and Action*, Oxford University Press, Oxford, London/New York, pp. 97–133.

Rabinowicz, W.: 2002, 'Does Practical Deliberation Crowd Out Self-Prediction?', *Erkenntnis* **57**, 91–122.

Savage, L. J.: 1954/1972, *The Foundations of Statistics*, Wiley, New York, Dover.

Schick, F.: 1979, 'Self-Knowledge, Uncertainty and Choice', *British Journal for the Philosophy of Science* **30**, 232–252. Reprinted in P. Gärdenfors and N.-E. Sahlin (eds.), *Decision, Probability, and Utility*, Cambridge University Press, Cambridge 1988, pp. 270–286.

Schick, F.: 1999, 'Surprise, Self-Knowledge and Commonality', in B. Hansson, S. Halldén, W. Rabinowicz and N.-E. Sahlin (eds.), *Spinning Ideas – An Electronic Festschrift for Peter Gärdenfors*, on-line at http://www.lucs.lu.se/spinning.

Shackle, G.L.S.: 1967, *Time in Economics*, North-Holland Publishing Company, Amsterdam.

Skyrms, B.: 1980, *Causal Necessity*, Yale University Press, New Haven, London.

Skyrms, B.: 1982, 'Causal Decision Theory', *Journal of Philosophy* **79**, 695–711.

Skyrms, B.: 1984, *Pragmatics and Empiricism*, Yale University Press, New Haven, London.

Skyrms, B.: 1990, 'Ratifiability and the Logic of Decision', in P. A. French, T. E. Uehling Jr. and H. K. Wettstein (eds.), *Midwest Studies in Philosophy, Vol. XV: Philosophy of the Human Sciences*, University of Notre Dame Press, Notre Dame, IND, pp. 44–56.

Sobel, J. H.: 1986, 'Notes on Decision Theory: Old Wine in New Bottles', *Australasian Journal of Philosophy* **64**, 407–437.

Spohn, W.: 1977, 'Where Luce and Krantz Do Really Generalize Savage's Decision Model', *Erkenntnis* **11**, 113–134.

Spohn, W.: 1978, Grundlagen der Entscheidungstheorie, in *Monographien Wissenschaftstheorie und Grundlagenforschung*, No. 8, Scriptor, Kronberg/Ts.

Spohn, W.: 1983, *Eine Theorie der Kausalität*, Habilitationsschrift, Ludwig-Maximilians-Universität München.

Spohn, W.: 1999, *Strategic Rationality*, Forschungsbericht der DFG-Forschergruppe 'Logik in der Philosophie', University of Konstanz.

University of Santa Cruz
Cowell College
Santa Cruz, CA 95064
U.S.A.
E-mail: mledwig@ucsc.edu

MANFRED JAEGER

A LOGIC FOR INDUCTIVE PROBABILISTIC REASONING

ABSTRACT. Inductive probabilistic reasoning is understood as the application of inference patterns that use statistical background information to assign (subjective) probabilities to single events. The simplest such inference pattern is direct inference: from "70% of As are Bs" and "a is an A" infer that a is a B with probability 0.7. Direct inference is generalized by Jeffrey's rule and the principle of cross-entropy minimization. To adequately formalize inductive probabilistic reasoning is an interesting topic for artificial intelligence, as an autonomous system acting in a complex environment may have to base its actions on a probabilistic model of its environment, and the probabilities needed to form this model can often be obtained by combining statistical background information with particular observations made, i.e., by inductive probabilistic reasoning. In this paper a formal framework for inductive probabilistic reasoning is developed: syntactically it consists of an extension of the language of first-order predicate logic that allows to express statements about both statistical and subjective probabilities. Semantics for this representation language are developed that give rise to two distinct entailment relations: a relation $\models$ that models strict, probabilistically valid, inferences, and a relation $\approx\!\!\!\mid$ that models inductive probabilistic inferences. The inductive entailment relation is obtained by implementing cross-entropy minimization in a preferred model semantics. A main objective of our approach is to ensure that for both entailment relations complete proof systems exist. This is achieved by allowing probability distributions in our semantic models that use non-standard probability values. A number of results are presented that show that in several important aspects the resulting logic behaves just like a logic based on real-valued probabilities alone.

1. INTRODUCTION

1.1. *Inductive Probabilistic Reasoning*

Probabilities come in two kinds: as statistical probabilities that describe relative frequencies, and as subjective probabilities that describe degrees of belief. To both kinds of probabilities the same rules of probability calculus apply, and notwithstanding a long and heated philosophical controversy over what constitutes the proper meaning of probability (de Finetti 1937; von Mises 1951; Savage

Synthese (2005) 144: 181–248

Knowledge, Rationality & Action 11–78

DOI 10.1007/s11229-004-6153-2

© Springer 2005

1954; Jaynes 1978), few conceptual difficulties arise when we deal with them one at a time.

However, in commonsense or inductive reasoning one often wants to use both subjective and statistical probabilities simultaneously in order to infer new probabilities of interest. The simplest example of such a reasoning pattern is that of *direct inference* (Reichenbach 1949, Section 72; Carnap 1950, Section 94), illustrated by the following example: from

(1)   2.7% of drivers whose annual mileage is between 10,000 and 20,000 miles will be involved in an accident within the next year

and

(2)   Jones is a driver whose annual mileage is between 10,000 and 20,000 miles

infer

(3)   The probability that Jones will be involved in an accident within the next year is 0.027.

The 2.7% in (1) is a statistical probability: the probability that a driver randomly selected from the set of all drivers with an annual mileage between 10,000 and 20,000 will be involved in an accident. The probability in (3), on the other hand, is attached to a proposition that, in fact, is either true or false. It describes a state of knowledge or belief, for which reason we call it a subjective probability.[1]

Clearly, the direct inference pattern is very pervasive: not only does an insurance company make (implicit) use of it in its computation of the rate it is willing to offer a customer, it also underlies some of the most casual commonsense reasoning ("In very few soccer matches did a team that was trailing 0:2 at the end of the first half still win the game. My team is just trailing 0:2 at halftime. Too bad".), as well as the use of probabilistic expert systems. Take a medical diagnosis system implemented by a Bayesian network (Pearl 1988; Jensen 2001), for instance: the distribution encoded in the network (whether specified by an expert or learned from data) is a statistical distribution describing relative frequencies in a large number of past cases. When using the system for the diagnosis of patient Jones, the symptoms that Jones exhibits are entered as evidence, and the (statistical) probabilities of various diseases conditioned on this evidence are identified with the probability of Jones having each of these diseases.

Direct inference works when for some reference class C and predicate P we are given the statistical probability of P in C, and for some singular object e all we know is that e belongs to C. If we have more information than that, direct inference may no longer work: assume in addition to (1) and (2) that

(4)   3.1% of drivers whose annual mileage is between 15,000 and 25,000 miles will be involved in an accident within the next year

and

(5)   Jones is a driver whose annual mileage is between 15,000 and 25,000 miles.

Now direct inference can be applied either to (1) and (2), or to (4) and (5), yielding the two conflicting conclusions that the probability of Jones having an accident is 0.027 and 0.031. Of course, from (1), (2), (4), and (5) we would infer neither, and instead ask for the percentage of drivers with an annual mileage between 15,000 and 20,000 that are involved in an accident. This number, however, may be unavailable, in which case direct inference will not allow us to derive any probability bounds for Jones getting into an accident. This changes if, at least, we know that

(6)   Between 2.7 and 3.1% of drivers whose annual mileage is between 15,000 and 20,000 miles will be involved in an accident within the next year.

From (1), (2), and (4)–(6) we will at least infer that the probability of Jones having an accident lies between 0.027 and 0.031. This no longer is direct inference proper, but a slight generalization thereof.

In this paper we will be concerned with inductive probabilistic reasoning as a very broad generalization of direct inference. By inductive probabilistic reasoning, for the purpose of this paper, we mean the type of inference where statistical background information is used to refine already existing, partially defined subjective probability assessments (we identify a categorical statement like (2) or (5) with the probability assessment: "with probability 1 is

[13]

Jones a driver whose... "). Thus, we here take a fairly narrow view of inductive probabilistic reasoning, and, for instance, do not consider statistical inferences of the following kind: from the facts that the individuals $jones_1, jones_2, \ldots, jones_{100}$ are drivers, and that $jones_1, \ldots, jones_{30}$ drive less and $jones_{31}, \ldots, jones_{100}$ more than 15,000 miles annually, infer that 30% of drivers drive less than 15,000 miles. Generally speaking, we are aiming at making inferences only in the direction from statistical to subjective probabilities, not from single-case observations to statistical probabilities.

Problems of inductive probabilistic reasoning that go beyond the scope of direct inference are obtained when the subjective input-probabilities do not express certainties

(7)   With probability 0.6 is Jones a driver whose annual
       mileage is between 10,000 and 20,000 miles.

What are we going to infer from (7) and the statistical probability (1) about the probability of Jones getting into an accident? There do not seem to be any sound arguments to derive a unique value for this probability; however, $0.6 \times 0.027 = 0.0162$ appears to be a sensible lower bound. Now take the subjective input probabilities

(8)   With probability 0.6 is Jones's annual mileage between
       10,000 and 20,000 miles, and with probability 0.8 between
       15,000 and 25,000 miles.

Clearly, it's getting more and more difficult to find the right formal rules that extend the direct inference principle to such general inputs.

In the guise of inductive probabilistic reasoning as we understand it, these generalized problems seem to have received little attention in the literature. However, the mathematical structure of the task we have set ourselves is essentially the same as that of *probability updating*: in probability updating we are given a *prior* (usually subjective) probability distribution representing a state of knowledge at some time t, together with new information in the form of categorical statements or probability values; desired is a new *posterior* distribution describing our knowledge at time $t + 1$, with the new information taken into account. A formal correspondence between the two problems is established by identifying the statistical and subjective probability distributions in inductive probabilistic inference

[14]

with the prior and posterior probability distribution, respectively, in probability updating.

The close relation between the two problems extends beyond the formal similarity, however: interpreting the statistical probability distribution as a canonical prior (subjective) distribution, we can view inductive probabilistic reasoning as a special case of probability updating. Methods that have been proposed for probability updating, therefore, also are candidates to solve inductive probabilistic inference problems.

For updating a unique prior distribution on categorical information, no viable alternative exists to *conditioning*: the posterior distribution is the prior conditioned on the stated facts. [2] Note that conditioning, seen as a rule for inductive reasoning, rather than probability updating, is just direct inference again.

As our examples already have shown, this basic updating/inductive reasoning problem can be generalized in two ways: first, the new information may come in the form of probabilistic constraints as in (7), not in the form of categorical statements; second, the prior (or statistical) information may be incomplete, and only specify a set of possible distributions as in (6), not a unique distribution. The problem of updating such partially defined beliefs has received considerable attention (e.g., Dempster 1967; Shafer 1976; Walley 1991; Gilboa and Schmeidler 1993; Moral and Wilson 1995; Dubois and Prade 1997; Grove and Halpern 1998). The simplest approach is to apply an updating rule for unique priors to each of the distributions that satisfy the prior constraints, and to infer as partial posterior beliefs only probability assignments that are valid for all updated possible priors. Inferences obtained in this manner can be quite weak, and other principles have been explored where updating is performed only on a subset of possible priors that are in some sense maximally consistent with the new information (Gilboa and Schmeidler 1993; Dubois and Prade 1997). These methods are more appropriate for belief updating than for inductive probabilistic reasoning in our sense, because they amount to a combination of prior and new information on a more or less symmetric basis. As discussed above, this is not appropriate in our setting, where the new single case information is not supposed to have any impact on the statistical background knowledge. Our treatment of incompletely specified priors, therefore, follows the first approach of taking every possible prior (statistical distribution) into account (see Section 4.1 for additional comments on this issue).

[15]

The main problem we address in the present paper is how to deal with new (single-case) information in the form of general probability constraints. For this various rules with different scope of application have previously been explored. In the case where the new constraints prescribe the probability values $p_1, \ldots, p_k$ of pairwise disjoint alternatives $A_1, \ldots, A_k$, *Jeffrey's rule* (Jeffrey 1965) is a straightforward generalization of conditioning: it says that the posterior should be the sum of the conditional distributions given the A_i, weighted with the prescribed values p_i. Applying Jeffrey's rule to (1) and (7), for instance, we would obtain $0.6 \times 0.027 + 0.4 \times r$ as the probability for Jones getting into an accident, where r is the (unspecified) statistical probability of getting into an accident among drivers who do less than 10,000 or more than 20,000 miles.

When the constraints on the posterior are of a more general form than permitted by Jeffrey's rule, there no longer exist updating rules with a similarly intuitive appeal. However, a number of results indicate that *cross-entropy minimization* is the most appropriate general method for probability updating, or inductive probabilistic inference (Shore and Johnson 1980; Paris and Vencovská 1992; Jaeger 1995b). Cross-entropy can be interpreted as a measure for the similarity of two probability distributions (originally in an information theoretic sense (Kullback and Leibler 1951)). Cross-entropy minimization, therefore, is a rule according to which the posterior (or the subjective) distribution is chosen so as to make it as similar as possible within the given constraints to the prior (resp. the statistical) distribution.

Inductive probabilistic reasoning as we have explained it so far clearly is a topic with its roots in epistemology and the philosophy of science rather than in computer science. However, it also is a topic of substantial interest in all areas of artificial intelligence where one is concerned with reasoning and decision making under uncertainty.

Our introductory example is a first case in point. The inference patterns described in this example could be part of a probabilistic expert system employed by an insurance company to determine the rate of a liability insurance for a specific customer.

As a second example, consider the case of an autonomous agent that has to decide on its actions based on general rules it has been programmed with, and observations it makes. To make things graphic, consider an unmanned spacecraft trying to land on some

[16]

distant planet. The spacecraft has been instructed to choose one of two possible landing sites: site A is a region with a fairly smooth surface, but located in an area subject to occasional severe storms; site B lies in a more rugged but atmospherically quiet area. According to the statistical information the spacecraft has been equipped with, the probabilities of making a safe landing are 0.95 at site A when there is no storm, 0.6 at site A under stormy conditions, and 0.8 at site B. In order to find the best strategy for making a safe landing, the spacecraft first orbits the planet once to take some meteorological measurements over site A. Shortly after passing over A it has to decide whether to stay on course to orbit the planet once more, and then land at A (20 h later, say), or to change its course to initiate landing at B. To estimate the probabilities of making a safe landing following either strategy, thus the probability of stormy conditions at A in 20 h time has to be evaluated. A likely method to obtain such a probability estimate is to feed the measurements made into a program that simulates the weather development over 20 h, to run this simulation, say, one hundred times, each time adding some random perturbation to the initial data and/or the simulation, and to take the fraction q of cases in which the simulation at the end indicated stormy conditions at A as the required probability. Using Jeffrey's rule, then $0.6q + 0.95(1-q)$ is the estimate for the probability of a safe landing at A.

This example illustrates why conditioning as the sole instrument of probabilistic inference is not enough: there is no way that the spacecraft could have been equipped with adequate statistical data that would allow it to compute the probability of storm at A in 20 h time simply by conditioning the statistical data on its evidence, consisting of several megabytes of meteorological measurements. Thus, even a perfectly rational, automated agent, operating on the basis of a well-defined finite body of input data cannot always infer subjective probabilities by conditioning statistical probabilities, but will sometimes have to engage in more flexible forms of inductive probabilistic reasoning.[3]

1.2. *Aims and Scope*

To make inductive probabilistic reasoning available for AI applications, two things have to be accomplished: first, a formal rule for this kind of probabilistic inference has to be found. Second, a formal representation language has to be developed that allows us to encode the kind of probabilistic statements we want to reason with, and on which inference rules for inductive probabilistic reasoning can be defined.

In this paper we will focus on the second of these problems, basically taking it for granted that cross-entropy minimization is the appropriate formal rule for inductive probabilistic reasoning (see Section 3.1 for a brief justification). The representation language that we will develop is first-order predicate logic with additional constructs for the representation of statistical and subjective probability statements. To encode both deductive and inductive inferences on this language, it will be equipped with two different entailment relations: a relation $\models$ that describes valid probabilistic inferences, and a relation $\approx$ that describes inductive probabilistic inferences obtained by cross-entropy minimization. For example, the representation language will be rich enough to encode all the example statements (1)–(8) in formal sentences $\phi_1, \ldots, \phi_8$.

If, furthermore, ψ_0 is a sentence that says that with probability 0.4 Jones drives less than 10,000 or more than 20,000 miles annually, then we will obtain in our logic

$$\phi_7 \models \psi_0,$$

because ψ_0 follows from ϕ_7 by the laws of probability theory. If, on the other hand, ψ_1 says that with probability at least 0.0162 Jones will be involved in an accident, then ψ_1 does not strictly follow from our premises, i.e.,

$$\phi_1 \wedge \phi_7 \not\models \psi_1.$$

However, for the inductive entailment relation we will obtain

$$\phi_1 \wedge \phi_7 \approx \psi_1.$$

Our probabilistic first-order logic with the two entailment relations $\models$ and $\approx$ will provide a principled formalization of inductive probabilistic reasoning in an expressive logical framework. The next problem, then, is to define inference methods for this logic. It is well known that for probabilistic logics of the kind we consider here no complete deduction calculi exist when probabilities are required to be real numbers (Abadi and Halpern 1994), but that completeness results can be obtained when probability values from more general algebraic structures are permitted (Bacchus 1990a). We will follow the approach of generalized probabilities and permit probabilities to take values in *logarithmic real-closed fields (lrc-fields)*, which provide a very good approximation to the real numbers. With the lrc-field based semantics we obtain a completeness result for our logic. It should be emphasized that with this approach we do not abandon real-valued probabilities: real numbers being an example for an lrc-field, they are, of course, not excluded by our generalized semantics. Moreover, a completeness result for lrc-field valued probabilities can also be read as a characterization of the degree of incompleteness of our deductive system for real-valued probabilities: the only inferences for real-valued probabilities that we are not able to make are those that are not valid in all other lrc-fields. By complementing the completeness result for lrc-field valued probabilities with results showing that core properties of real-valued probabilities are actually shared by all lrc-field valued probabilities, we obtain a strong and precise characterization of how powerful our deductive system is for real-valued probabilities.

The main part of this paper (Sections 2 and 3) contains the definition of our logic $\mathscr{L}_{ip}$ consisting of a probabilistic representation language L_p, a strict entailment relation $\models$ (both defined in Section 2), and an inductive entailment relation $\approx$ (defined in Section 3). The basic design and many of the properties of the logic $\mathscr{L}_{ip}$ do not rely on our use of probability values from logarithmic real-closed fields, so that Sections 2 and 3 can also be read ignoring the issue of generalized probability values, and thinking of real-valued probabilities throughout. Only the key properties of $\mathscr{L}_{ip}$ expressed in Corollary 2.11 and Theorem 2.12 are not valid for real-valued probabilities.

To analyze in detail the implications of using lrc-fields we derive a number of results on cross-entropy and cross-entropy minimization in logarithmic real-closed fields. The basic technical results here have been collected in Appendix A. These results are used in Section 3 to show that many important inference patterns for inductive probabilistic reasoning are supported in $\mathscr{L}_{ip}$. The results of Appendix A also are of some independent mathematical interest, as they constitute an alternative derivation of basic properties of cross-entropy minimization in (real-valued) finite probability spaces only from elementary algebraic properties of the logarithmic function. Previous derivations of these properties required more powerful analytic methods (Kullback 1959; Shore and Johnson 1980).

[19]

This paper is largely based on the author's PhD thesis (Jaeger 1995a). A very preliminary exposition of the logic $\mathscr{L}_{ip}$ was given in Jaeger (1994a). A statistical derivation of cross-entropy minimization as the formal model for inductive probabilistic reasoning was given in Jaeger (1995b).

1.3. *Previous Work*

Clearly, the work here presented is intimately related to a sizable body of previous work on combining logic and probability, and on the principles of (probabilistic) inductive inference.

Boole (1854) must probably be credited for being the first to combine logic and probability. He saw events to which probabilities are attached as formulas in a (propositional) logic, and devised probabilistic inference techniques that were based both on logical manipulations of the formulas and algebraic techniques for solving systems of (linear) equations (see Hailperin (1976) for a modern exposition of Boole's work).

The work of Carnap (1950, 1952) is of great interest in our context in more than one respect: Carnap was among the first to acknowledge the existence of two legitimate concepts of probability, (in Carnap's terminology) expressing degrees of confirmation and relative frequencies, respectively. The main focus in Carnap's work is on probability as degree of confirmation, which he considers to be defined on logical formulas. His main objective is to find a canonical probability distribution c on the algebra of (first-order) formulas, which would allow to compute the degree of confirmation $c(h/e)$ of some hypothesis h, given evidence e in a mechanical way, i.e., from the syntactic structure of h and e alone. Such a confirmation function c would then be seen as a normative rule for inductive reasoning. While eventually abandoning the hope to find such a unique confirmation function (Carnap 1952), Carnap (1950) proves that for a general class of candidate functions c a form of the direct inference principle can be derived: if e is a proposition that says that the relative frequency of some property M in a population of n objects is r, and h is the proposition that one particular of these n objects has property M, then $c(h/e) = r$.

Carnap's work was very influential, and many subsequent works on probability and logic (Gaifman 1964; Scott and Krauss 1966; Fenstad 1967; Gaifman and Snir 1982) were more or less directly spawned by Carnap (1950). They are, however, more concerned with purely logical and mathematical questions arising out of the study of probabilistic interpretations for logical language, than with the foundations of probabilistic and inductive reasoning.

In none of the works mentioned so far were probabilistic statements integrated into the logical language under consideration. Only on the semantic level were probabilities assigned to (non-probabilistic) formulas. This changes with Kyburg (1974), who, like Carnap, aims to explain the meaning of probability by formalizing it in a logical framework. In doing so, he develops within the framework of first-order logic special syntactic constructs for statistical statements. These statistical statements, in conjunction with a body of categorical knowledge, then are used to define subjective probabilities via direct inference.

Keisler (1985) and Hoover (1978) developed first-order and infinitary logics in which the standard quantifiers $\forall x$ and $\exists x$ are replaced by a probability quantifier $Px \geq r$, standing for "for x with probability at least r". The primary motivation behind this work was to apply new advances in infinitary logics to probability theory.

In AI, interest in probabilistic logic started with Nilsson's (1986) paper, which, in many aspects, was a modern reinvention of (Boole 1854) (see Hailperin (1996) for an extensive discussion).

Halpern's (1990) and Bacchus's (1990a,b) seminal works introduced probabilistic extensions of first-order logic for the representation of both statistical and subjective probabilities within the formal language. The larger part of Halpern's and Bacchus's work is concerned with coding strict probabilistic inferences in their logics. A first approach towards using the underlying probabilistic logics also for inductive probabilistic reasoning is contained in Bacchus (1990b), where an axiom schema for direct inference is presented. Much more general patterns of inductive (or default) inferences are modeled by the *random worlds* method by Bacchus, Grove, Halpern, and Koller (Bacchus et al. 1992, 1997; Grove et al. 1992a,b). By an approach very similar to Carnap's definition of the confirmation function c, in this method a degree of belief $\Pr(\phi|\psi)$ in ϕ given the knowledge ψ is defined. Here ϕ and ψ now are formulas in the statistical probabilistic languages of Halpern and Bacchus. As ψ, thus, cannot encode prior constraints on the subjective probabilities (or degrees of belief), the reasoning patterns supported by this method are quite different from what we have called inductive probabilistic reasoning in Section 1.1, and what forms the subject of the current paper. A more detailed discussion of

[21]

the random worlds method and its relation to our framework is deferred to Section 4.1.

2. THE LOGIC OF STRICT INFERENCE

2.1. *Outline*

In this section, we introduce the logic $\mathscr{L}_p = (L_p, \models)$ consisting of a language L_p for the representation of statistical and subjective probabilities, and an entailment relation $\models$ capturing inferences that are validated by probability calculus. Thus, the nature of the logic $\mathscr{L}_p$ will be very similar to that of the logics of Halpern (1990) and Bacchus (1990b), and we will follow in our presentation of $\mathscr{L}_p$ these previously defined formalisms as far as possible.

The main difference between our logic $\mathscr{L}_p$ and the logics of Halpern and Bacchus lies in the definition of terms expressing subjective probabilities. Here our approach is guided by the goal to later extend the logic $\mathscr{L}_p$ to a logic $\mathscr{L}_{ip} = (L_p, \models, \mathrel{\vHorbar\kern-6pt\approx})$ with an additional entailment relation $\mathrel{\vHorbar\kern-6pt\approx}$ for inductive probabilistic inferences. This inductive entailment relation will be obtained by implementing cross-entropy minimization between the statistical and subjective probability distribution in the semantic structures for the language. As we can only speak of the cross-entropy of two probability distributions that are defined on the same probability space, we cannot follow Bacchus and Halpern in interpreting statistical and subjective probability terms by probability distributions over the domains of semantical structures, and distributions over sets of semantic structures, respectively. Instead, we choose to interpret both statistical and subjective probability terms over the domain of semantic structures. To make this feasible for subjective probability terms, we have to impose a certain restriction on their formulation: it will be required that subjective probability terms always refer to some specific objects or events about which there is some uncertainty. In our introductory example, for instance, all the uncertainty expressed in the subjective probability statements was attached to the object "Jones" about whose exact properties we have incomplete information. In a somewhat more complicated example, a subjective probability statement may be about the probability that in an accident "crash010899Madison/5th", involving drivers "Jones" and "Mitchell", driver "Jones" was to be blamed for the accident. This statement, then, would express uncertainty about the exact relations between the elements of the tuple (crash010899Madison/5th,Jones,Mitchell) of objects and events.

Considering only subjective probability expressions that fit this pattern allows us to interpret them by probability distributions over the domain of a semantic structure: we interpret the concrete objects and events appearing in the subjective probability expression as randomly drawn elements of the domain. This approach stands in the tradition of frequentist interpretations of subjective probabilities (Reichenbach 1949; Carnap 1950). For the denotation of such random domain elements we will use a special type of symbols, called *event symbols*, that are used syntactically like constants, but are interpreted by probability measures.

Another point where we will deviate from the previous approaches by Halpern and Bacchus is in the structure of the probability measures appearing as part of the semantic structures. In Halpern (1990) and Bacchus (1990b) these measures were assumed to come from the very restricted class of real-discrete measures (cf. Example 2.7). Halpern (1990) states that this restriction is not essential and briefly outlines a more general approach, perhaps somewhat understating the technical difficulties arising in these approaches (as exemplified by our Theorem 2.8). In Bacchus (1990a) a more general concept of probability distributions is used, allowing arbitrary finitely additive field-valued probabilities. We will use a closely related approach, requiring probabilities to take values in *lrc-fields* (Definition 2.1).

2.2. *Syntax*

The syntax of our logic is that of first-order predicate logic with three extensions: first, the language of logarithmic, ordered fields is integrated as a fixed component into the language; second, a term-forming construction (taken directly from Bacchus (1990b)) is introduced that allows us to build terms denoting statistical probabilities; and third, a term-forming construction is introduced for building terms denoting subjective probabilities.

We use two sets of variables in the language: *domain variables* ranging over the elements of the domain of discourse, and *field variables* ranging over numbers, especially probability values. The vocabulary

$$S_{LOF} = \{0, 1, +, \cdot, \leqslant, Log\}$$

of ordered fields with a logarithmic function is considered to belong to the logical symbols of the language. The non-logical symbols consist of a set $S = \{R, Q, \ldots, f, g, \ldots, c, d, \ldots\}$ of relation, function, and constant symbols, as in first-order logic, and a tuple $\mathbf{e} = (e_1, \ldots, e_N)$ of *event symbols*.

The language $L_p(S, \mathbf{e})$ now is defined by the following rules. Since in part (f) of the formation rule for field terms a condition on the free variables of a formula is required, we have to define simultaneously with the construction of terms and formulas the set of free variables they contain. Except for the non-standard syntactic constructions we omit these obvious declarations.

A *domain-term* is constructed from domain-variables $v_0, v_1, \ldots,$ constant and function symbols from S according to the syntax rules of first-order logic.

Atomic domain formulas are formulas of the form

$$Rt_1 \ldots t_k \quad \text{or} \quad t_1 = t_2,$$

where R is a k-ary relation symbol from S, and the t_i are domain-terms.

Boolean operations. If ϕ and ψ are formulas, then so are $(\phi \wedge \psi)$ and $\neg \phi$.

Quantification. If ϕ is a formula and $v(x)$ is a domain-variable (field-variable), then $\exists v \phi (\exists x \phi)$ is a formula.

Field-terms:

(a) Every field-variable $x_0, x_1, \ldots$ is a field-term.
(b) 0 and 1 are field-terms.
(c) If t_1 and t_2 are field-terms, then so are $(t_1 \cdot t_2)$ and $(t_1 + t_2)$.
(d) If t is a field term, then so is $Log(t)$.
(e) If ϕ is a formula, and $\mathbf{w}$ a tuple of domain variables, then

$$[\phi]_{\mathbf{w}}$$

is a field-term. The free variables of $[\phi]_{\mathbf{w}}$ are the free variables of ϕ not appearing in $\mathbf{w}$. A field term of this form is called a *statistical probability term*.

(f) If $\phi(\mathbf{v})$ is a formula whose free variables are among the domain variables $\mathbf{v}$, ϕ does not contain any terms of the form prob($\ldots$), and if $\mathbf{v}/\mathbf{e}$ is an assignment that maps every $v \in \mathbf{v}$ to some $e \in \mathbf{e}$, then

$$prob(\phi[\mathbf{v}/\mathbf{e}])$$

is a field-term (without free variables). A field term of this form is called a *subjective probability term*.

Atomic field formulas. If t_1, t_2 are field-terms, then $t_1 \leqslant t_2$ is an atomic field formula.

Rule (f) for field terms essentially says that event symbols $e_1, \ldots,$ e_N are used syntactically like constant symbols, but are restricted to only appear within the scope of a prob()-operator. Moreover, subjective probability terms may not be nested or contain free variables. These are fairly serious limitation that are not essential for the definition of $\mathscr{L}_p$, but will be crucially important for the definition of $\approx$ in $\mathscr{L}_{ip}$.

We may freely use as definable abbreviations (in)equalities like $t_1 > t_2, t_1 = t_2, t_1 \geqslant t_2$, and conditional probability expressions like $[\phi|\psi]_w$ or $\mathrm{prob}(\phi[e]|\psi[e])$. These conditional probability expressions are interpreted by the quotients $[\phi \wedge \psi]_w/[\psi]_w$, respectively, $\mathrm{prob}(\phi[e] \wedge \psi[e])/\mathrm{prob}(\psi[e])$, provided the interpretations of $[\psi]_w$, respectively, $\mathrm{prob}(\psi[e])$, are positive. Several conventions may be employed to interpret conditional probability terms when the conditioning expressions are assigned probability zero. We will not explore this issue here and refer the reader to Bacchus (1990b), Halpern (1990), and Jaeger (1995a) for alternative proposals.

To illustrate the use of the language L_p, we encode some of the example sentences of Section 1.1. We use a vocabulary that contains two unary predicate symbols D and M that partition the domain into elements of the sorts driver and mileage, respectively. Another unary predicate symbol IIA stands for "involved in accident", and a unary function am maps drivers to their annual mileage. Also we use constants $10, 15, \ldots$ for specific mileages (in thousands), and a binary order relation $\preceq$ on mileages (this relation $\preceq$ defined on the domain is to be distinguished from the relation $\leqslant$ defined on probability values). Finally, there is a single event symbol jones. Statement (1) can now be formalized as

(9) $\phi_1 :\equiv [\mathrm{IIA}(d)|\mathrm{D}(d) \wedge 10 \preceq \mathrm{am}(d) \preceq 20]_d = 0.027.$

Statement (3) becomes

(10) $\phi_3 :\equiv \mathrm{prob}(\mathrm{IIA}(jones)) = 0.027.$

2.3. *Semantics*

Key components of the semantic structures that we will use to interpret L_p are finitely additive probability measures with values in logarithmic real-closed fields. We briefly review the concepts we require.

DEFINITION 2.1. *An* S_{LOF}*-structure* $\mathfrak{J} = (\mathbb{F}, 0, 1, +, \cdot, \leqslant, \text{Log})$ *over a domain* $\mathbb{F}$ *is a logarithmic real-closed field (lrc-field) if it satisfies the axioms* **LRCF** *consisting of*

  (i) *The axioms of ordered fields.*
  (ii) *An axiom for the existence of square roots*

$$\forall x \exists y (0 \leqslant x \rightarrow y^2 = x).$$

  (iii) *A schema demanding that every polynomial of uneven degree has a root*

$$\forall y_0 \ldots y_{n-1} \exists x (y_0 + y_1 \cdot x + \cdots + y_{n-1} \cdot x^{n-1} + x^n = 0),$$
$$n = 1, 3, 5, \ldots$$

  (iv) $\forall x, y > 0 \quad \text{Log}(x \cdot y) = \text{Log}(x) + \text{Log}(y).$
  (v) $\forall x > 0 \quad x \neq 1 \rightarrow \text{Log}(x) < x - 1.$
  (vi) *The approximation schema*

$$\forall x \in (0, 1] \; q_n(x) \leqslant \text{Log}(x) \leqslant p_n(x), \quad n = 1, 2, \ldots$$

where

$$q_n(x) :\equiv (x-1) - \frac{(x-1)^2}{2} + \frac{(x-1)^3}{3} - \cdots + (-1)^{n-1}\frac{(x-1)^n}{n}$$
$$+ (-1)^n \frac{(x-1)^{n+1}}{x},$$
$$p_n(x) :\equiv (x-1) - \frac{(x-1)^2}{2} + \frac{(x-1)^3}{3} - \cdots + (-1)^{n-1}\frac{(x-1)^n}{n}.$$

A structure over the vocabulary $S_{OF} := \{+, \cdot, \leqslant, 0, 1\}$ that satisfies the axioms **RCF** consisting of (i)–(iii) alone is called a *real-closed field*. By classic results in model theory, **RCF** is a complete axiomatization of the S_{OF}-theory of the real numbers. In other words, every first-order S_{OF}-sentence ϕ that is true in $\mathbb{R}$ also is true in every other real-closed field (see Rabin (1977) for an overview). To what extent

[26]

similar results hold for logarithmic real closed fields is a long-standing open problem in model theory (there studied w.r.t. (real-closed) fields augmented by an exponential, rather than a logarithmic, function (see e.g., Dahn and Wolter (1983)).

DEFINITION 2.2. *Let M be a set. An algebra over M is a collection $\mathfrak{A}$ of subsets of M that contains M, and is closed under complementation and finite unions. If M is also closed under countable unions, it is called a σ-algebra. If $\mathfrak{A}$ is an algebra on M, and $\mathfrak{A}'$ an algebra on M', then the product algebra $\mathfrak{A} \times \mathfrak{A}'$ is the algebra on $M \times M'$ generated by the sets $A \times A'$ $(A \in \mathfrak{A},\ A' \in \mathfrak{A}')$.*

DEFINITION 2.3. *Let $\mathfrak{A}$ be an algebra over M, $\mathfrak{F}$ an lrc-field. Let $\mathbb{F}^+ := \{x \in \mathbb{F} \mid 0 \leqslant x\}$. A function*

$$P: \mathfrak{A} \to \mathbb{F}^+$$

is an $\mathbb{F}$-probability measure iff $P(\emptyset) = 0$, $P(M) = 1$, and $P(A \cup B) = P(A) + P(B)$ for all $A, B \in \mathfrak{A}$ with $A \cap B = \emptyset$. The elements of $\mathfrak{A}$ also are called the measurable sets. The set of all probability measures with values in $\mathbb{F}$ on the algebra $\mathfrak{A}$ is denoted by

$$\Delta_{\mathbb{F}}\mathfrak{A}.$$

Thus, even when the underlying algebra is a σ-algebra, we do not require σ-additivity, because this would usually make no sense in arbitrary lrc-fields, where infinite sums of non-negative numbers need not be defined. If $\mathfrak{A}$ is a finite algebra with n atoms, then $\Delta_{\mathbb{F}}\mathfrak{A}$ can be identified with

$$\Delta_{\mathbb{F}}^n := \{(x_1, \ldots, x_n) \in \mathbb{F}^n \mid x_i \geqslant 0, \sum_i x_i = 1\}.$$

If $\mathfrak{A}'$ is a subalgebra of $\mathfrak{A}$, and $P \in \Delta_{\mathbb{F}}\mathfrak{A}$, then $P \restriction \mathfrak{A}'$ denotes the restriction of P to $\mathfrak{A}'$, i.e., a member of $\Delta_{\mathbb{F}}\mathfrak{A}'$. By abuse of notation we also use $P \restriction \mathfrak{A}'$ to denote the marginal distribution on $\mathfrak{A}'$ when $\mathfrak{A}'$ is a factor, rather than a subalgebra, of $\mathfrak{A}$, i.e., $\mathfrak{A} = \mathfrak{A}' \times \mathfrak{A}''$ for some $\mathfrak{A}''$.

Semantic structures for the interpretation of $L_p(S, \mathbf{e})$ are based on standard model theoretic structures for the vocabulary S, augmented by probability measures for the interpretation of probability terms.

The basic form of a probabilistic structure will be

$$\mathfrak{M} = (M, I, \mathfrak{F}, (\mathfrak{A}_n, P_n)_{n \in \mathbb{N}}, Q_{\mathbf{e}}),$$

where (M, I) is a standard S-structure consisting of domain M and interpretation function I for S, $\mathfrak{F}$ is a logarithmic real closed field, the $(\mathfrak{A}_n, P_n)$ are probability measure algebras on M^n, and $Q_{\mathbf{e}}$ is a probability measure on $\mathfrak{A}_{|\mathbf{e}|}$ (we use $|\mathbf{e}|$, $|\mathbf{v}|$, etc., to denote the number of elements in a tuple of event symbols $\mathbf{e}$, variables $\mathbf{v}$, etc.).

Statistical probability terms $[\phi]_{\mathbf{w}}$ will be interpreted by $P_{|\mathbf{w}|}(A)$ where A is the set defined by ϕ in $M^{|\mathbf{w}|}$. The measure P_n, thus, is intended to represent the distribution of a sample of n independent draws from the domain, identically distributed according to P_1 (an "iid sample of size n"). In the case of real-valued σ-additive measures this would usually be achieved by defining P_n to be the n-fold product of P_1, defined on the product σ-algebra $\mathfrak{A}_1 \times \cdots \times \mathfrak{A}_1$ (n factors). A corresponding approach turns out to be infeasible in our context, because the product algebra $\mathfrak{A}_1 \times \cdots \times \mathfrak{A}_1$ usually will not be fine-grained enough to give semantics to all statistical probability terms $[\phi]_{\mathbf{w}}$. In order to ensure that the sequence $(\mathfrak{A}_1, P_1), (\mathfrak{A}_2, P_2), \ldots$, nevertheless, behaves in several essential aspects like a sequence of product algebras and product measures, we explicitly impose three *coherence conditions: homogeneity,* the *product property,* and the *Fubini property.* These are essentially the same conditions as can be found in Hoover (1978), there summarily called Fubini property. Bacchus (1990a) requires homogeneity and the product property only.

Homogeneity. For all n, $A \in \mathfrak{A}_n$ and permutations π of $\{1, \ldots, n\}$:

$$\pi(A) := \{\pi \mathbf{a} \mid \mathbf{a} \in A\} \in \mathfrak{A}_n, \quad \text{and} \quad P_n(\pi(A)) = P_n(A).$$

Homogeneity expresses the permutation invariance of iid samples: if we sample two drivers from our example domain, for instance, then the probability that the first one drives a Toyota, and the second one a Ford is the same as the probability that the first one drives a Ford, and the second one a Toyota.

Product property. For all $k, l \in \mathbb{N}$: $A \in \mathfrak{A}_k$ and $B \in \mathfrak{A}_l$ implies $A \times B \in \mathfrak{A}_{k+l}$, and $P_{k+l}(A \times B) = P_k(A) \cdot P_l(B)$.

The product property expresses independence of samples. For an example let $k = l = 1$, A comprise the set of Toyota drivers, and B comprise the set of Ford drivers. Then $P_1(A)(P_1(B))$ is the probability of sampling a Toyota (Ford) driver in a single draw. $P_2(A \times B)$ is the probability of first drawing a Toyota driver, then a Ford driver, in a two-element sample. When sampling is iid, $P_2(A \times B)$ must be equal to $P_1(A)P_1(B)$.

For the formulation of the last coherence condition we first introduce some notation for sections of sets: Let $I \subset \{1, \ldots, n\}$ with $I \neq \emptyset$ and $I' := \{1, \ldots, n\} \setminus I$. Let $A \subseteq M^n$ and $\boldsymbol{a} \in M^I$. Then the *section* of A in the coordinates I along $\boldsymbol{a}$ is defined as

$$\sigma_{\boldsymbol{a}}^I(A) := \{\boldsymbol{b} \in M^{I'} \,|\, (\boldsymbol{a}, \boldsymbol{b}) \in A\}.$$

Fubini property. For all $n \in \mathbb{N}, I \subset \{1, \ldots, n\}$ with $1 \leqslant |I| =: k$, $A \in \mathfrak{A}_n$, and $\boldsymbol{a} \in M^I$:

(11) $\sigma_{\boldsymbol{a}}^I(A) \in \mathfrak{A}_{n-k}$

for all $r \in [0, 1]$:

(12) $A_{I, \geqslant r} := \{\boldsymbol{a} \in M^I \,|\, P_{n-k}(\sigma_{\boldsymbol{a}}^I(A)) \geqslant r\} \in \mathfrak{A}_k$

and

(13) $P_n(A) \geqslant r P_k(A_{I, \geqslant r}).$

Furthermore, we require (13) to hold with strict inequality for the set $A_{I, >r}$ defined by replacing $\geqslant$ by $>$ in (12).

The Fubini property expresses a fundamental "commensurability" property of product measures in different dimensions. For standard σ-additive measures it plays a vital role in the theory of integration. It is best illustrated by a geometric example: obviously, if a geometric figure A in the plane contains a rectangle with sides of lengths s and r, then the area of A must be at least $r \cdot s$. This is essentially the defining property of area as the product measure of one-dimensional lengths. Furthermore, the lower bound $r \cdot s$ also holds when A only contains a "distorted rectangle" of dimensions $r \times s$, as illustrated in Figure 1. The Fubini property establishes the lower bound of $r \cdot s$ for the measure of A from a condition that further generalizes the property of containing a "distorted rectangle".

We are now ready to define our semantic structures.

DEFINITION 2.4. *Let* S *be a vocabulary,* e *a tuple of event symbols. A probabilistic structure for* (S, e) *is a tuple*

$$\mathfrak{M} = (M, I, \mathfrak{F}, (\mathfrak{A}_n, P_n)_{n \in \mathbb{N}}, Q_e),$$

[29]

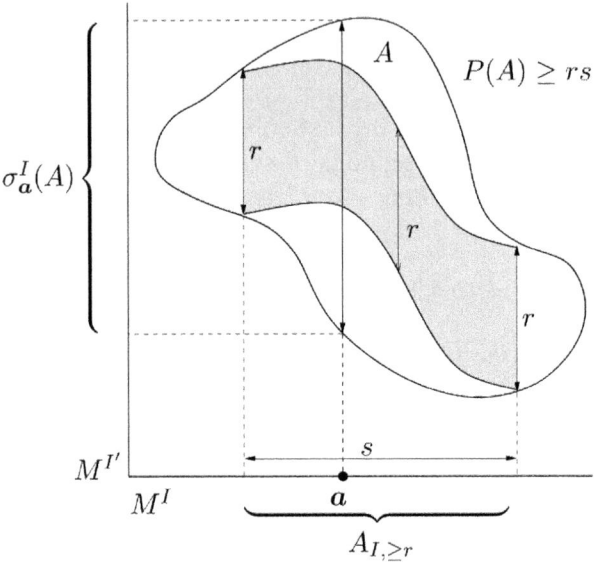

Figure 1. The Fubini property.

where M is a set (the domain), I is an interpretation function for S over M, $\mathfrak{F}$ is a lrc-field, $(\mathfrak{A}_n, P_n)$ is a measure algebra on M^n ($n \in \mathbb{N}$), such that the sequence $(\mathfrak{A}_n, P_n)_{n \in \mathbb{N}}$ satisfies homogeneity, the product property, and the Fubini property, and Q_e is a probability measure on $\mathfrak{A}_{|e|}$.

Now let a probabilistic structure $\mathfrak{M}$ for (S, e) be given, let γ be a variable assignment that maps domain-variables into M and field-variables into $\mathbb{F}$. The notation $\gamma[v/a, x/r]$ is used for the variable assignment that maps v to a, x to r, and for all other variables is the same as γ.

We now need to define the satisfaction relation between $(\mathfrak{M}, \gamma)$ and L_p-formulas. Due to the possible non-measurability of L_p-definable sets, this relation may only be partial. In detail, we define a partial interpretation that maps an (S, e)-term t to its interpretation $(\mathfrak{M}, \gamma)(t)$ in M (if it is a domain term), or in $\mathbb{F}$ (if it is a field term). In parallel, a relation $\models$ is defined between $(\mathfrak{M}, \gamma)$ and $L_p(S, e)$-formulas ϕ. This relation, too, may be only partial in the sense that it is possible that neither $(\mathfrak{M}, \gamma) \models \phi$, nor $(\mathfrak{M}, \gamma) \models \neg\phi$.

Domain-terms. For a domain-term t, the interpretation $(\mathfrak{M}, \gamma)(t)$ is defined just as in first-order logic. Note that t cannot contain any field-terms as subterms.

Atomic domain formulas. If ϕ is an atomic domain formula then the relation $(\mathfrak{M}, \gamma) \models \phi$ is defined as in first-order logic.

Boolean operations. The definition of $(\mathfrak{M}, \gamma) \models \phi$ for $\phi = \psi \vee \chi$ and $\phi = \neg \psi$ is as usual, provided that $\models$ is defined between $(\mathfrak{M}, \gamma)$ and the subformulas ψ, χ. Otherwise $\models$ is undefined between $(\mathfrak{M}, \gamma)$ and ϕ.

Quantification. Let $\phi(v, x) \equiv \exists w \psi(v, w, x)$. Then

$$(\mathfrak{M}, \gamma) \models \phi(v, x) \quad \text{iff} \quad \exists a \in M (\mathfrak{M}, \gamma[w/a]) \models \psi(v, w, x).$$

Similarly for quantification over field variables and universal quantification.

Field-terms. Let t be a field-term.

(a) $t \equiv x$. Then $(\mathfrak{M}, \gamma)(t) = \gamma(x)$.
(b) $t \equiv 0$. Then $(\mathfrak{M}, \gamma)(t) = 0$. Similarly for $t \equiv 1$.
(c) $t \equiv t_1 + t_2$. Then $(\mathfrak{M}, \gamma)(t) = (\mathfrak{M}, \gamma)(t_1) + (\mathfrak{M}, \gamma)(t_2)$ if $(\mathfrak{M}, \gamma)(t_1)$ and $(\mathfrak{M}, \gamma)(t_2)$ are defined. $(\mathfrak{M}, \gamma)(t)$ is undefined otherwise. Similarly for $t \equiv t_1 \cdot t_2$.
(d) $t \equiv \mathrm{Log}(t')$. Then $(\mathfrak{M}, \gamma)(t) = \mathrm{Log}((\mathfrak{M}, \gamma)(t'))$ if $(\mathfrak{M}, \gamma)(t')$ is defined. $(\mathfrak{M}, \gamma)(t)$ is undefined otherwise.
(e) $t \equiv [\phi(v, w, x)]_w$. Then

$$(\mathfrak{M}, \gamma)(t) = P_{|w|}(\{a | (\mathfrak{M}, \gamma[w/a]) \models \phi(v, w, x)\})$$

if $\{a | (\mathfrak{M}, \gamma[w/a]) \models \phi(u, w, x)\} \in \mathfrak{A}_{|w|}$; $(\mathfrak{M}, \gamma)(t)$ is undefined otherwise.
(f) $t \equiv \mathrm{prob}(\phi[v/e])$. Then

$$(\mathfrak{M}, \gamma)(t) = Q_e(\{a | (\mathfrak{M}, \gamma[v/a]) \models \phi(v)\})$$

if $\{a | (\mathfrak{M}, \gamma[v/a]) \models \phi(v)\} \in \mathfrak{A}_{|e|}$; $(\mathfrak{M}, \gamma)(t)$ is undefined otherwise.
Atomic field formulas. Let $\phi \equiv t_1 \leqslant t_2$. Then $(\mathfrak{M}, \gamma) \models \phi$ iff $(\mathfrak{M}, \gamma)$ (t_1) and $(\mathfrak{M}, \gamma)(t_2)$ are defined, and $(\mathfrak{M}, \gamma)(t_1) \leqslant (\mathfrak{M}, \gamma)(t_2)$.

DEFINITION 2.5. *A probabilistic structure $\mathfrak{M}$ is sufficient if the relation $(\mathfrak{M}, \gamma) \models \phi$ is defined for all γ and all $\phi \in L_p$.*

In other words, $\mathfrak{M}$ is sufficient if all L_p-definable sets are measurable. We define semantic entailment with respect to sufficient structures only.

DEFINITION 2.6. *For $\Phi \subseteq L_p$, $\psi \in L_p$ we write $\Phi \models \psi$ if for all sufficient probabilistic structures $\mathfrak{M}: (\mathfrak{M}, \gamma) \models \Phi$ implies $(\mathfrak{M}, \gamma) \models \psi$.*

Because of the importance of definability, we introduce a somewhat more compact notation for sets defined by formulas: if ϕ is
an $L_p(S, \mathbf{e})$-formula, $\mathfrak{M}$ a probabilistic structure, γ a variable assignment, and $\mathbf{v}$ a tuple of n distinct domain variables, then we write

(14) $(\mathfrak{M}, \gamma, \mathbf{v})(\phi) := \{\mathbf{a} \in M^n | (\mathfrak{M}, \gamma[\mathbf{v}/\mathbf{a}]) \models \phi\}.$

Furthermore, when $\phi \equiv \phi(\mathbf{v}, \mathbf{w}, \mathbf{x})$, $\gamma(\mathbf{w}) = \mathbf{b}$, and $\gamma(\mathbf{x}) = \mathbf{r}$, then we
also denote (14) by $(\mathfrak{M}, \mathbf{v})(\phi(\mathbf{v}, \mathbf{b}, \mathbf{r}))$.

It can be very difficult to verify sufficiency for a given structure
$\mathfrak{M}$. In fact, the only class of examples of probabilistic structures for
which sufficiency is easily proved is the following.

EXAMPLE 2.7. Let S be a vocabulary, $\mathbf{e} = (e_1, \ldots, e_N)$ a tuple of
event symbols. Let (M, I) be a standard S-structure; for $i \in \mathbb{N}$ let $a_i \in
M, b_i \in M^N$, $p_i, q_i \in \mathbb{R}$ with $\sum p_i = \sum q_i = 1$. Let $\mathfrak{A}_n = 2^{M^n}$ for all $n \in
\mathbb{N}$, and define

$$P_n(A) = \sum_{(a_{i_1}, \ldots, a_{i_n}) \in A} p_{i_1} \cdots \cdots p_{i_n} \ (A \subseteq M^n)$$

and

$$Q_{\mathbf{e}}(A) = \sum_{b_i \in A} q_i \ (A \subseteq M^N).$$

It is easy to see that $(\mathfrak{A}_n, P_n)_{n \in \mathbb{N}}$ satisfies the coherency conditions.
Moreover, sufficiency is trivially satisfied, because every subset of
M^n is measurable. We refer to structures of this form as *real-discrete
structures*.

2.4. *Probabilistic Reasoning in* $\mathscr{L}_p$

The logic $\mathscr{L}_p$ supports reasoning with statistical and subjective probabilities as two separate entities, and thus has much in common
with Halpern's (1990) logic $\mathscr{L}_3$. However, due to the domain distribution semantics of subjective probabilities, $\mathscr{L}_p$ exhibits some distinguishing properties. In this section, we will discuss some of these
properties. First, however, we turn to purely statistical reasoning,
and illustrate by an example the role of the coherence conditions.

Let $\{D, M, \ldots\}$ be the vocabulary introduced in Section 2.2 for
encoding our introductory example. To provide the basis for some
inferences in $\mathscr{L}_p$, we first axiomatize some aspects of the intended

[32]

meaning of the given symbols. Notably, we want $\preceq$ to be an order relation on M, which we can formalize in L_p by a (standard first-order) sentence $\phi_{\preceq}$. Also, according to the intended meaning of am, this function takes values in M

$$\forall vw(\text{am}(v) = w \to M(w)) \equiv: \phi_{\text{am}}.$$

Now consider the statistical probability term

$$[\text{am}(d) \prec \text{am}(d')]_{d,d'}$$

(where $\prec$, naturally, is shorthand for "$\preceq$ and not $=$"), which represents the statistical probability that of two randomly chosen drivers d and d', d has a lower annual mileage than d'. We want to derive that $1/2$ is an upper bound for this probability. For this let $\mathfrak{M}$ be a sufficient probabilistic structure for the given vocabulary. Then

(15) $A := (\mathfrak{M}, (d, d'))(\text{am}(d) \prec \text{am}(d'))$
$$= \{(a, b) \in M \times M \,|\, \text{am}(a) \prec \text{am}(b)\} \in \mathfrak{A}_2.$$

Also, the permutation of A

(16) $A' := \{(a, b) \in M \times M \,|\, \text{am}(b) \prec \text{am}(a)\}$

belongs to $\mathfrak{A}_2$. If $\mathfrak{M}$ is a model of $\phi_{\preceq} \wedge \phi_{\text{am}}$, then A and A' are disjoint, and by homogeneity $P_2(A) = P_2(A')$. It follows that $P_2(A) \leqslant 1/2$. Hence, we can infer in $\mathscr{L}_p$:

(17) $\phi_{\preceq} \wedge \phi_{\text{am}} | = [\text{am}(d) \prec \text{am}(d')]_{d,d'} \leqslant 1/2.$

Next, we show that from $\phi_{\preceq} \wedge \phi_{\text{am}}$ we can derive

(18) $\exists d [\text{am}(d')] \preceq [\text{am}(d)]_{d'} \geqslant 1/2,$

i.e., there exists a driver whose annual mileage is at least as great as that of 50% of all drivers (an "at least median mileage"-driver). To derive (18) we have to appeal to the Fubini property: let $\mathfrak{M}$ be a model of $\phi_{\preceq} \wedge \phi_{\text{am}}$, and assume that

(19) $\mathfrak{M} | = \forall d [\text{am}(d') \preceq \text{am}(d)]_{d'} < 1/2$, i.e.,
(20) $\mathfrak{M} | = \forall d [\text{am}(d) \prec \text{am}(d')]_{d'} > 1/2.$

Now consider again the set A defined by (15). Then, according to (20),

$$A_{1, > 1/2} = \{a \in M \,|\, P_1(\{b \in M \,|\, a \prec b\}) > 1/2\} = M.$$

[33]

By the Fubini property this leads to

$$P_2(A) > 1/2 P_1(M) = 1/2,$$

a contradiction to (17). Hence (20) cannot hold, and (18) follows from $\phi_{\leq} \wedge \phi_{am}$.

We now turn to reasoning with subjective probabilities. To simplify notation, we assume in the following that there is only one event symbol e in our vocabulary, i.e. $|e| = 1$.

Even though e is interpreted by a probability distribution over the domain, the logic does support the intuition that e, in fact, stands for a unique domain element, because

(21) $prob(\exists^{=1} w(e = w)) = 1$

is a tautology in $\mathcal{L}_p$ (here $\exists^{=1}$ is an abbreviation for 'there exists exactly one'). To see that (21) is indeed valid, it only must be realized that the interpretation of the formula $\exists^{=1} w(v = w)$ is always M, and so must be assigned probability 1 by Q_e.

Now let $\phi(w)$ be a formula. Then

(22) $\forall w(\phi(w) \vee \neg\phi(w))$

is a tautology. It might now appear as though from (21) and (22) one should be able to infer

(23) $\phi(e) \vee \neg\phi(e),$

and hence

(24) $prob(\phi(e)) = 0 \vee prob(\phi(e)) = 1.$

This would mean that reasoning with subjective probabilities reduces to trivial 0–1 valued probability assignments that simply mirror truth value assignments. This is not the case, however, because (23) is an expression that is not allowed by the syntax of $\mathcal{L}_p$, and hence cannot be used for deriving (24). This changes if we introduce a standard constant symbol e as an alternative name for e via the axiom

(25) $prob(e = e) = 1.$

Since $\forall w(w = e \rightarrow (\phi(w) \leftrightarrow \phi(e)))$ is a tautology, we have

(26) $prob(e = e \rightarrow (\phi(e) \leftrightarrow \phi(e))) = 1$

and (24) becomes an immediate consequence of (25) and (26).

[34]

We thus see that $\mathscr{L}_p$ in this way supports two views on single case probabilities: as long as individual events are only represented by event symbols, the probabilities of their properties can be identified with frequencies obtained by repeated sampling according to Q_e, which means that they are only constrained by the conditions of a coherent domain distribution. If the single case nature of e is made explicit by an axiom of the form (25), the logic enforces the view that the probability for a proposition relating to a single case event can only be 0 or 1, according to whether the proposition is true or false. Both these views are shades of frequentist interpretations of single case probabilities: the latter is the strict frequentist view of von Mises (1957), whereas the former is a less dogmatic frequentist perspective in which single case probabilities are admitted as meaningful, but are given an empirical interpretation (Reichenbach 1949, Jaeger 1995b).

Limitations on possible subjective probability assignments can be imposed in $\mathscr{L}_p$ also by restricting the sampling distribution Q_e in less obvious ways than the axiom (25). Consider the sentence

(27) $\exists^{=1} v \mathtt{President}(v) \wedge \mathrm{prob}(\mathtt{President}(e)) = 1$

$\qquad \wedge \forall v(\mathtt{President}(v) \rightarrow (\mathtt{Republican}(v) \leftrightarrow \neg\mathtt{Democrat}(v))).$

The first two conjuncts of this sentence tie the interpretation of e to the one element interpretation of the predicate President in very much the same way as (25) tied it to the one element interpretation of e. As before, we thus obtain that properties of e can only have 0–1 probabilities, and hence (27) is inconsistent with

(28) $\mathrm{prob}(\mathtt{Republican}(e)) = 1/2 \wedge \mathrm{prob}(\mathtt{Democrat}(e)) = 1/2.$

This may seem counterintuitive at first sight, as (27) and (28) seem to express a meaningful subjective probability assessment. On the other hand, however, it also seems natural to demand that for any formula $\phi(x)$ the implication

(29) $\mathrm{prob}(\phi(e)) > 0 \models \exists v \phi(v)$

should be valid, since we should not be able to assign a nonzero probability to e having the impossible property ϕ. If, now, (27) and (28) were jointly consistent, then (29) would be violated in some model with either $\phi(v) = \mathtt{President}(v) \wedge \mathtt{Democrat}(v)$, or

[35]

$\phi(v) = \text{President}(v) \wedge \text{Republican}(v)$. Thus, the minimal consistency requirement between domain knowledge and subjective probability assessment as expressed by (29) already forces the joint inconsistency of (27) and (28).

A somewhat more careful modeling resolves the apparent conflict: by introducing a time parameter into our representation, we can make the more accurate statement that there only exists a single president at any given point in time, and that e refers to the next president

$$\forall t \ \text{Time}(t) \rightarrow \exists^{=1} v \text{President}(v, t)$$

(30) $\wedge \text{prob}(\text{President}(e, \ \text{next})) = 1.$

Here 'next' must be another event, not a constant symbol. Now (28) is consistent with our premises since $Q_{e,\text{next}}$ can be any distribution that samples presidents at different points in time.

2.5. Sufficient Structures

So far, the only type of sufficient probabilistic structures we have encountered are the real-discrete structures of Example 2.7. For many interesting theories one can find models that belong to this class. For instance, all our example sentences (1),(3), etc. have real-discrete models. This is not always the case, though. Consider the sentence

$$\phi^{\text{cont}} :\equiv \forall v [v = w]_w = 0,$$

which explicitly states that no single element carries a positive probability mass. Clearly ϕ^{cont} does not have a real-discrete model. Probabilistic structures that do satisfy ϕ^{cont} we call *continuous structures*. Do sufficient continuous structures exist? The answer is yes. An explicit construction of sufficient continuous structures for the special case that S only contains unary relation symbols is given in Jaeger (1995a). For more expressive vocabularies it becomes extremely difficult to verify sufficiency in an explicit construction. In particular, as the following theorem shows, we cannot follow the example of real-discrete structures, and try to obtain sufficiency simply by making every set measurable.

THEOREM 2.8. *There does not exist a sufficient continuous probability structure* $\mathfrak{M}$ *with* $\mathfrak{A}_n = 2^{M^n}$ *for all n.*

[36]

Proof. We show the stronger result that we cannot even construct the first two elements $(2^M, P_1), (2^{M^2}, P_2)$ of a sequence $(2^{M^n}, P_n)_{n \in \mathbb{N}}$ such that the coherency conditions hold for these two measure algebras.

For this let M be a set, P_1 a continuous probability measure on 2^M, P_2 a permutation invariant probability measure on 2^{M^2} such that P_1 and P_2 satisfy the product property. We show that there exists an $A \subseteq M^2$ with $P_1(\sigma_a^1(A)) = 0$ for all $a \in M$, and $P_2(A) > 0$, thus providing a counterexample to the Fubini property.

Let λ be the cardinality of M. Let Γ be the set of ordinals $\kappa \leqslant \lambda$ that have the following property: there exists a sequence of pairwise disjoint subsets $\{E_\nu \subset M | \nu \text{ ordinal}, \nu < \kappa\}$ with

(31) $\forall \nu < \kappa : P_1(E_\nu) = 0$ and $P_1(\cup_{\nu < \kappa} E_\nu) > 0.$

Γ is non-empty, because $\lambda \in \Gamma$.

Let ρ be the minimal element in Γ; let $\{E_\nu | \nu < \rho\}$ be a sequence for ρ with (31). For each ordinal $\nu < \rho$ let

$$\tilde{E}_\nu := \cup_{\theta < \nu} E_\theta.$$

By the minimality of ρ in Γ, we have $P_1(\tilde{E}_\nu) = 0$ for all $\nu < \rho$. Now define

$$A_0 := \cup_{\nu < \rho} (E_\nu \times \tilde{E}_\nu),$$
$$A_1 := \cup_{\nu < \rho} (E_\nu \times E_\nu),$$
$$B := \cup_{\nu < \rho} E_\nu.$$

Let $a \in M$ be arbitrary. If $a \notin B$, then $\sigma_a^1(A_0) = \sigma_a^1(A_1) = \emptyset$. For $a \in B$ there exists exactly one $\nu < \rho$ with $a \in E_\nu$, so that $\sigma_a^1(A_0) = \tilde{E}_\nu$ and $\sigma_a^1(A_1) = E_\nu$. Thus, for all $a \in M$, $P_1(\sigma_a^1(A_0)) = P_1(\sigma_a^1(A_1)) = 0$.

Now consider any $(a, b) \in B \times B$ where $a \in E_\nu, b \in E'_\nu$. If $\nu > \nu'$ then $(a, b) \in A_0$. For $\nu = \nu'$ we have $(a, b) \in A_1$, and if $\nu < \nu'$, then (a, b) belongs to the permutation $\pi A_0 := \cup_{\nu < \rho} (\tilde{E}_\nu \times E_\nu)$ of A_0. Thus,

$$B \times B = A_0 \cup \pi A_0 \cup A_1.$$

Since $r := P_1(B) > 0$, and therefore $P_2(B \times B) = r^2 > 0$, by the permutation invariance of P_2, it follows that $P_2(A_0) > 0$, or $P_2(A_1) > 0$. Hence, at least one of A_0 and A_1 violates the Fubini property.   □

[37]

2.6. *Reduction to First-Order Logic*

The previous section has highlighted the difficulties in the model theory of $\mathscr{L}_p$. In this section, we provide results that, on the other hand, provide powerful tools for the analysis of $\mathscr{L}_p$. These tools are obtained by showing that $\mathscr{L}_p$ can be reduced to standard first-order logic. This reduction is based on the observation that a statistical probability term $[\phi(v, w, x)]_w$ maps tuples $(a, r) \in M^{|v|} \times \mathbb{F}^{|x|}$ to elements $s \in \mathbb{F}$, and thus behaves essentially like a standard function term $f(v, x)$ over a domain $M \cup \mathbb{F}$. A similar observation applies to subjective probability terms. To reduce $\mathscr{L}_p$ to first-order logic, one can define a translation from $\mathscr{L}_p$ into the language $L_I(S^*)$ of first-order logic over an expanded (infinite) vocabulary $S^* \supset S$. In this translation, probability terms are inductively replaced by standard function terms using new function symbols. This syntactic translation is complemented by a transformation between sufficient probabilistic structures and standard first-order structures. Finally, the class of standard first-order structures that correspond to sufficient probabilistic structures under such a transformation can be axiomatized by a first-order theory AX. We then obtain the following result.

THEOREM 2.9. Let S be a vocabulary. There exist

- a vocabulary $S^* \supset S$,
- a recursively enumerable set of axioms $\mathrm{AX} \subset L_I(S^*)$,
- computable mappings

$$t: L_p(S) \rightarrow L_I(S^*),$$
$$t^{-1}: t(L_p(S)) \rightarrow L_p(S),$$

  such that $t^{-1}(t(\phi)) = \phi$,
- transformations

$$T: \mathfrak{M} \mapsto \mathfrak{M}^* \quad (\mathfrak{M} \text{ a sufficient probabilistic } S\text{-structure,}$$
$$\mathfrak{M}^* \text{ a } S^*\text{-structure with } \mathfrak{M}^* \models AX),$$
$$T^{-1}: \mathfrak{N} \rightarrow \mathfrak{N}^{-1} \quad (\mathfrak{N} \text{ a } S^*\text{-structure with } \mathfrak{N} \models AX,$$
$$\mathfrak{N}^{-1} \text{ a sufficient probabilistic } S\text{-structure}),$$

  such that $T^{-1}(T(\mathfrak{M})) = \mathfrak{M}$,

so that for all $\phi \in L_p(S)$, all sufficient probabilistic S-structures $\mathfrak{M}$, and all S^*-structures $\mathfrak{N} \models AX$:

(32) $\mathfrak{M} \models \phi$ iff $T(\mathfrak{M}) \models t(\phi)$ and $\mathfrak{N} \models t(\phi)$ iff $T^{-1}(\mathfrak{N}) \models \phi$.

For the detailed proof of this theorem the reader is referred to Jaeger (1995a). We obtain several useful corollaries. The first one reduces semantic implication in $\mathscr{L}_p$ to first-order entailment.

COROLLARY 2.10. *For all* $\Phi \cup \{\phi\} \subseteq L_p(S)$:

$$\Phi \models \phi \quad iff \ t(\Phi) \cup AX \models t(\phi).$$

Using this corollary, one can easily transfer compactness of first-order logic to $\mathscr{L}_p$.

COROLLARY 2.11. *$\mathscr{L}_p$ is compact.*

As an application of compactness consider the L_p-theory

$$\Phi := \{\delta_n \mid n \in \mathbb{N}\} \cup \exists x > 0 \forall v[v = w]_w = x,$$

where δ_n is a standard first-order sentence that says that the domain contains at least n elements. A model of Φ thus is an infinite structure in which every singleton has the same positive probability mass. Since every finite subset of Φ is satisfiable (by a finite domain real-discrete structure), we know by Corollary 2.11 that Φ is satisfiable. However, Φ is clearly not satisfiable by a structure with real-valued probabilities: the probability of the singletons in a model of Φ must be some infinitesimal. Thus, Φ also provides an example of what we lose in terms of semantic strength by allowing probabilities to be lrc-field-valued, not necessarily real-valued, and shows that Corollary 2.11 cannot hold when we limit ourselves to real-valued probability structures.

Finally, we obtain as a corollary to Theorem 2.9 a completeness result.

THEOREM 2.12. *There exists a sound and complete proof system for $\mathscr{L}_p$.*

[39]

Again, this corollary is in marked contrast to what one obtains when probabilities are required to be real-valued, in which case no complete proof system can exist (Abadi and Halpern 1994).

3. THE LOGIC OF INDUCTIVE INFERENCE

3.1. *Inductive Reasoning by Cross-Entropy Minimization*

The statistical knowledge expressed in our example sentences (1),(4) and (6) can be expressed by the L_p-sentences

(33) $\phi_1 := \equiv [\text{IIA}(d)|\text{D}(d) \wedge 10 \preceq \text{am}(d) \preceq 20]_d = 0.027,$

(34) $\phi_4 := \equiv [\text{IIA}(d)|\text{D}(d) \wedge 15 \preceq \text{am}(d) \preceq 25]_d = 0.031,$

(35) $\phi_6 := \equiv [\text{IIA}(d)|\text{D}(d) \wedge 15 \preceq \text{am}(d) \preceq 20]_d \in [0.027, 0.031].$

The belief about Jones expressed in (2) can be expressed by

(36) $\phi_2 := \equiv \text{prob}(\text{D}(jones) \wedge 10 \preceq \text{am}(jones) \preceq 20) = 1.$

As discussed in Section 1, it seems reasonable to infer from $\phi_1 \wedge \phi_2$

(37) $\phi_3 := \equiv \text{prob}(\text{IIA}(jones)) = 0.027.$

However, this inference is not valid in $\mathscr{L}_p$, i.e.,

$$\phi_1 \wedge \phi_2 \not\models \phi_3.$$

This is because in a probabilistic structure the statistical and subjective probability terms are interpreted by the measures P_1 and Q_{jones}, respectively, and the constraint ϕ_1 on admissible statistical measures does not constrain the possible choices for Q_{jones}. Moreover, it would clearly not be desirable to have that $\phi_1 \wedge \phi_2$ strictly implies ϕ_3, because then $\phi_1 \wedge \phi_2$ would be inconsistent with $\text{prob}(\neg\text{IIA}(jones)) = 1$, i.e., the knowledge that Jones will, in fact, not be involved in an accident. Hence, if we wish to infer ϕ_3 from $\phi_1 \wedge \phi_2$, this can only have the character of a *non-monotonic*, or *defeasible*, inference, which may become invalid when additional information becomes available. Our aim, then, will be to augment the logic $\mathscr{L}_p$ with an additional non-monotonic entailment relation $\models\!\!\sim$ for which

$$\phi_1 \wedge \phi_2 \models\!\!\sim \phi_3, \quad \text{but} \quad \phi_1 \wedge \phi_2 \wedge \text{prob}(\neg\text{IIA}(jones)) = 1 \not\models\!\!\sim \phi_3.$$

[40]

As a second example for the intended inference relation $\approx$ consider the formula

(38) $\phi_{2,5} :\equiv \text{prob}(D(jones) \wedge 15 \preceq \text{am}(jones) \preceq 20) = 1.$

As argued in Section 1, our inductive inference relation then should satisfy

$$\phi_6 \wedge \phi_{2,5} \approx \text{prob}(\text{IIA}(jones)) \in [0.027, 0.031].$$

Adding that these should be the sharpest bounds that $\approx$ allows us to derive for $\text{prob}(\text{IIA}(jones))$, this example illustrates an important aspect of the intended relation $\approx$: it will not be used to make any default assumptions about the statistical distribution in the sense that, for example, we could derive

$$\phi_6 \approx [\text{IIA}(d)|D(d) \wedge 15 \preceq \text{am}(d) \preceq 20]_d = 0.029$$

(i.e., assuming without further information that the correct statistical probability is given by the center point of the admissible interval, or else, maybe, by 0.031 as the value closest to 0.5). Only inferring the bounds [0.027, 0.031] for $\text{prob}(\text{IIA}(jones))$ means that we take every admissible statistical distribution into consideration, and apply the inductive inference relation $\approx$ to the subjective distribution alone with respect to each of the statistical possibilities.

As an example where the given information on Jones is not deterministic consider the sentence

(39) $\phi_{39} :\equiv \text{prob}(D(jones) \wedge 10 \preceq \text{am}(jones) \preceq 15) = 0.4$
$\wedge \text{prob}(D(jones) \wedge 15 \preceq \text{am}(jones) \preceq 20) = 0.6.$

Here Jeffrey's rule is applicable, because the two constraints in (39) are on disjoint subsets. Jeffrey's rule, now, leads to the inductive inference

(40) $\phi_{39} \approx \text{prob}(\text{IIA}(jones)) = 0.4[\text{IIA}(d)|D(d) \wedge 10 \preceq \text{am}(d)15]_d$
$+ 0.6[\text{IIA}(d)|D(d) \wedge 15 \preceq \text{am}(d) \preceq 20]_d.$

As the statistical information $\phi_1 \wedge \phi_6$ implies the bounds [0, 0.027] and [0.027, 0.031] for the two conditional probabilities on the right-hand side of (40), we obtain

(41) $\phi_1 \wedge \phi_6 \wedge \phi_{39} \approx \text{prob}(\text{IIA}(jones)) \in [0.6 \cdot 0.027, 0.4 \cdot 0.027$
$+ 0.6 \cdot 0.031]$
$= [0.0162, 0.0294].$

While the step from direct inference to Jeffrey's rule is very easy, the step to the general case where subjective probability constraints can be on arbitrary, non-disjoint, sets is rather non-trivial. The guiding principle both in direct inference and Jeffrey's rule can be seen as the attempt to make the subjective probability distribution as similar as possible to the statistical distribution. To follow this principle in general requires to be able to measure the similarity, or distance, between probability distributions. A very prominent distance measure for probability distributions is *cross-entropy*: if $P = (p_1, \ldots, p_n)$ and $Q = (q_1, \ldots, q_n)$ are two probability measures on an n-element probability space, and $p_i = 0$ implies $q_i = 0$ for $i = 1, \ldots, n$ (i.e., Q is *absolutely continuous* with respect to Q, written $Q \ll P$), then the cross-entropy of Q with respect to P is defined by

$$(42) \quad CE(Q, P) := \sum_{\substack{i=1 \\ p_i > 0}}^{n} q_i \operatorname{Log} \frac{q_i}{p_i}.$$

Given a measure $P \in \Delta\mathfrak{A}$ with $\mathfrak{A}$ a finite algebra, and a subset $J \subseteq \Delta\mathfrak{A}$, we can define the $CE-projection$ of P onto J

$$(43) \quad \Pi_J(P) := \{Q \in J \mid Q \ll P, \forall Q' \in J : CE(Q', P) \geqslant CE(Q, P)\}.$$

The set $\Pi_J(P)$ can be empty (either because J does not contain any Q with $Q \ll P$, or because the infimum of $\{CE(Q', P) \mid Q' \in J\}$ is not attained by any $Q \in J$), can be a singleton, or contain more than one element.

To use CE in modeling inductive probabilistic reasoning, we identify the distributions P and Q in (42) with the statistical and subjective probability distributions, respectively. We can then formalize the process of inductive probabilistic reasoning as follows: if K is the set of statistical measures consistent with our knowledge, J is the set of subjective measures consistent with our already formed, partial beliefs, then we will sharpen our partial beliefs by going from J to

$$\Pi_J(K) := \cup\{\Pi_J(P) \mid P \in K\} \subseteq J,$$

i.e., by discarding all subjective distributions that are not as close as possible to at least one feasible statistical distribution.

Is this an adequate formalization of inductive probabilistic reasoning? Clearly, this question, being non-mathematical in nature,

[42]

does not admit of an affirmative answer in the form of a strict correctness proof. However, it is arguable that, short of such a proof, the justification for using cross-entropy minimization is as strong as it possibly can be.

A first justification consists in the observation that cross-entropy minimization does indeed generalize Jeffrey's rule: if J is defined by prescribing values for the elements of a partition, then $\Pi_J(P)$ is obtained by applying Jeffrey's rule to P and these values. This property, however, is not unique to cross-entropy minimization (Diaconis and Zabell 1982). Justifications that identify cross-entropy minimization as the unique method satisfying certain desirable properties can be brought forward along two distinct lines: the first type of argument consists of formal conditions on the input/output relation defined by a method, and a proof that cross-entropy minimization is the only rule that will satisfy these conditions. This approach underlies the well-known works both by Shore and Johnson (1980, 1983) and of Paris and Vencovská (1990, 1992). A typical condition that will be postulated in derivations of this type can be phrased in terms of inductive inference in L_p as follows: if the input consists of separate constraints on two event variables, e.g.,

(44) $\mathrm{prob}(10 \preceq \mathrm{am}(jones) \preceq 15) \leqslant 0.7 \wedge \mathrm{prob}(\mathrm{IIA}(mitchell)) \leqslant 0.1,$

then the output, i.e., the selected joint subjective distribution for Jones and Mitchell, should make the two variables independent, and therefore satisfy e.g.,

(45) $\mathrm{prob}(\mathrm{IIA}(jones) \wedge 10 \preceq \mathrm{am}(mitchell))$
$= \mathrm{prob}(\mathrm{IIA}(jones)) \cdot \mathrm{prob}(10 \preceq \mathrm{am}(mitchell)).$

Abstracting from such particular examples, this independence principle becomes a general property of the inductive entailment operator $\mathrel{\vicon}$, which can be formally stated as in Theorem 3.8 (and which corresponds to the *system independence* property of (Shore and Johnson 1980), respectively, the principle of independence of Paris (1994)). A second condition, or desideratum, for an inductive inference rule is the conditional reasoning property, expressed in Theorem 3.9 (which is closely related to the *subset independence* property of Shore and Johnson (1980)). Variants of these two properties form the core of axiomatic derivations of CE-minimization as the formal rule for inductive probabilistic inference.

A second type of justification for the minimum CE-principle has been developed in Jaeger (1995a,b). This justification follows the tradition of frequentist interpretations for single case probabilities as predicted frequencies in a sequence of trials (Reichenbach 1949, Section 72; Carnap 1950, p. 189ff).

Since single case probabilities often cannot be associated with observable frequencies in actual, repeated, physical experiments, such trials may only take an imaginary form, i.e., be carried out as a thought experiment (Jaeger 1995b). For example, to assess the probability that the driver of the car, the wreckage of which we have just seen at the roadside, has survived the crash, we may mentally reenact the accident several times, and take a mental count of how often the driver comes away alive. We now make two assumptions about how the thought experiment is performed. The first assumption is that the sampling in the thought experiment is according to our statistical knowledge of the domain. If, for example, we happen to know exact statistics on the average speed of vehicles on this road, the prevalence of seat-belt use, the frequency of drunk driving, etc., then our mental sampling will be in accordance with these known statistics. The second assumption is that already existing constraints on the subjective probability being assessed are used to condition the statistical distribution over possible samples on frequencies consistent with these constraints. If, for example, we happen to believe that with probability at least 0.7 the driver in the accident was drunk (this being well above the statistical probability of drunk driving), then we condition the distribution over possible samples of repeated accidents on the event of containing at least 70% incidences of drunk driving. More loosely speaking, we perform the mental sampling according to the underlying statistical distribution, but bias the result so as to contain at least 70% drunk drivers.

This semi-formal thought experiment model can be translated into a precise statistical model, and it can then be proven that according to this model the predicted frequencies must be exactly those that are obtained by CE-minimization (Jaeger 1995b).

As an example for a result obtained by CE-minimization in a situation where Jeffrey's rule no longer applies, consider the sentence

(46) $\phi_{46} :\equiv \mathrm{prob}(10 \preceq \mathrm{am}(jones) \preceq 20) = 0.5$
        $\wedge \mathrm{prob}(15 \preceq \mathrm{am}(jones) \preceq 25) = 0.7.$

This sentence imposes probability constraints on the two non-disjoint sets defined by $10 \preceq \text{am}(v) \preceq 20$ and $15 \preceq \text{am}(v) \preceq 25$. As usual, we want to derive a probability estimate for IIA(*jones*). It is another distinctive feature of *CE*-minimization that this estimate can be derived in two steps as follows: in the first step probability estimates for Jones belonging to the elements of the partition generated by the sets appearing in (46) are computed (by *CE*-minimization). In the second step the probability assignments found for the partition are extended to other sets by Jeffrey's rule, which now is applicable. For example ϕ_{46} the relevant partition consists of four different sets of possible annual mileages, for which we might have the following statistical information:

(47) $[10 \preceq \text{am}(d) \preceq 15]_d = 0.4,$

(48) $[15 \preceq \text{am}(d) \preceq 20]_d = 0.3,$

(49) $[20 \prec \text{am}(d) \preceq 25]_d = 0.1,$

(50) $[\text{am}(d) \prec 10 \vee 25 \prec \text{am}(d)]_d = 0.2.$

To obtain the probability estimates for Jones's membership in the elements of the partition, we have to compute the distribution $Q = (q_1, q_2, q_3, q_4)$ that minimizes $CE(\cdot, P)$ with respect to $P = (0.4, 0.3, 0.1, 0.2)$ under the constraints $q_1 + q_2 = 0.5$ and $q_2 + q_3 = 0.7$. This computation is a non-linear optimization problem, and yields the (approximate) solution

(51) $Q = (0.128 \ldots, 0.37 \ldots, 0.329 \ldots, 0.171 \ldots),$

meaning that in the first step we have made, for example, the inductive inference

(52) $\text{prob}(10 \preceq \text{am}(jones) \preceq 15) \in (0.128, 0.129).$

Given the probabilities for the four disjoint reference classes we can now apply Jeffrey's rule, and obtain bounds for $\text{prob}(\text{IIA}(jones))$ in the same way as (41) was derived from (39) and the relevant statistical information.

3.2. *Preferred Models*

Having identified cross-entropy minimization as the formal rule we want to employ for inductive reasoning, we want to use it as the basis for inductive entailment $\approx$ in $\mathscr{L}_p$.

[45]

Our plan is to implement CE-minimization by developing a *preferred model semantics* (Shoham 1987) for L_p: for a given L_p-sentence ϕ we will single out from the set of all models of ϕ a subset of preferred models. A model $\mathfrak{M} = (M, \ldots, (\mathfrak{A}_n, P_n)_n, Q_\mathbf{e})$ is going to be a preferred model if the subjective probability measure $Q_\mathbf{e}$ minimizes cross-entropy with respect to the measure $P_{|\mathbf{e}|}$ that describes the statistical distribution of a random sample of $|\mathbf{e}|$ domain elements. An inductive entailment relation $\phi \mathrel{\vert\!\approx} \psi$ then holds if ψ is true in all preferred models of ϕ.

Several difficulties arise when we put this plan into practice, because we have defined cross-entropy by (42) only for real-valued measures on finite algebras. As we are now dealing with lrc-field valued measures on infinite algebras, the concepts of cross-entropy and CE-minimization have to be generalized. Furthermore, we have to ascertain that this generalization retains those essential properties of cross-entropy in $\mathbb{R}$ on which the justification of the minimum CE-principle is based. For instance, we will have to check that the generalized minimum CE-principle still has the independence property, so that the inductive inference of (45) from (44) remains valid with our lrc-field based semantics.

We tackle this complex of questions in two stages: first we define cross-entropy for lrc-field valued measures on finite spaces, and prove that here generalized cross-entropy exhibits the same essential properties as cross-entropy on the reals. In a second step we show that for our purpose it is already sufficient to define cross-entropy on finite algebras, because a suitable notion of CE-minimization for measures on the infinite algebra $\mathfrak{A}_{|\mathbf{e}|}$ can be obtained by "lifting" cross-entropy minimal measures from finite subalgebras of $\mathfrak{A}_{|\mathbf{e}|}$ to $\mathfrak{A}_{|\mathbf{e}|}$.

To begin, we have to define cross-entropy and CE-projections for lrc-field valued measures on finite algebras. This, however, is immediate, and is done by (42) and (43) just as for real-valued measures simply by interpreting the function *Log* now as an arbitrary logarithmic function in an lrc-field.

This leads us to the question of what properties of cross-entropy in the reals carry over to the generalized CE function. We give a fairly comprehensive answer to this question in Appendix A: first we show that CE-projections in lrc-fields retain the key structural properties of CE-projections in the reals, namely those properties on which Shore and Johnson (1980) base their derivation of the minimum CE-principle. From these results it follows, for example, that the inductive inference from (44) to (45) also is warranted on the basis of lrc-field valued probabilities. Second, it is shown in Appendix A that generalized *CE*-minimization also behaves numerically essentially as *CE*-minimization in the reals. This means, for example, that the numerical result (52) also is obtained with lrc-field valued probabilities. In summary, the results developed in Appendix A constitute a collection of far-reaching completeness results that show that for finite algebras we retain for *CE*-minimization in lrc-fields most of the salient features of *CE*-minimization for real-valued measures. In some of the proofs of theorems in this section references are made to results of Appendix A. It should be noted that all these references are to facts that are long established for real-valued probabilities, and therefore are inessential as long as one follows the main development thinking of real-valued probabilities alone.

It remains to find a suitable notion of *CE*-minimization for measures defined on $\mathfrak{A}_{|\mathbf{e}|}$ by a reduction to *CE*-minimization on finite algebras. Although the following construction contains some technicalities, the underlying idea is extremely simple, and consists essentially of the same two-step procedure used in the example (46)–(52) of the preceding section. To be able to carry out the first step of that procedure, it is necessary that the given constraints on the subjective distribution only refer to finitely many sets, which will generate a finite partition on which we know how to conduct *CE*-minimization. In the following we give a precise semantic definition for what it means that constraints only refer to finitely many sets. Later (Lemma 3.6) we will see that constraints expressible in $\mathscr{L}_p$ are guaranteed to have this semantic property.

DEFINITION 3.1. *Let* $\mathfrak{A}$ *be an algebra over M. Let* $J \subseteq \Delta_{\mathbb{F}}\mathfrak{A}$, *and* $\mathfrak{A}'$ *a finite subalgebra of* $\mathfrak{A}$. *Let* $J \upharpoonright \mathfrak{A}' := \{P \upharpoonright \mathfrak{A}' | P \in J\}$. *We say that* J *is defined by constraints on* $\mathfrak{A}'$, *iff*

$$\forall P \in \Delta_{\mathbb{F}}\mathfrak{A}: \quad P \in J \quad \text{iff} \quad P \upharpoonright \mathfrak{A}' \in J \upharpoonright \mathfrak{A}'.$$

Given a set $J \subseteq \Delta_{\mathbb{F}}\mathfrak{A}$ defined by constraints on some finite $\mathfrak{A}' \subseteq \mathfrak{A}$, we can apply the two-step process of first computing $\Pi_{J \upharpoonright \mathfrak{A}'}(P \upharpoonright \mathfrak{A}')$, and then extend the result to $\mathfrak{A}$ by Jeffrey's rule as formally described in the following definition.

DEFINITION 3.2. *Let* $\mathfrak{A}$ *be an algebra,* $P \in \Delta_{\mathbb{F}}\mathfrak{A}$. *Let* $\mathfrak{A}' \subseteq \mathfrak{A}$ *a finite subalgebra with atoms* $\{A_1, \ldots, A_L\}$, *and* $Q \in \Delta_{\mathbb{F}}\mathfrak{A}'$ *such that*

$Q \ll P \restriction \mathfrak{A}'$. Let P^h be the conditional distribution of P on $A_h (h = 1, \ldots, L; P(A_h) > 0)$. The extension Q^* of Q to $\mathfrak{A}$ defined by

$$Q^* := \sum_{\substack{h=1 \\ P(A_h)>0}}^{L} Q(A_h) P^h$$

is called the Jeffrey-extension of Q to $\mathfrak{A}$ by P, denoted by $\mathscr{J}(Q, P, \mathfrak{A})$.

The following lemma says that if J is defined by constraints on $\mathfrak{A}'$, then Jeffrey extensions realize cross-entropy minimization on all finite algebras that refine $\mathfrak{A}'$.

LEMMA 3.3. *Let $\mathfrak{A}$ be an algebra, $P \in \Delta_{\mathbb{F}} \mathfrak{A}$. Let $J \subseteq \Delta_{\mathbb{F}} \mathfrak{A}$ be defined by constraints on a finite subalgebra $\mathfrak{A}' \subseteq \mathfrak{A}$. Then for all finite $\mathfrak{A}'' \supseteq \mathfrak{A}'$:*

(53) $\Pi_{J \restriction \mathfrak{A}''}(P \restriction \mathfrak{A}'') = \{Q \restriction \mathfrak{A}'' \mid Q = \mathscr{J}(Q', P, \mathfrak{A}), Q' \in \pi_{J \restriction \mathfrak{A}'}(P \restriction \mathfrak{A}')\}.$

Conversely, for $Q \in \Delta_{\mathbb{F}} \mathfrak{A}$, if

(54) $Q \restriction \mathfrak{A}'' \in \Pi_{J \restriction \mathfrak{A}''}(P \restriction \mathfrak{A}'')$

for all finite $\mathfrak{A}'' \supseteq \mathfrak{A}'$, then $Q = \mathscr{J}(Q \restriction \mathfrak{A}', P, \mathfrak{A})$.

Proof. Let $\{A_1, \ldots, A_p\}$ be the set of atoms of $\mathfrak{A}''$. Let $Q'' \in \Delta_{\mathbb{F}} \mathfrak{A}'', Q'' \ll P \restriction \mathfrak{A}''$. By Lemma A.2 then

$$CE(Q'', P \restriction \mathfrak{A}'') \geqslant CE(Q'' \restriction \mathfrak{A}', P \restriction \mathfrak{A}')$$

with equality iff

(55) $(Q'')^h = (P \restriction \mathfrak{A}'')^h, \quad h = 1, \ldots, p,$

where $(\cdot)^h$ is the conditional distribution on A_h. Equivalent to (55) is

$$Q'' = \mathscr{J}(Q'' \restriction \mathfrak{A}', P \restriction \mathfrak{A}'', \mathfrak{A}'').$$

Since J is defined by constraints on $\mathfrak{A}'$, we have for all $Q' \in J \restriction \mathfrak{A}'$ that

$$\mathscr{J}(Q', P \restriction \mathfrak{A}'', \mathfrak{A}'') \in J \restriction \mathfrak{A}''$$

and therefore

(56) $\Pi_{J \restriction \mathfrak{A}''}(P \restriction \mathfrak{A}'') = \{\mathscr{J}(Q', P \restriction \mathfrak{A}'', \mathfrak{A}'') | Q' \in \Pi_{J \restriction \mathfrak{A}'}(P \restriction \mathfrak{A}')\}.$

With

$$\mathscr{J}(Q', P \restriction \mathfrak{A}'', \mathfrak{A}'') = \mathscr{J}(Q', P, \mathfrak{A}) \restriction \mathfrak{A}''$$

this proves (53).

Conversely, assume that (54) holds for Q and all finite $\mathfrak{A}''$. Then, in particular, $Q \restriction \mathfrak{A}' \in \Pi_{J \restriction \mathfrak{A}'}(P \restriction \mathfrak{A}')$, and, again by Lemma A.2,

$$Q \restriction \mathfrak{A}'' = \mathscr{J}(Q \restriction \mathfrak{A}', P \restriction \mathfrak{A}'', \mathfrak{A}'')$$

for all finite $\mathfrak{A}'' \supseteq \mathfrak{A}'$. Thus, also $Q = \mathscr{J}(Q \restriction \mathfrak{A}', P, \mathfrak{A})$.   □

Lemma 3.3 suggests to define for $J \subseteq \Delta\mathfrak{A}$ that is defined by constraints on the finite subalgebra $\mathfrak{A} \subseteq \mathfrak{A}$:

(57) $\Pi_J(P) := \{\mathscr{J}(Q', P, \mathfrak{A}) | Q' \in \Pi_{J \restriction \mathfrak{A}'}(P \restriction \mathfrak{A}')\}.$

However, there is still a slight difficulty to overcome: the algebra $\mathfrak{A}'$ is not uniquely determined, and (57) would be unsatisfactory if it depended on the particular choice of $\mathfrak{A}'$. We therefore show, next, that this is not the case, which is basically due to the fact that there is a unique smallest algebra $\mathfrak{A}'$ by constraints on which J is defined.

LEMMA 3.4. *Let $\mathfrak{A}$ be an algebra, $\mathfrak{A}'$ and $\mathfrak{A}''$ finite subalgebras of $\mathfrak{A}$. Assume that $J \subseteq \Delta\mathfrak{A}$ is defined by constraints on $\mathfrak{A}'$, and also by constraints on $\mathfrak{A}''$. Then J also is defined by constraints on*

$$\mathfrak{A}^\cap := \mathfrak{A}' \cap \mathfrak{A}''.$$

Proof. Let $\mathfrak{A}^\cup$ be the subalgebra of $\mathfrak{A}$ generated by $\mathfrak{A}'$ and $\mathfrak{A}''$. Then J also is defined by constraints on $\mathfrak{A}^\cup$, and it suffices to show that for all $Q \in \Delta\mathfrak{A}$

(58) $Q \restriction \mathfrak{A}^\cup \in J \restriction \mathfrak{A}^\cup \Leftrightarrow Q \restriction \mathfrak{A}^\cap \in J \restriction \mathfrak{A}^\cap.$

To obtain a more economical notation, we may therefore work within a completely finitary context, and assume that $\mathfrak{A} = \mathfrak{A}^\cup$ and $J \subseteq \Delta_{\mathbb{F}}\mathfrak{A}^\cup$.

With $\{A_i' = 1, \ldots, p\}$ the atoms of $\mathfrak{A}'$, and $\{A_j'' | j = 1, \ldots, q\}$ the atoms of $\mathfrak{A}''$, atoms of $\mathfrak{A}^\cup$ are the non-empty intersections

$$B_{ij} := A_i' \cap A_j'', \qquad i = 1, \ldots, p, \quad j = 1, \ldots, q.$$

Elements of $\mathfrak{A}^\cap$ are just the unions of atoms of $\mathfrak{A}'$ that simultaneously can be represented as a union of atoms of $\mathfrak{A}'$, i.e.,

$$A = \bigcup_{i \in I} A'_i \in \mathfrak{A}'$$

with $I \subseteq \{1, \ldots, p\}$ belongs to $\mathfrak{A}^\cap$ iff there exists $K \subseteq \{1, \ldots, q\}$, such that also

$$A = \bigcup_{k \in K} A''_k.$$

Now assume that there exist $Q, Q' \in \Delta \mathfrak{A}^\cup$ with

(59) $Q \upharpoonright \mathfrak{A}^\cap = Q' \upharpoonright \mathfrak{A}^\cap$

and $Q \in J$, but $Q' \notin J$. Furthermore, assume that Q, Q' are minimal with these properties in the sense that the number of atoms of $\mathfrak{A}^\cup$ to which Q and Q' assign different probabilities is minimal.

From $Q \neq Q'$ and (59) it follows that there exists an atom C of $\mathfrak{A}^\cap$, and atoms $B_{hk}, B_{h'k'} \subset C$ of $\mathfrak{A}^\cup$, such that

$$Q(B_{hk}) = Q'(B_{hk}) + r,$$
$$Q(B_{h'k'}) = Q'(B_{h'k'}) - s$$

for some $r, s > 0$. Assume that $r \leqslant s$ (the argument for the case $s < r$ proceeds similarly). We show that there exists a sequence

(60) $(i_0, j_0), (i_1, j_1), \ldots, (i_n, j_n)$

in $\{1, \ldots, p\} \times \{1, \ldots, q\}$ such that

(61) $(i_0, j_0) = (h, k), \quad (i_n, j_n) = (h', k')$

and for all $h = 1, \ldots, n$:

(62) $i_h = i_{h-1}$ or $j_h = j_{h-1}$, and $B_{i_h, j_h} \neq \emptyset$.

Once we have such a sequence, we derive a contradiction to the minimality assumption for Q, Q' as follows: we construct a sequence

$$Q = Q_0, Q_1, \ldots, Q_n$$

[50]

by defining for all atoms B of $\mathfrak{A}^{\cup}$ and for $h = 1, \ldots, n$:

$$Q_h(B) := \begin{cases} Q_{h-1}(B), & B \notin \{B_{i_{h-1} j_{h-1}}, B_{i_h j_h}\}, \\ Q_{h-1}(B) - r, & B = B_{i_{h-1} j_{h-1}}, \\ Q_{h-1}(B) + r, & B = B_{i_h j_h} \end{cases}$$

(i.e., we just "shift" probability mass r from B_{hk} to $B_{h'k'}$ via the $B_{i_h j_h}$). For all $h = 1, \ldots, n$ then $Q_h \in J$, because $Q_0 \in J$, and $Q_h \upharpoonright \mathfrak{A}' = Q_{h-1} \upharpoonright \mathfrak{A}'$ (if $i_h = i_{h-1}$), or $Q_h \upharpoonright \mathfrak{A}'' = Q_{h-1} \upharpoonright \mathfrak{A}''$ (if $j_h = j_{h-1}$). Thus, $Q_n \in J$, $Q_n \upharpoonright \mathfrak{A}^{\cap} = Q' \upharpoonright \mathfrak{A}^{\cap}$, and Q_n agrees with Q' on one atom more than does Q, a contradiction.

It remains to show the existence of the sequence (60). For this we define a relation $(h, k) \to \cdot$ on $\{1, \ldots, p\} \times \{1, \ldots, q\}$ by: $(h, k) \to (i, j)$ iff there exists a sequence (60) with $(i_0, j_0) = (h, k)$ and $(i_n, j_n) = (i, j)$ so that (62) holds. Now consider

$$A := \bigcup_{(i,j):(h,k) \to (i,j)} B_{ij}.$$

As $(h, k) \to (i, j)$ and $B_{i'j} \neq \emptyset$ implies $(h, k) \to (i', j)$ (respectively, $B_{ij'} \neq \emptyset$ implies $(h, k) \to (i, j')$), we obtain

$$A = \bigcup_{i : \exists j (h,k) \to (i,j)} A'_i = \bigcup_{j : \exists i (h,k) \to (i,j)} A''_j,$$

which means that $A \in \mathfrak{A}^{\cap}$ (in fact, $A = C$). From $A \in \mathfrak{A}^{\cap}$, $B_{hk} \subseteq A$, and $B_{h'k'}$ belonging to the same atom of $\mathfrak{A}^{\cap}$ as B_{hk}, it follows that $B_{h'k'} \subseteq A$, i.e., $(h, k) \to (h'k')$.   □

From Lemmas 3.3 and 3.4 it follows that the set $\Pi_J(P)$ defined in (57) does not depend on the choice of $\mathfrak{A}'$: by Lemma 3.4 there exists a unique smallest algebra $\mathfrak{A}^*$ by constraints on which J is defined, and by Lemma 3.3 we have for every $\mathfrak{A}' \supseteq \mathfrak{A}^*$:

$$\{\mathscr{I}(Q', P, \mathfrak{A}) | Q' \in \Pi_{J \upharpoonright \mathfrak{A}'}(P \upharpoonright \mathfrak{A}')\}$$
$$= \{\mathscr{I}(Q^*, P, \mathfrak{A}) | Q^* \in \Pi_{J \upharpoonright \mathfrak{A}^*}(P \upharpoonright \mathfrak{A}^*)\}.$$

DEFINITION 3.5. *Let $\mathfrak{A}$ be an algebra over M, $\mathfrak{A}'$ a finite subalgebra of $\mathfrak{A}$. Let $J \subseteq \Delta_{\mathbb{F}} \mathfrak{A}$ be defined by constraints on $\mathfrak{A}'$, and $P \in \Delta_{\mathbb{F}} \mathfrak{A}$. The set $\Pi_{J \upharpoonright \mathfrak{A}'}(P \upharpoonright \mathfrak{A}')$ is defined by (43). The cross-entropy projection of P onto J then is defined by (57).*

We are now ready to define the preferred model semantics for L_p. Recall that it is our aim to identify those models $\mathfrak{M}$ of a L_p-formula ϕ for which the subjective probability measure Q_e minimizes cross-entropy with respect to the statistical measure $P_{|e|}$, and that this minimization is to be effected only by choosing suitable Q_e for every possible given $P_{|e|}$, not by selecting any preferred $P_{|e|}$.

For a probabilistic structure $\mathfrak{M} = (M, \ldots, \mathfrak{F}, \ldots, Q_e)$ and $Q \in \Delta_{\mathbb{F}}\mathfrak{A}_{|e|}$ we denote by $\mathfrak{M}[Q_e/Q]$ the structure $\mathfrak{M}'$ that is obtained by replacing Q_e with Q. For a sufficient probabilistic structure $\mathfrak{M}$, and an L_p-sentence ϕ we define

(63) $\Delta_{\mathbb{F}}(\phi, \mathfrak{M}) := \{Q \in \Delta_{\mathbb{F}}\mathfrak{A}_{|e|} | M[Q_e/Q] \models \phi\}.$

Thus, $\Delta_{\mathbb{F}}(\phi, \mathfrak{M})$ is the set of subjective probability measures that will turn the non-subjective part $(M, I, \mathfrak{F}, (\mathfrak{A}_n, P_n)_{n \in \mathbb{N}})$ of $\mathfrak{M}$ into a model of ϕ (it is not difficult to show that such a substitution cannot destroy sufficiency).

The following lemma is the main reason for the syntactic restrictions that were imposed on subjective probability terms.

LEMMA 3.6. For all $\mathfrak{M}$ and $\phi: \Delta_{\mathbb{F}}(\phi, \mathfrak{M})$ is defined by constraints on a finite subalgebra $\mathfrak{A}'$ of $\mathfrak{A}_{|e|}$.

Proof. ϕ contains a finite number of subjective probability terms $\text{prob}(\psi_1(e)), \ldots, \text{prob}(\psi_k(e))$. Membership of $Q \in \Delta\mathfrak{A}_{|e|}$ in $\Delta(\phi, \mathfrak{M})$ only depends on the values $Q((\mathfrak{M}, v)(\psi_i(v)))(i = 1, \ldots, k)$. By the condition that the ψ_i do not contain any occurrences of $\text{prob}(\cdot)$, the sets $(\mathfrak{M}, v)(\psi_i(v))$ do not depend on the component Q_e of $\mathfrak{M}$. Let $\mathfrak{A}'$ be the finite subalgebra of $\mathfrak{A}_{|e|}$ generated by the sets $(\mathfrak{M}, v)(\psi_i(v))$. Then $\mathfrak{A}'$ is a finite algebra so that for every $Q \in \Delta\mathfrak{A}_{|e|}$ the validity of $\mathfrak{M}[Q_e/Q] \models \phi$ is determined by the values of Q on $\mathfrak{A}'$.   □

No analogue of Lemma 3.6 would hold if we dropped either the prohibition of nested subjective probability terms, or of free variables in subjective probability terms. Together, Definition 3.5 and Lemma 3.6 permit the following final definition of the inductive entailment relation $\approx$ for $\mathcal{L}_{ip}$.

DEFINITION 3.7. Let $\phi \in L_p(S, e)$, $\mathfrak{M} = (M, \ldots, Q_e)$ a sufficient probabilistic structure for (S, e). $\mathfrak{M}$ is called a *preferred model of ϕ*, written $\mathfrak{M} \mathrel{|\!\approx} \phi$, iff

(64) $Q_{\mathbf{e}} \in \Pi_{\Delta_{\mathbb{F}}(\phi, \mathfrak{M})}(P_{|\mathbf{e}|})$.

For, $\phi, \psi \in L_p(S, \mathbf{e})$ we define: $\phi \mathrel{|\!\approx} \psi$ iff $\mathfrak{M} \models \psi$ for every preferred model $\mathfrak{M}$ of ϕ.

3.3. *Inductive Reasoning in* $\mathscr{L}_{ip}$

Having formally defined our inductive entailment relation $\mathrel{|\!\approx}$, we now investigate some of its logical properties. Our first goal is to verify that the relation $\mathrel{|\!\approx}$ indeed supports the patterns of inductive inference described in Sections 1.1 and 3.1, which motivated the approach we have taken. This is established in the following using the structural properties of CE-projections described in Theorems A.5 (system independence) and A.6 (subset independence).

At the very outset we stipulated that the relation $\mathrel{|\!\approx}$ should implement direct inference, where direct inference is applicable. From Corollary A.7 one immediately obtains that the inductive inference

(65) $[\psi(\boldsymbol{v})]_v > 0 \wedge [\phi(\boldsymbol{v})|\psi(\boldsymbol{v})]_v = r \wedge \mathrm{prob}(\psi[\mathbf{e}]) = 1 \mathrel{|\!\approx} \mathrm{prob}(\phi[\mathbf{e}]) = r$

is valid in $\mathscr{L}_{ip}$ for all formulas ϕ, ψ. Usually, however, our total knowledge does not have the form of the premise of (65): one does not only know that $\psi[\mathbf{e}]$ is true for a single property ψ, but rather that $\psi_1[\mathbf{e}], \ldots, \psi_n[\mathbf{e}]$ are true. Assuming the necessary statistical knowledge as well, our premise then is

(66) $\wedge_{i=1}^n ([\psi_i(\boldsymbol{v})]_v > 0 \wedge [\phi(\boldsymbol{v})|\psi_i(\boldsymbol{v})]_v = r_i \wedge \mathrm{prob}(\psi_i[\mathbf{e}]) = 1)$.

The question of what to inductively infer from this body of knowledge is essentially the problem of the choice of the best *reference class* for direct inference (Pollock 1983; Kyburg 1983). The original prescription by Reichenbach (1949) was to take the smallest reference class for which reliable statistics exist. We cannot follow this principle in $\mathscr{L}_{ip}$, because, first, in our framework we do not have the means to distinguish the reliabilities of two statistical statements $[\phi(\boldsymbol{v})|\psi_i(\boldsymbol{v})]_v = r_i$ and $[\phi(\boldsymbol{v})|\psi_k(\boldsymbol{v})]_v = r_k$, and second, from the logical equivalence of (66) and

(67) $\wedge_{i=1}^n ([\psi_i(\boldsymbol{v})]_v > 0 \wedge [\phi(\boldsymbol{v})|\psi_i(\boldsymbol{v})]_v = r_i) \wedge \mathrm{prob}(\wedge_{i=1}^n \psi_i[\mathbf{e}]) = 1$,

[53]

it follows with (65) that from (66) we will always have to infer

(68) $[\wedge_{i=1}^{n} \psi_i(v)]_v > 0 \rightarrow \text{prob}(\phi[\mathbf{e}]) = [\phi(v) | \wedge_{i=1}^{n} \psi_i(v)]_v.$

Thus, we always base direct inference on the smallest reference class that **e** belongs to, whether or not the statistics for this reference class can be deemed reliable – or even are available. In extreme cases this leads to inferences that may seem overly conservative: consider

$$\phi_1 \equiv [\text{IIA}(d)|\neg\text{Drinks}(d)]_d = 0.01$$
$$\wedge \text{prob}(\neg\text{Drinks}(jones)) = 1,$$
$$\phi_2 \equiv [\text{IIA}(d)|\text{Drives}(\text{Toyota}, d)]_d = 0.01$$
$$\wedge \text{prob}(\text{Drives}(\text{Toyota}, jones)) = 1.$$

Then $\phi_1 \approx \text{prob}(\text{IIA}(jones)) = 0.01$, and $\phi_2 \approx \text{prob}(\text{IIA}(jones)) = 0.01$, but not

(69) $\phi_1 \wedge \phi_2 \approx \text{prob}(\text{IIA}(jones)) = 0.01.$

This is because we will infer

(70) $\phi_1 \wedge \phi_2 \approx \text{prob}(\text{IIA}(jones)) = [\text{IIA}(d)|\neg\text{Drinks}(d)$
$$\wedge \text{Drives}(\text{Toyota}, d)]_d.$$

Going from (70) to (69) amounts to an implicit default inference about statistical probabilities

$$[\text{IIA}(d)| \neg\text{Drinks}(d)]d = 0.01 \wedge [\text{IIA}(d)|\text{Drives}(\text{Toyota}, d)]_d = 0.01$$
$$\approx [\text{IIA}(d)|\neg\text{Drinks}(d) \wedge \text{Drives}(\text{Toyota}, d)]_d = 0.01,$$

which $\mathcal{L}_{ip}$ is not designed to do.

Basing direct inference on the narrowest possible reference class can lead to difficulties when the subject of the direct inference (**e** in our case) is referenced in the definition of the reference class (see e.g., Pollock (1983, Section 6)). In particular, one then might consider the single point reference class {**e**}. and argue that direct inference in $\mathcal{L}_{ip}$ must always identify $\text{prob}(\phi(\mathbf{e}))$ with $[\phi(v)|v = \mathbf{e}]_v$. Since this statistical probability can only assume the values 0 or 1 (according to whether $\phi(\mathbf{e})$ holds), it might therefore appear as though

(71) $\text{prob}(\phi(\mathbf{e})) = 0 \vee \text{prob}(\phi(\mathbf{e})) = 1$

is valid in $\mathcal{L}_{ip}$ with respect to $\approx$-entailment. As in the derivation of (24), however, this argument is based on incorrectly using

$\mathbf{e}$ in the expression $[\phi(v)|v=\mathbf{e}]_v$ like a standard constant symbol. The syntactic condition that $\mathbf{e}$ must always appear within the scope of a prob()-operator prevents the construction of reference classes involving $\mathbf{e}$.

When our knowledge base is of a form that makes Jeffrey's rule applicable, then we derive from Corollary A.7 that $\approx$ coincides with Jeffrey's rule.

Leaving the elementary cases of direct inference and Jeffrey's rule behind, we next consider some logical properties of $\mathscr{L}_{ip}$ that in a more general way reflect the system- and subset-independence properties of CE-projections. First, we use system-independence to derive the general (logical) independence property of $\approx$, an instance of which was illustrated by (44) and (45).

THEOREM 3.8. *Let S be a vocabulary, $\mathbf{e}$ and $\mathbf{f}$ two disjoint tuples of event symbols. Let $\phi_{\mathbf{e}}, \psi_{\mathbf{e}}(v) \in L_p(S, \mathbf{e}), \phi_{\mathbf{f}}, \psi_{\mathbf{f}}(w) \in L_p(S, \mathbf{f})$, with $|v| = |\mathbf{e}|$ and $|w| = |\mathbf{f}|$. Then*

(72) $\phi_{\mathbf{e}} \wedge \phi_{\mathbf{f}} \approx prob(\psi_{\mathbf{e}}[\mathbf{e}] \wedge \psi_{\mathbf{f}}(\mathbf{f})) = prob(\psi_{\mathbf{e}}[\mathbf{e}])prob(\psi_{\mathbf{f}}(\mathbf{f})).$

Proof. Consider a probabilistic structure $\mathfrak{M}$ for $(S, (\mathbf{e}, \mathbf{f}))$. The set $\Delta(\phi_{\mathbf{e}} \wedge \phi_{\mathbf{f}}, \mathfrak{M})$ is defined by constraints on a finite algebra $\mathfrak{A}^{\times} = \mathfrak{A} \times \mathfrak{A}' \subset \mathfrak{A}_{|\mathbf{e}, \mathbf{f}|}$, and its restriction $J^{\times}$ to $\mathfrak{A}^{\times}$ has the form

$$\{Q \in \Delta \mathfrak{A}^{\times} | Q \upharpoonright \mathfrak{A} \in J_{\mathbf{e}}, Q \upharpoonright \mathfrak{A}' \in J_{\mathbf{f}}\}$$

for $J_{\mathbf{e}} \subseteq \Delta \mathfrak{A}, J_{\mathbf{f}} \subseteq \Delta \mathfrak{A}'$. The restriction $P^{\times}$ of the statistical distribution $P_{|\mathbf{e}, \mathbf{f}|}$ to $\mathfrak{A}^{\times}$ is a product measure, so that every

$$Q \in \Pi_{J^{\times}}(P^{\times})$$

also is a product measure on $\mathfrak{A}^{\times}$. The theorem now follows from Theorem A.5, and by observing (using Lemma 3.3) that the Jeffrey-extension $\mathscr{J}(Q, P_{|\mathbf{e}, \mathbf{f}|}, \mathfrak{A}_{|\mathbf{e}, \mathbf{f}|})$ preserves the product property for sets of the form $A \times B$ with $A \in \mathfrak{A}_{|\mathbf{e}|}, B \in \mathfrak{A}_{|\mathbf{f}|}$. $\square$

The next theorem transforms subset-independence (Theorem A.6) into a statement about the coherency of conditional reasoning in $\mathscr{L}_{ip}$.

THEOREM 3.9. Let $\phi_{|\gamma}, \psi_{|\gamma} \in L_p$ only contain subjective probability terms of the form $prob(\phi[\mathbf{e}]|\gamma[\mathbf{e}])$ for some fixed $\gamma \in L_p$. Let

[55]

ϕ, ψ be the sentences obtained from $\phi_{|\gamma}, \psi_{|\gamma}$ by replacing each term $\mathrm{prob}(\phi[\mathbf{e}]|\gamma[\mathbf{e}])$ with the corresponding unconditional term $\mathrm{prob}(\phi[\mathbf{e}])$. Then

$$(73) \qquad \phi_{|\gamma} \wedge \mathrm{prob}(\gamma[\mathbf{e}]) > 0 \mathrel{|\!\approx} \psi_{|\gamma}$$

iff

$$(74) \qquad \phi \wedge \mathrm{prob}(\gamma[\mathbf{e}]) = 1 \mathrel{|\!\approx} \psi.$$

Note that adding the conjunct $\mathrm{prob}(\gamma[\mathbf{e}]) > 0$ to the premise of (73) means that there is no ambiguity in the interpretations of the conditional probability terms in $\phi_{|\gamma}$ and $\psi_{|\gamma}$, so that the theorem holds independent from the conventions adopted for dealing with conditioning events of probability zero. The proof of the theorem is similar to that of the previous one, by first noting that the structure of the set $\Delta(\phi_{|\gamma} \wedge \mathrm{prob}(\gamma[\mathbf{e}]) > 0, \mathfrak{M})$ is a special case of the form described in Theorem A.6, then applying that theorem, and finally observing that the structural property expressed in (A.15) is preserved under Jeffrey extensions.

In Section 1.1 we said that $\mathscr{L}_{ip}$ is not intended to model any inductive inferences about statistical probabilities, based on (even numerous) single case observations. By defining preferred models in terms of the condition (64) on the subjective distribution $Q_{\mathbf{e}}$ for any given statistical distribution $P_{|\mathbf{e}|}$ this goal is essentially realized, but with the following caveat: statistical distributions $P_{|\mathbf{e}|}$ for which $\Pi_{\Delta_{\mathrm{F}}}(\phi, \mathfrak{M})(P_{|\mathbf{e}|})$ is empty are ruled out. This means, in particular, that distributions $P_{|\mathbf{e}|}$ are ruled out for which $\Delta_{\mathrm{F}}(\phi, \mathfrak{M})$ does not contain any $Q_{\mathbf{e}}$ with $Q_{\mathbf{e}} \ll P_{|\mathbf{e}|}$ (cf. (43) and Definition 3.7). In consequence, for example the following is a valid inference pattern in $\mathscr{L}_{ip}$:

$$(75) \qquad \mathrm{prob}(\phi(\mathbf{e})) > 0 \mathrel{|\!\approx} [\phi(v)]_v > 0.$$

While, in principle, this is a default inference about statistical probabilities from subjective probabilities, (75) may still be considered unproblematic even from our conservative point of view, because it just amounts to the reasonable constraint that in preferred models we cannot assign nonzero probabilities to events $\mathbf{e}$ having some statistically impossible property ϕ. Observe that (75) means that for $\mathrel{|\!\approx}$ we obtain a strengthening of (29).

[56]

The set $\Pi_{\Delta_{\mathbb{F}}(\phi, \mathfrak{M})}(P_{|e|})$ can also be empty because the infimum is not attained in CE-minimization. Consider, for example, the sentence

(76) $\phi_{76} = ([\psi(v)]_v = 0.3 \vee [\psi(v)]_v = 0.5) \wedge \mathrm{prob}(\psi(e)) > 0.4.$

For any model $\mathfrak{M}$ of ϕ_{76} with $P_{|e|}((\mathfrak{M}, v)(\psi)) = 0.3$ then $\Pi_{\Delta_{\mathbb{F}}(\phi, \mathfrak{M})}$ $(p_{|e|}) = \emptyset$, because $CE(\cdot, P_{|e|})$ is not minimized over the open interval $[0.4, 1]$ defining $\Delta_{\mathbb{F}}(\psi, \mathfrak{M})$. When $P_{|e|}((\mathfrak{M}, v)(\psi)) = 0.5$, on the other hand, the infimum is attained for $Q \in \Delta_{\mathbb{F}}(\psi, M)$ with $Q((\mathfrak{M}, v)(\psi)) = 0.5$. Thus, ϕ_{76} only has preferred models in which the statistical probability of ψ is 0.5, i.e.,

$$\phi_{76} \approx [\psi(v)]_v = 0.5.$$

Thus, some potentially undesired inferences can occur when constraints on the subjective distribution define non-closed sets $\Delta_{\mathbb{F}}(\phi, \mathfrak{M})$. This is a typical limitation of methods based on minimizing distance measures, and often circumvented by prohibiting non-closed constraint sets. In the very general language L_p it is difficult to enforce closedness of $\Delta_{\mathbb{F}}(\phi, \mathfrak{M})$ by a simple syntactic condition on ϕ. Such a condition, therefore, has not been imposed in the basic definitions. However, in practical modeling with L_p some attention should be paid to the question whether the sets $\Delta_{\mathbb{F}}(\phi, \mathfrak{M})$ will be closed (see also Section 4.2).

3.4. *Axiomatization*

In this section, we obtain a completeness result for the inductive entailment relation $\approx$. The result is derived by showing that for a given L_p-sentence ϕ there exists a recursively enumerable set Min $CE(\phi) \subseteq L_p$ that axiomatizes inductive entailment, i.e.,

(77) $\phi \approx \psi$ iff $\mathrm{MinCE}(\phi) \models \psi,$ $\psi \in L_p.$

By the completeness result for strict inference we then obtain a completeness result for $\approx$. This approach of capturing the preferred models of ϕ by adjoining to ϕ a set of axioms dependent on ϕ is closely related to the *circumscription* framework (McCarthy 1980) in non-monotonic reasoning.

To establish (77) it is sufficient to find a set $\mathrm{MinCE}(\phi)$ that axiomatizes the class of preferred models of ϕ up to elementary equivalence, i.e., to obtain that a probabilistic structure $\mathfrak{M}$ is a model of MinCE(ϕ) iff it is elementarily equivalent to a structure $\mathfrak{M}'$ with $\mathfrak{M}' \approx \phi$ (recall that two structures are called elementarily equivalent iff they satisfy the same sentences). For a structure $\mathfrak{M} = (\dots, (\mathfrak{A}_n, P_n)_{n\in\mathbb{N}}, Q_\mathbf{e})$ to be a preferred model of ϕ, by definition, is equivalent for $\mathfrak{M}$ to satisfy the condition

(78) $Q_\mathbf{e} \in \Pi_{\Delta_\mathbb{F}(\phi,\mathfrak{M})}(P_{|\mathbf{e}|}).$

Elementary equivalence to a preferred model, on the other hand, is guaranteed by the weaker condition

(79) $Q_\mathbf{e} \restriction \mathfrak{A}^* \in \Pi_{\Delta_\mathbb{F}(\phi,\mathfrak{M}) \restriction \mathfrak{A}^*}(P_{|\mathbf{e}|} \restriction \mathfrak{A}^*),$

where $\mathfrak{A}^* \subseteq \mathfrak{A}_{|\mathbf{e}|}$ is the subalgebra consisting of those sets that are definable by an L_p-formula without parameters, i.e., $A \in \mathfrak{A}^*$ iff there exists $\psi(\boldsymbol{v}) \in L_p$ with $A = (\mathfrak{M}, \boldsymbol{v})(\psi)$. That (79) implies elementary equivalence to a preferred model follows from the fact that any two structures $\mathfrak{M}$ and $\mathfrak{M}'$ that differ only with respect to $Q_\mathbf{e}$-values for elements $A \in \mathfrak{A}_{|\mathbf{e}|} \backslash \mathfrak{A}^*$ are elementarily equivalent, and that any structure $\mathfrak{M}$ that satisfies (79) can be modified into a preferred model of ϕ by only changing $Q_\mathbf{e}$-values on $\mathfrak{A}_{|\mathbf{e}|} \backslash \mathfrak{A}^*$. Thus, it will be sufficient to capture with MinCE(ϕ) the class of models that satisfy (79).

Using that we have defined CE-projections on infinite algebras via the two steps (43) and (57), we can split (79) into two parts: abbreviating $\Delta_\mathbb{F}(\phi, \mathfrak{M})$ by J, and letting $\mathfrak{A}'$ be a finite subalgebra by constraints on which J is defined, we obtain out of (43) the condition

(80) $Q_\mathbf{e} \restriction \mathfrak{A}' \in \Pi_{J|\mathfrak{A}'}(P_{|\mathbf{e}|} \restriction \mathfrak{A}').$

When (80) is fulfilled, and $A_1, \dots, A_L$ are the atoms of $\mathfrak{A}'$, then the defining Equation (57) can be expressed by

(81) $\displaystyle Q_\mathbf{e}(B) = \sum_{\substack{h=1 \\ P_{|\mathbf{e}|(A_h)}>0}}^{L} Q_\mathbf{e}(A_h)P_{|\mathbf{e}|}(B|A_h), \quad B \in \mathfrak{A}^*.$

We now axiomatize (80) by a single L_p-formula, and (81) by a schema, ranging over the B. Our first task is to identify a suitable algebra $\mathfrak{A}'$, and its atoms $A_1, \dots, A_L$. As in the proof of lemma 3.6 let

$$\text{prob}(\psi_1[\mathbf{e}]), \dots, \text{prob}(\psi_n[\mathbf{e}])$$

be the subjective probability terms contained in ϕ. Then $\Delta_{\mathbb{F}}(\phi, \mathfrak{M})$ is defined by constraints on the algebra $\mathfrak{A}'$ generated by the extensions of the ψ_i. The atoms of $\mathfrak{A}'$ are the non-empty extensions of the formulas

$$\alpha_j(\boldsymbol{v}) := \wedge_{i=1}^n \tilde{\psi}_i(\boldsymbol{v}) \quad (\tilde{\psi}_i(\boldsymbol{v}) \in \{\psi_i(\boldsymbol{v}), \neg\psi_i(\boldsymbol{v})\}, j = 1, \ldots, 2^n).$$

As a first building block for the formalization of (80) we can now formulate an L_p-formula that defines as a subset of $\mathbb{F}^{2^n}$ the set of all probability measures on $\mathfrak{A}'$:

$$(82) \quad \delta(x_1, \ldots, x_{2^n}) :\equiv \bigwedge_{j=1}^{2^n} x_j \geqslant 0 \wedge \sum_{j=1}^{2^n} x_j = 1$$

$$\wedge \bigwedge_{j=1}^{2^n} (\neg \exists \boldsymbol{v} \alpha_j(\boldsymbol{v}) \to x_j = 0).$$

Now let $\phi[\text{prob}/\boldsymbol{x}]$ denote the formula that is obtained from ϕ by substituting for every term $\text{prob}(\psi_i[\boldsymbol{e}])$ the term $x_{j_1} + \cdots + x_{j_k}$ where $k = 2^{n-1}$, and $\{j_1, \ldots, j_k\} \subset \{1, \ldots, 2^n\}$ is the collection of indices j_h for which the atom α_{j_h} is contained in ψ_i (i.e., α_{j_h} is a conjunction in which ψ_i appears un-negated). For the formula

$$(83) \quad \iota(\boldsymbol{x}) := \delta(\boldsymbol{x}) \wedge \phi[\text{prob}/\boldsymbol{x}]$$

and a probabilistic structure $\mathfrak{M}$ we then have

$$(84) \quad (\mathfrak{M}, \boldsymbol{x})(\iota(\boldsymbol{x})) = \Delta_{\mathbb{F}}(\phi, \mathfrak{M}) \restriction \mathfrak{A}'.$$

The formula

$$(85) \quad \zeta(\boldsymbol{x}) :\equiv \bigwedge_{j=1}^{2^n} ([\alpha_j(\boldsymbol{v})]_v = 0 \to x_j = 0)$$

encodes the condition of absolute continuity with respect to the statistical distribution on the algebra $\mathfrak{A}'$. In particular, the sentence

$$(86) \quad \zeta[\text{prob}] :\equiv \bigwedge_{j=1}^{2^n} ([\alpha_j(\boldsymbol{v})]_v = 0 \to \text{prob}(\alpha_j[\boldsymbol{e}]) = 0)$$

[59]

says that $Q_{\mathbf{e}} \upharpoonright \mathfrak{A}' \ll P_{|\mathbf{e}|} \upharpoonright \mathfrak{A}'$. We now can axiomatize (80) by the L_p-sentence

$$
\begin{aligned}
(87) \quad & \zeta[\text{prob}] \wedge \forall x (\iota(x) \wedge \zeta(x)) \\
& \rightarrow \sum_{j:[\alpha_j(v)]_v > 0} x_j \operatorname{Log} \frac{x_j}{[\alpha_j(v)]_v} \\
& \geqslant \sum_{j:[\alpha_j(v)]_v > 0} \text{prob}(\alpha_j[\mathbf{e}]) \operatorname{Log} \frac{\text{prob}(\alpha_j[\mathbf{e}])}{[\alpha_j(v)]_v}
\end{aligned}
$$

(we are here taking some liberties with the syntax of L_p, but one can easily expand this formula so as to eliminate the sum-expressions, and obtain a proper L_p-sentence).

To encode (81), let B be defined by the formula $\beta(v)$. Then (81) can be written in L_p as

$$
(88) \quad \text{prob}(\beta[\mathbf{e}]) = \sum_{j:[\alpha_j(v)]_v > 0} \text{prob}(\alpha_j[\mathbf{e}])[\beta(v)|\alpha_j(v)]_v.
$$

Taking the union over all L_p-formulas $\beta(v)$ with $|v| = |\mathbf{e}|$ turns (88) into a recursively enumerable sentence schema.

Finally, let $\text{MinCE}(\phi)$ consist of ϕ, of (87), and all instances of (88). Clearly there exists an algorithm that for any given sentence ϕ enumerates $\text{MinCE}(\phi)$ (we only need a uniform method to generate the atoms α_j determined by ϕ, and then simply list (87) and all instances of (88)). Also, by our derivation of $\text{MinCE}(\phi)$, clearly (77) is satisfied. Thus, the enumeration algorithm for $\text{MinCE}(\phi)$, together with a complete inference system for $\models$, constitutes a complete inference system for $\approx$.

4. RELATED WORK AND CONCLUSION

4.1. *Related Work*

Closely related to our logic of strict inference, $\mathscr{L}_p$, are the probabilistic first order logics of Bacchus (1990a,b) and Halpern (1990). Our logic of inductive inference, $\mathscr{L}_{ip}$, on the other hand, has to be compared with the random worlds method of Bacchus, Grove, Halpern, and Koller (Bacchus et al. 1992, 1997; Grove et al. 1992a,b).

There are two main differences between our logic $\mathscr{L}_p$ and the combined subjective and statistical probability logic $\mathscr{L}_3$ of Halpern

(1990). The first difference lies in basing the semantics of $\mathscr{L}_p$ on arbitrary lrc-field valued measures, whereas the semantics of $\mathscr{L}_3$ is based on real-discrete measures alone. As a result, a completeness result corresponding to our Theorem 2.12 cannot be obtained for $\mathscr{L}_3$ (Abadi and Halpern 1994). However, measures in more general algebraic structures were also already used by Bacchus (1990a) to obtain a completeness result for his statistical probability logic, and the same approach could clearly also be directly applied to Halpern's $\mathscr{L}_3$. The second difference between $\mathscr{L}_p$ and $\mathscr{L}_3$, therefore, is the much more significant one: in $\mathscr{L}_3$ statistical and subjective probabilities are interpreted by probability measures on the domains of first-order structures, and probability measures on sets of such structures (or possible worlds), respectively (leading to *type-3 probability structures*). As a result, the logic does not enforce any connections between statistical and subjective probabilities, or, more generally, domain knowledge and subjective probabilities. For example, the sentence

(89) $\neg \exists v \phi(v) \wedge \mathrm{prob}(\phi(e)) = 0.5$

is satisfiable in $\mathscr{L}_3$ by a type-3 structure containing a possible world that does not have any elements with property ϕ, and also containing possible worlds in which $\phi(e)$ is true (when interpreting (89) as a sentence in $\mathscr{L}_3$, the symbol e is considered as a standard constant). Halpern (1990) also shows that some dependencies between statistical and subjective probabilities are obtained in $\mathscr{L}_3$ when the semantics is restricted to type-3 structures in which all relation and function symbols are *rigid*, i.e., have the same interpretation in all possible worlds, and only the interpretations of some constant symbols are allowed to vary over the possible worlds. These dependencies are very weak, however, and do "not begin to settle the issue of how to connect statistical information with degrees of belief" (Halpern 1990). Our probabilistic structures are closely related to these rigid type-3 structures. In fact, we can view a probabilistic structure in our sense as a superimposition of the possible worlds in a rigid type-3 structure, where non-rigid constant symbols now become our event symbols, and the distribution Q_e represents their distribution in the different possible worlds. This collapsing of the possible worlds into a single structure gives us the crucial technical advantage that subjective and statistical probabilities are defined on the same space, and their discrepancy can be measured by cross-entropy.

[61]

The statistical probability logics of Bacchus and Halpern serve as the foundation for the random-worlds method of Bacchus, Grove, Halpern, and Koller (Bacchus et al. 1992, 1997; Grove et al. 1992a, b). Aim of this approach is to assign to pairs ϕ, ψ of formulas in the statistical representation language a degree of belief $\Pr(\phi|\psi)$ in the proposition ϕ, given the knowledge ψ. The definition of $\Pr(\phi|\psi)$ proceeds by considering for fixed $n \in \mathbb{N}$ the fraction $\Pr_n(\phi|\psi)$ of models of over domain $\{1, \ldots, n\}$ that also satisfy ϕ, and to define $\Pr(\phi|\psi) := \lim_{n \to \infty} \Pr_n(\phi|\psi)$, provided that limit exists.

Like our logic $\mathscr{L}_{ip}$, the random worlds method derives much of its motivation from direct inference. A typical example to which the method would be applied is

(90) $\psi \equiv [\mathrm{IIA}(d)|\neg\mathrm{Drinks}(d)]_d = 0.01 \wedge \neg\mathrm{Drinks}(jones)$,

(91) $\phi \equiv \mathrm{IIA}(jones)$

for which the random-worlds method yields the direct inference $\Pr(\phi|\psi) = 0.01$. The similarity of motivation, and a connection of the random-worlds method with entropy maximization (Grove et al. 1992b), at first sight suggests a fairly close relationship between that method and $\mathscr{L}_{ip}$. On closer examination it turns out, however, that the two frameworks differ substantially with respect to fundamental mathematical properties. The first major difference between the two approaches is that the random-worlds method does not permit to include in the input information ψ any prior constraints on degrees of belief. A second difference lies in the fact that the random-worlds method leads to inferences that go very much beyond the type of inductive probabilistic inferences supported by $\mathscr{L}_{ip}$. In particular, the random-worlds method also leads to default inferences about the statistical distribution, and give, e.g., the degree of belief $\Pr([\mathrm{Drinks}(d)]_d = 0.5|[\mathrm{Drinks}(d)]_d \geqslant 0.3) = 1$. One sees that, thus, the random-worlds method does not model inductive probabilistic reasoning as we understand it – as an inference pattern that is strictly directed from general (statistical) knowledge to beliefs about a particular case – but leads to a much stronger form of probabilistic default inferences.

Another vital difference arises out of the random-worlds method's commitment to finite domains: if ϕ is a sentence that is not satisfiable on finite domains, and ψ is any sentence, then we obtain $\Pr(\phi|\psi) = 0$; no corresponding phenomenon occurs in $\mathscr{L}_{ip}$. Finally, the random-worlds method differs from $\mathscr{L}_{ip}$ greatly with respect to computational properties. As shown in Grove et al. (1992a), the set of pairs (ϕ, ψ) for which $\Pr(\phi|\psi)$ is defined, i.e., the limit $\lim_{n \to \infty} \Pr_n(\phi|\psi)$ exists, is not recursively enumerable. Thus, there exists no complete proof system for the random-worlds method (a solution to this problem by a move to generalized probabilities here is infeasible, as the very definition of the degrees of belief as limits of sequences of rational numbers is tied to the real number system).

In Section 1.1 we argued that for our inductive inference problem a conservative approach is appropriate for combining partial prior information with new information: we simply combine each possible exact prior (i.e., statistical distribution) with the new information (i.e., subjective probability constraints). It is instructive, though, to compare this to some more adventurous rules that have been considered in the literature. A very natural possibility is to perform CE-minimization over both the statistical and the subjective probability distribution, i.e., preferred models will be those in which $CE(Q_\mathbf{e}, P_{|\mathbf{e}|})$ is minimal for all feasible choices of $Q_\mathbf{e}$ and $P_{|\mathbf{e}|}$ (given the non-probabilistic part of the model). This is an instance of *revising based on similarity relationships* (Moral and Wilson 1995). This approach is also closely related to the *classical (maximum likelihood) update rule* of Gilboa and Schmeidler (1993): according to that rule a set C of priors is updated based on the observation of an event A by selecting from C those distributions that assign maximal likelihood to A. If we again identify the categorical observation A with a probability constraint $\text{prob}(A) = 1$, then this means that we select all distributions $q \in C$ with $\min_{p:p(A)=1} CE(p, q) = \min_{q':q' \in C} \min_{p:p(A)=1} CE(p, q)$. Thus, the rule by Gilboa and Schmeidler can also be understood as CE-minimization in two arguments (though originally restricted to categorical observations); however, here the result of the updating consists of distributions selected from the set of priors, not from the set defined by the new constraints.

To compare such stronger update rules with our conservative rule, consider the following example:

(92) $[\psi(v)]_v \geqslant [\phi(v)]_v \wedge \text{prob}(\phi(e)) = 1.$

According to our conservative inference rule, we apply direct inference to every statistical distribution satisfying the statistical constraint in (92). These include for every $q \in [0, 1]$ distributions with $[\psi(v)|\phi(v)]_v = q$. Consequently, we will not derive any non-trivial bounds on $\text{prob}(\psi(e))$. If we perform CE-minimization in both arguments, then we will effectively only consider statistical distributions with $[\psi(v)]_v = [\phi(v)]_v = 1$, and derive $\text{prob}(\psi(e)) = 1$. This may not seem unreasonable based on the abstract formulation (92), but consider e.g. the case where $\psi(v) = \text{Drives}(\text{Toyota}, v)$ and $\phi(v) = \text{Drives}(\text{RollsRoyce}, v)$.

4.2. *Conclusion*

To formalize the process of inductive probabilistic reasoning within an expressive logical framework we have defined the logic $\mathcal{L}_{ip}$ with its inductive entailment relation $\approx$. Three design principles have largely guided the definition of $\mathcal{L}_{ip}$: expressiveness, completeness, and epistemic justifiability. Expressiveness: the logic provides a rich first-order representation language that enables the encoding of complex probabilistic information. Completeness: the expressiveness of the language should be complemented with a powerful deductive system. We have obtained a complete deductive system for lrc-field valued probabilities, and have furthermore established a strong agreement between the behaviors of real-valued and lrc-field valued probabilities (especially with regard to cross-entropy minimization). Combined these results entail a strong characterization of the deductive power of our inference system also with respect to real-valued probabilities. Epistemic justifiability: it was our aim to model with the inductive entailment relation $\approx$ only a well-justified pattern of defeasible probabilistic reasoning – how statistical information enables us to refine an already partially formed subjective probability assignment. For this particular inference pattern we argue that cross-entropy minimization relative to every possible statistical distribution is the adequate formal model (more fully than in the present paper this argument is given in Jaeger (1995a,b)). The resulting relation $\approx$ is necessarily weak when only little statistical information is available. However, in typical applications one can expect the statistical background information to be much more specific than the partial subjective probability assignments made in the observation of a single event, in which case $\approx$ will lead to strong conclusions.

The full logic $\mathcal{L}_{ip}$ should be regarded as a rich reference logic for the theoretical analysis of the formal rules of inductive probabilistic reasoning. For practical applications and implementations one should consider suitable fragments of this logic, e.g., the probabilistic description logics described in Jaeger (1994b). Such fragments can reduce the complexities of reasoning in $\mathscr{L}_{ip}$ in several ways: they can enforce the closure of the sets $\Delta_{\mathbb{F}}(\phi, \mathfrak{M})$, so that some of the difficulties described in Section 3.3 are avoided; they can further reduce the discrepancy between real-valued and lrc-field valued probabilities, and thereby become complete also for real-valued probabilities; finally, and most importantly, fragments will give rise to specialized inference techniques that can make automated reasoning more effective.

APPENDIX A. CROSS ENTROPY IN LOGARITHMIC REAL-CLOSED FIELDS

In this appendix we prove that the most important properties of CE and CE-minimization in the reals carry over to the general case of CE in arbitrary lrc-fields. We partition these results into two groups: the first group describes qualitative properties that can be derived on the basis of the axioms LRCF without the approximation schema (viii). The second group deals with the numerical agreement between CE in the reals and in other lrc-fields, and is essentially based on the schema LRCF (viii).

A.1. Qualitative Properties

LEMMA A.1. *The following sentences are derivable from LRCF:*

(A.1)	$Log(1) = 0,$
(A.2)	$\forall x > 0 \; Log(1/x) = -Log(x),$
(A.3)	$\forall x \in (0, 1) \; Log(x) < 0,$
(A.4)	$\forall x > 1 \; Log(x) > 0,$
(A.5)	$\forall x, y > 0 \; x < y \rightarrow Log(x) < Log(y),$
(A.6)	$0 \cdot Log(0) = 0.$

The proofs for (A.1)–(A.5) are straightforward from the axioms LRCF. For (A.6) note that in every model $\mathfrak{F}$ for S_{LOF} a value $Log(0) \in \mathbb{F}$ has to be defined, and that by the field axioms $0 \cdot Log(0) = 0$ must hold.[4]

The following property of the logarithm is the basis for all the subsequent results in this section.

LEMMA A.2. *In every lrc-field the following holds:*

$$\forall x_1, y_1, x_2, y_2 > 0: \quad x_1 \mathrm{Log}\left(\frac{x_1}{y_1}\right) + x_2 \mathrm{Log}\left(\frac{x_2}{y_2}\right)$$

(A.7)
$$\geqslant (x_1 + x_2)\mathrm{Log}\left(\frac{x_1 + x_2}{y_1 + y_2}\right),$$

where equality holds iff

(A.8) $$\frac{x_1}{x_1 + x_2} = \frac{y_1}{y_1 + y_2}.$$

Proof. Let $\mathbb{F}$ be an lrc-field, and $x_1, y_1, x_2, y_2 \in \mathbb{F}$ be positive. Defining

$$x := x_1 + x_2, \quad \lambda_x := \frac{x_1}{x_1 + x_2},$$
$$y := y_1 + y_2, \quad \lambda_y := \frac{y_1}{y_1 + y_2},$$

we can write

$$x_1 = \lambda_x x, \quad x_2 = (1 - \lambda_x)x, \quad y_1 = \lambda_y y, \quad y_2 = (1 - \lambda_y)y$$

and the left-hand side of (A.7) may be rewritten as

$$\lambda_x x \mathrm{Log}\left(\frac{\lambda_x x}{\lambda_y y}\right) + (1 - \lambda_x)x \mathrm{Log}\left(\frac{(1 - \lambda_x)x}{(1 - \lambda_y)y}\right)$$

(A.9) $$= x \mathrm{Log}\left(\frac{x}{y}\right) + x\left(\lambda_x \mathrm{Log}\left(\frac{\lambda_x}{\lambda_y}\right) + (1 - \lambda_x)\mathrm{Log}\left(\frac{1 - \lambda_x}{1 - \lambda_y}\right)\right).$$

If (A.8) holds, i.e., $\lambda_x = \lambda_y$, then the second term of (A.9) vanishes by (A.1), so that (A.7) holds with equality.

Now suppose that $\lambda_x \neq \lambda_y$, Then $\lambda_y / \lambda_x \neq 1$ and $(1 - \lambda_y)/(1 - \lambda_x) \neq 1$. By LRCF($v$), $-\mathrm{Log}(x) > 1 - x$ for $x \neq 1$, so that

$$\lambda_x \mathrm{Log}\left(\frac{\lambda_x}{\lambda_y}\right) + (1 - \lambda_x)\mathrm{Log}\left(\frac{1 - \lambda_x}{1 - \lambda_y}\right)$$

$$= \lambda_x\left(-\mathrm{Log}\left(\frac{\lambda_y}{\lambda_x}\right)\right) + (1 - \lambda_x)\left(-\mathrm{Log}\left(\frac{1 - \lambda_y}{1 - \lambda_x}\right)\right)$$

$$> \lambda_x\left(1 - \frac{\lambda_y}{\lambda_x}\right) + (1 - \lambda_x)\left(1 - \frac{1 - \lambda_y}{1 - \lambda_x}\right) = 0.$$

Since $x > 0$, this means that the second term of (A.9) is strictly greater than 0. This proves the lemma. $\square$

[66]

LEMMA A.3. (positivity). *Let $\mathfrak{F}$ be an lrc-field, $n \geq 2$, $Q, P \in \Delta_{\mathbb{F}}^{n}$ with $Q \ll P$. Then $CE(Q, P) \geq 0$, with equality iff $Q = P$.*

Proof. By induction on n. Let $n = 2$, $Q = (Q_1, Q_2)$, $P = (P_1, P_2) \in \Delta_{\mathbb{F}}^{2}$, $Q \ll P$. If one of the P_i equals 0, then so does the corresponding Q_i, in which case $Q = P$ and $CE(Q, P) = 1 \ \mathrm{Log}(1) = 0$. Suppose, then, that $P_i > 0$ ($i = 1, 2$). If $Q_i = 0$ for one i, say $i = 1$, then $Q \neq P$ and $CE(Q, P) = \mathrm{Log}(1/P_2) > 0$ by (A.4).

For the case that $Q_i, P_i > 0$ ($i = 1, 2$), we have

$$CE(Q, P) = Q_1 \mathrm{Log}\left(\frac{Q_1}{P_1}\right) + Q_2 \mathrm{Log}\left(\frac{Q_2}{P_2}\right)$$
$$\geq (Q_1 + Q_2)\mathrm{Log}\left(\frac{Q_1 + Q_2}{P_1 + P_2}\right)$$
$$= 1 \ \mathrm{Log}(1) = 0$$

by Lemma A.2, with equality iff $Q_1/(Q_1 + Q_2) = P_1/(P_1 + P_2)$, i.e., $Q = P$.

Now let $n > 2$, and assume that the lemma has been shown for $n - 1$. For $Q = P$ we again obtain $CE(Q, P) = 1 \ \mathrm{Log}(1) = 0$. Suppose, then, that $Q \neq P$. Without loss of generality, $Q_1 \neq P_1$. Define $\bar{Q}, \bar{P} \in \Delta_{\mathbb{F}}^{n-1}$ by

$$\bar{Q}_i := Q_i, \quad \bar{P}_i := P_i, \quad i = 1, \ldots, n - 2,$$

and

$$\bar{Q}_{n-1} := Q_{n-1} + Q_n, \quad \bar{P}_{n-1} := P_{n-1} + P_n.$$

Then $\bar{Q} \ll \bar{P}$, $\bar{Q} \neq \bar{P}$, so that by induction hypothesis $CE(\bar{Q}, \bar{P}) > 0$. By Lemma A.2 we have $CE(Q, P) \geq CE(\bar{Q}, \bar{P})$, which proves the lemma. □

LEMMA A.4. (convexity). *Let $\mathfrak{F}$ be an lrc-field, $n \geq 2$, $Q, Q', P \in \Delta_{\mathbb{F}}^{n}$, $Q \neq Q'$ with $Q, Q' \ll P$. Let $0 < \lambda < 1$. Then*

$$CE(\lambda Q + (1 - \lambda)Q', P) < \lambda CE(Q, P) + (1 - \lambda)CE(Q', P).$$

Proof. For the proof of the lemma it is sufficient to show that for fixed $y \in \mathbb{F}$, $y > 0$, the function

$$c_y : x \mapsto x \ \mathrm{Log}\left(\frac{x}{y}\right)$$

[67]

defined for $x \geqslant 0$ is strictly convex, because then

$$CE(\lambda Q + (1-\lambda)Q', P) = \sum_{P_i > 0} c_{P_i}(\lambda Q_i + (1-\lambda)Q'_i)$$
$$< \sum_{P_i > 0} \lambda c_{P_i}(Q_i) + (1-\lambda)c_{P_i}(Q'_i)$$
$$= \lambda CE(Q, P) + (1-\lambda)CE(Q', P),$$

where the strict inequality holds because $Q_i \neq Q'_i$ for at least one $i \in \{1, \ldots, n\}$ with $P_i > 0$.

For the proof of the convexity of c_y, let $y > 0, x_1, x_2 \geqslant 0, x_1 \neq x_2, 0 < \lambda < 1$. Abbreviate $\lambda x_1 + (1-\lambda)x_2$ by $\bar{x}$.

We distinguish two cases: first assume that one of the x_i is equal to 0, e.g., $x_1 = 0$. Then

$$c_y(\bar{x}) = (1-\lambda)x_2 \, \mathrm{Log}\left(\frac{(1-\lambda)x_2}{y}\right)$$
$$< (1-\lambda)x_2 \, \mathrm{Log}\left(\frac{x_2}{y}\right)$$
$$= \lambda c_y(x_1) + (1-\lambda)c_y(x_2),$$

where the inequality is due to (A.5), and the final equality holds because $c_y(0) = 0$ by (A.6).

Now suppose that $x_1, x_2 > 0$. By Lemma A.2 we obtain

$$(A.10) \quad c_y(\bar{x}) \leqslant \lambda x_1 \, \mathrm{Log}\left(\frac{\lambda x_1}{y/2}\right) + (1-\lambda)x_2 \, \mathrm{Log}\left(\frac{(1-\lambda)x_2}{y/2}\right)$$

with equality iff $\lambda x_1/\bar{x} = 1/2$, i.e.,

$$(A.11) \quad \lambda x_1 = (1-\lambda)x_2.$$

The right side of (A.10) may be rewritten as

$$\lambda x_1 \, \mathrm{Log}\left(\frac{x_1}{y}\right) + \lambda x_1 \, \mathrm{Log}(2\lambda) + (1-\lambda)x_2 \, \mathrm{Log}\left(\frac{x_2}{y}\right)$$
$$+ (1-\lambda)x_2 \mathrm{Log}(2(1-\lambda)).$$

Without loss of generality, assume that $\lambda x_1 \geqslant (1-\lambda)x_2$, so that we obtain

$$(A.12) \quad c_y(\bar{x}) \leqslant \lambda c_y(x_1) + (1-\lambda)c_y(x_2) + \lambda x_1 \, \mathrm{Log}(4\lambda(1-\lambda)),$$

still with equality iff (A.11) holds.

[68]

First consider the case that (A.11) in fact is true. Then, because $x_1 \neq x_2$, we have that $\lambda \neq 1/2$. By the completeness of RCF, and the fact that

$$\mathbb{R} \models \forall \lambda \in (0, 1) \quad \lambda \neq \frac{1}{2} \to \lambda \cdot (1 - \lambda) < \frac{1}{4},$$

we infer that $4\lambda(1 - \lambda) < 1$, which (with (A.3)) entails that λx_1 Log $(4\lambda(1-\lambda)) < 0$, thus proving that

(A.13) $c_y(\bar{x}) < \lambda c_y(x_1) + (1 - \lambda)c_y(x_2)$.

In almost the same manner (A.13) is derived for the case that (A.11) does not hold: the last term in (A.12) then is found to be $\leqslant 0$, which suffices to prove (A.13) because we have strict inequality in (A.12). □

So far we have established properties of CE as a function. Next we turn to the process of CE-minimization. The following two theorems state two key structural properties of cross-entropy minimization. These properties are the cornerstones of Shore's and Johnson's (1980) axiomatic justification of cross-entropy minimization, and, in a somewhat different guise, also of Paris's and Vencovská's (1990) derivation of the maximum entropy principle.

THEOREM A.5. (system independence). *Let $\mathfrak{A}, \mathfrak{A}'$ be finite algebras. Let $\mathbb{F}$ be an lrc-field, $J \cup \{P\} \subseteq \Delta_{\mathbb{F}}\mathfrak{A}$, $J' \cup \{P'\} \subseteq \Delta_{\mathbb{F}}\mathfrak{A}'$. Define*

$$\mathfrak{A}^{\times} := \mathfrak{A} \times \mathfrak{A}', \quad P^{\times} := P \otimes P'$$

and let $J^{\times} \subseteq A^{\times}$ be defined as the set of measures with marginal distribution on $\mathfrak{A}$ in J and marginal distribution on $\mathfrak{A}'$ in J', i.e.,

$$J^{\times} = \{Q^{\times} \in \Delta_{\mathbb{F}}\mathfrak{A}^{\times} \mid Q^{\times} \upharpoonright \mathfrak{A} \in J, Q^{\times} \upharpoonright \mathfrak{A}' \in J'\}.$$

Then

$$\Pi_{J^{\times}}(P^{\times}) = \Pi_J(P) \otimes \Pi_{J'}(P')$$
(A.14) $$:= \{Q \otimes Q' \mid Q \in \Pi_J(P), Q' \in \Pi_{J'}(P')\}.$$

Having established Lemmas A.1–A.4, the proof of this theorem and the following can be carried out for lrc-field valued probabilities just as for real-valued probabilities. We will therefore omit the proofs here, and refer the reader to Shore and Johnson (1980) and Jaeger (1995a).

THEOREM A.6. (subset independence). *Let $\mathfrak{A}$ be a finite algebra on M, $A = \{A_1, \ldots, A_L\} \subseteq \mathfrak{A}$ a partition of M, and $\mathfrak{F}$ an lrc-field. Let $P \in \Delta_{\mathbb{F}}\mathfrak{A}$.*

Denote by $\bar{\mathfrak{A}}$ the subalgebra of $\mathfrak{A}$ generated by A, and by $\mathfrak{A}^h$ the relative algebra of $\mathfrak{A}$ with respect to A_h ($h = 1, \ldots, L$). For $Q \in \Delta_{\mathbb{F}}\mathfrak{A}$ let $\bar{Q}$ denote the restriction $Q \restriction \bar{\mathfrak{A}}$, and Q^h the conditional of Q on $\mathfrak{A}^h$ ($h = 1, \ldots, L$; $Q(A_h) > 0$).

Let $J \subseteq \Delta_{\mathbb{F}}\mathfrak{A}$ be of the form

$$J = \bar{J} \cap J_1 \cap \cdots \cap J_L$$

with $\bar{J}$ a set of constraints on $\bar{Q}$, and J_h a set of constraints on Q^h ($h = 1, \ldots, L$). Precisely:

$$\bar{J} = \{Q \in \Delta_{\mathbb{F}}\mathfrak{A} \mid \bar{Q} \in \bar{J}^*\} \quad \text{for some } \bar{J}^* \subseteq \Delta_{\mathbb{F}}\bar{\mathfrak{A}},$$
$$J_h = \{Q \in \Delta_{\mathbb{F}}\mathfrak{A} \mid Q(A_h) = 0 \vee Q^h \in J_h^*\} \quad \text{for some } J_h^* \subseteq \Delta_{\mathbb{F}}\mathfrak{A}^h.$$

Let $Q \in \Pi_J(P)$. For all $h \in \{1, \ldots, L\}$ with $Q(A_h) > 0$ then

(A.15) $Q^h \in \Pi_{J_h^*}(P^h)$.

An important consequence of Theorem A.6 is that in the special case where J is defined by prescribing fixed probability values for the elements of a partition of M, then cross-entropy minimization reduces to *Jeffrey's rule* (Jeffrey 1965).

COROLLARY A.7. (Jeffrey's rule) *Let $\mathfrak{A}$ be a finite algebra on M, $P \in \Delta_{\mathbb{F}}\mathfrak{A}$, $\{A_1, \ldots, A_L\} \subset \mathfrak{A}$ a partition of M, and $(r_1, \ldots, r_L) \in \Delta_{\mathbb{F}}^L$ with $r_h > 0 \Rightarrow P(A_h) > 0$ for $h = 1, \ldots, L$. For*

$$J := \{Q \in \Delta_{\mathbb{F}}\mathfrak{A} \mid Q(A_h) = r_h, \quad h = 1, \ldots, L\}$$

then $\Pi_J(P) = \{Q\}$ *for*

$$(A.16) \quad Q = \sum_{\substack{h=1 \\ r_h > 0}}^{L} r_h P^h,$$

where P^h is the conditional of P on A_h.

A.2. *Numerical Approximations*

To motivate the results in this Ssection, reconsider the example of Section 3.1 given by (46)–(50). Here (47)–(50) defined a unique statistical probability measure $P = (0.4, 0.3, 0.1, 0.2)$ on a four-element algebra. The components of P being rational, P can be interpreted as an element $P(\mathbb{F})$ of $\Delta_{\mathbb{F}}^4$ for any lrc-field $\mathbb{F}$. Similarly, the constraint (46) defines a subset

$$J(\mathbb{F}) := \left\{ (x_1, \ldots, x_4) \in \Delta_{\mathbb{F}}^4 \mid x_1 + x_2 = 0.5,\, x_2 + x_3 = 0.7 \right\}$$

of $\Delta_{\mathbb{F}}^4$ for every $\mathbb{F}$. For the inductive inference relation of $\mathscr{L}_{ip}$ we now have to consider the CE-projections $\Pi_{J(\mathbb{F})}(P(\mathbb{F}))$ for arbitrary $\mathbb{F}$. For $\mathbb{F} = \mathbb{R}$ we know that $\Pi_{J(\mathbb{F})}(P(\mathbb{F}))$ contains a unique element Q, and, using an iterative non-linear optimization algorithm, we can determine the value of Q approximately, as stated in (51). More precisely, the meaning of (51) is

(A.17) $\Pi_{J(\mathbb{R})}(P(\mathbb{R})) \subseteq \left\{ (q_1, \ldots, q_4) \in \Delta_{\mathbb{R}}^4 \mid q_1 \in (0.128, 0.129), \ldots, \right.$
$$\left. q_4 \in (0.171, 0.172) \right\}.$$

In order to use this numerical result obtained for the reals for showing that certain inductive entailment relations hold in $\mathscr{L}_{ip}$ – e.g., that (52) follows from (46)–(50) – we have to ascertain that (A.17) implies

(A.18) $\Pi_{J(\mathbb{F})}(P(\mathbb{F})) \subseteq \left\{ (q_1, \ldots, q_4) \in \Delta_{\mathbb{F}}^4 \mid q_1 \in (0.128, 0.129), \ldots, \right.$
$$\left. q_4 \in (0.171, 0.172) \right\}$$

for every $\mathbb{F}$. Theorem A.10 will show that this is indeed the case. We obtain this result by showing successively that the bounds given for Log by LRCF (viii) are sufficient to determine uniform bounds (i.e., valid in every $\mathbb{F}$) for the function $x \operatorname{Log}(x/q) (q \in \mathbb{Q}$ fixed), for $CE(Q, P)(P \in \Delta_{\mathbb{Q}}^n$ fixed), and finally for $\Pi_{J(\mathbb{F})}(P(\mathbb{F}))$. The first lemma gives a piecewise approximation of $x \operatorname{Log}(x/q)$.

LEMMA A.8. *Let $\epsilon > 0$ and $P \in (0, 1]$ be rational numbers[5], let p_n and q_n be as defined in* LRCF (viii). *There exists a rational number $r(\epsilon) > 0$ and an $m \in \mathbb{N}$ such that the following S_{LOF}-sentences hold in* *all lrc-fields:*

(A.19) $\forall x \in (0, r(\epsilon)] x \operatorname{Log} \left(\dfrac{x}{P} \right) \in (-\epsilon, 0)$,

(A.20) $\forall x \in [r(\epsilon), P] x \operatorname{Log} \left(\dfrac{x}{P} \right) \in \left[x q_m \left(\dfrac{x}{P} \right), x p_m \left(\dfrac{x}{P} \right) \right]$,

(A.21) $\forall x \in [r(\epsilon), P] x p_m \left(\dfrac{x}{P} \right) - x q_m \left(\dfrac{x}{P} \right) \in [0, \epsilon)$,

(A.22) $\forall x \in [P, 1] x \operatorname{Log} \left(\dfrac{x}{P} \right) \in \left[-x p_m \left(\dfrac{P}{x} \right), -x q_m \left(\dfrac{P}{x} \right) \right]$,

(A.23) $\forall x \in [P, 1] - x q_m \left(\dfrac{P}{x} \right) + x p_m \left(\dfrac{P}{x} \right) \in [0, \epsilon)$.

Proof. We first determine a number $r(\epsilon)$ such that the approximation (A.19) holds. We then choose a sufficiently large n such that the bounds (A.21) and (A.23) hold. Properties (A.20) and (A.22) directly follow from LRCF (viii).

By elementary calculus we find that in $\mathbb{R} \lim_{x \to 0} x \operatorname{Log}(x/P) = 0$, and that $x \operatorname{Log}(x/P)$ attains its absolute minimum at $x = P/e > 0$. We choose an arbitrary rational $r(\epsilon) \in (0, P/e)$ that satisfies

$$ r(\epsilon) \operatorname{Log} \left(\frac{r(\epsilon)}{P} \right) > \max \left\{ -\epsilon, \frac{P}{e} \operatorname{Log} \left(\frac{P}{e} \right) \right\}. $$

Also, choose a rational $r' \in (r(\epsilon), (P/e))$. By the strict convexity of $x \longmapsto x \operatorname{Log}(x/P)$ then $r' \operatorname{Log}(r'/P) < r(\epsilon) \operatorname{Log}(r(\epsilon)/P)$. For sufficiently large $n \in \mathbb{N}$

$$ r(\epsilon) q_m \left(\frac{r(\epsilon)}{P} \right) > r' p_m \left(\frac{r'}{P} \right) \quad \text{and} \quad r(\epsilon) q_m \left(\frac{r(\epsilon)}{P} \right) > -\epsilon $$

now holds in $\mathbb{R}$, and hence in every lrc-field. It follows that in every lrc-field we have

$$ r(\epsilon) \operatorname{Log} \left(\frac{r(\epsilon)}{P} \right) > r' \operatorname{Log} \left(\frac{r'}{P} \right) \quad \text{and} \quad r(\epsilon) \operatorname{Log} \left(\frac{r(\epsilon)}{P} \right) > -\epsilon. $$

By the strict convexity of the function $x \longmapsto x \operatorname{Log}(x/P)$ (Lemma A.4) we can now infer

$$ \forall x \in (0, r(\epsilon)] x \operatorname{Log} \left(\frac{x}{P} \right) > r(\epsilon) \operatorname{Log} \left(\frac{r(\epsilon)}{P} \right) $$

[72]

and thus

$$\forall x \in (0, r(\epsilon)] x \, \mathrm{Log}\left(\frac{x}{P}\right) > -\epsilon.$$

Also, because $r(\epsilon) < P$, by (A.2) and (A.4) we get

$$\forall x \in (0, r(\epsilon)] x \, \mathrm{Log}\left(\frac{x}{P}\right) < 0,$$

proving (A.19).

For the approximation of $x \, \mathrm{Log}(x/P)$ on $[r(\epsilon), 1]$ choose an $m \in \mathbb{N}$ such that

$$\max\left\{\frac{(r(\epsilon) - 1)^{m+1}}{r(\epsilon)}, \frac{(P - 1)^{m+1}}{P}\right\} < \epsilon.$$

For such m then (A.21) and (A.23) are satisfied. $\qquad\square$

The next lemma combines bounds for $Q_i \, \mathrm{Log}(Q_i/P_i)$ to find bounds for $CE(Q, P)$. In the formulation of the lemma we employ the notations introduced in Section 2.3 for the interpretations of terms in a structure, and for the sets defined in a structure by a formula.

LEMMA A.9. *Let* $n \geqslant 1$, $P \in \Delta_{\mathbb{Q}}^n$, *and* $\epsilon \in \mathbb{Q}, \epsilon > 0$. *There exist* $L_I(\mathrm{S_{OF}})$-*formulas* $\alpha_1(x), \ldots, \alpha_k(x)$ *and* $L_I(\mathrm{S_{OF}})$-*terms* $l_1(x)$, $u_1(x), \ldots, l_k(x), u_k(x)$ *with* $x = (x_1, \ldots, x_n)$, *such that the following holds in all lrc-fields* $\mathfrak{F}$:

(i) $\Delta_{\mathbb{F}}^n \cap \{Q \mid Q \ll P\} = \cup_{i=1}^k (\mathfrak{F}, x)(\alpha_i)$,

(ii) $\forall i \in \{1, \ldots, k\} \forall Q \in (\mathfrak{F}, x)(\alpha_i) : \mathfrak{F}(l_i(Q)) \leqslant CE(Q, P) \leqslant \mathfrak{F}(u_i(Q))$, and $\mathfrak{F}(u_i(Q)) - \mathfrak{F}(l_i(Q)) < \epsilon$.

Proof. Let $P \in \Delta_{\mathbb{Q}}^n$. Assume, first, that $P_i > 0$ for all $i = 1, \ldots, n$, so that $Q \ll P$ for all $Q \in \Delta_{\mathbb{F}}^n$. Applying Lemma A.8 to the P_i and ϵ/n, we find rational constants $r_1(\epsilon/n), \ldots, r_n(\epsilon/n)$, such that $Q_i \, \mathrm{Log}(Q_i/P_i)$ can be bounded for $Q_i \in (0, r_i(\epsilon/n)]$ by the constants $-\epsilon/n$ and 0, and for $Q_i \in [r_i(\epsilon/n), 1]$ by the terms $Q_i q_m(Q_i/P_i), Q_i q_m(P_i/Q_i), Q_i p_m(Q_i/P_i), Q_i p_m(P_i/Q_i)$ as described in Lemma A.8.

We now let the formulas α_j run over all conjunctions of the form

$$\wedge_{i=1}^n (x_i \in I_i),$$

where I_i is either $(0, r_i(\epsilon/n)], [r_i(\epsilon/n), P_i]$, or $[P_i, 1]$. The lower bound $l_j(x)$ on $CE(Q, P)$ for elements Q of $\alpha_j(x)$ then is given by the sum of the lower bounds $-\epsilon/n$, $Q_i q_m(Q_i/P_i)$, respectively, $-Q_i p_m(P_i/Q_i)$, obtained for each component Q_i $\mathrm{Log}(Q_i/P_i)$ of $CE(Q, P)$. Similarly for the upper bounds $u_j(x)$.

If $P_i = 0$ for some $i \in \{1, \dots, n\}$ we proceed in the same way, simply using a conjunct $x_i = 0$ instead of a conjunct $x_i \in I_i$ in the definition of the α_j.   □

Now the desired theorem can be formulated. Roughly speaking, it says that approximations of the CE-projection $\Pi_J(P)$ that are expressible by a S_{OF}-formula, and that are valid in $\mathbb{R}$, also are valid in arbitrary $\mathbb{F}$.

THEOREM A.10. *Let* $\phi(x_1, \dots, x_n)$ *and* $\psi(x_1, \dots, x_n)$ *be* $L_I(S_{OF})$-*formulas. Let* $P \in \Delta^n_{\mathbb{Q}}$. *Define*

$$\chi(\phi, \psi) :\equiv \exists x > 0\, \exists z (\phi(z) \wedge \forall y (\phi(y) \wedge \neg\psi(y)$$
$$\rightarrow CE(z, P) < CE(y, P) - x)).$$

If $\mathfrak{R} \models \chi(\phi, \psi)$, *then* $\mathrm{LRCF} \models \chi(\phi, \psi)$.

To connect this theorem with our introductory example, think of ϕ as the formula defining the set $J(\mathbb{F})$ and of ψ as the formula defining the right-hand side of (A.18). Then $\chi(\phi, \psi)$ essentially is the general statement whose interpretation over $\mathbb{R}$ is (A.17), and whose interpretation over $\mathbb{F}$ is (A.18). The theorem now says that (A.17) implies (A.18).

Proof. Assume that $\mathbb{R} \models \chi(\phi, \psi)$, and let $0 < \epsilon \in \mathbb{Q}$ be such that $\mathbb{R}$ is a model of

$$(A.24)\quad \exists z(\phi(z) \wedge \forall y(\phi(y) \wedge \neg\psi(y) \rightarrow CE(z, P) < CE(y, P) - \epsilon)).$$

Let $\alpha_1(x), \dots, \alpha_k(x)$ and $l_1(x), u_1(x), \dots, l_k(x), u_k(x)$ be as given by Lemma A.9 for P and $\epsilon/3$. Then, for some $j \in \{1, \dots, k\}$, $\mathbb{R}$ also is a model of

$$(A.25)\quad \exists z(\phi(z) \wedge \alpha_j(z) \wedge \forall y \ll P \exists i \in \{1, \dots, k\}(\alpha_i(y) \wedge (\phi(y) \wedge \neg\psi(y)$$
$$\rightarrow u_j(z) < l_i(y) - \epsilon/3))),$$

which, some abuse of first-order syntax notwithstanding, is a pure $L_I(S_{OF})$-sentence. Thus (A.25) holds in every lrc-field $\mathfrak{F}$.

Furthermore, by Lemma A.9, we have for arbitrary $\mathfrak{F}$:

$$(A.26)\quad \mathfrak{F} \models \forall y \forall i \in \{1, \dots, k\}(\alpha_i(y) \rightarrow CE(y, P) - l_i(y) \in [0, \epsilon/3]$$
$$\wedge u_i(y) - CE(y, P) \in [0, \epsilon/3]).$$

Combining the bounds $l_i(y) - u_j(z) > \epsilon/3, CE(y, P) - l_i(y) \leqslant \epsilon/3$, and $u_j(z) - CE(z, P) \leqslant \epsilon/3$, one obtains $CE(y, P) - CE(z, P) > \epsilon/3$, so that (A.24) with ϵ replaced by $\epsilon/3$ holds in arbitrary $\mathfrak{F}$, and hence also $\mathfrak{F} \vDash \chi(\phi, \psi)$. □

The following corollary mediates between the rather abstract formulation of Theorem A.10 and our introductory example.

COROLLARY A.11. *Let* $J \subseteq \Delta_{\mathbb{R}}^n$ *be closed and defined by an* $L_I(S_{OF})$-*formula* $\phi(x_1, \ldots, x_n)$. *Let* $H \subseteq \Delta_{\mathbb{R}}^n$ *be open and defined by an* $L_I(S_{OF})$-*formula* $\psi(x_1, \ldots, x_n)$. *Let* $P \in \Delta_{\mathbb{Q}}^n$, *and assume that* $\Pi_J(P) \subset H$. *For an arbitrary lrc-field* $\mathfrak{F}$, *and the sets* $\bar{J}, \bar{H}$ *defined in* $\mathfrak{F}$ *by* ϕ *and* ψ, *respectively, then* $\Pi_{\bar{J}}(P) \subset \bar{H}$.

Proof. According to the assumptions the set $H^c \cap J$ is closed. Let $Q \in \Pi_J(P)$. From $\Pi_J(P) \subset H$ and the compactness of $H^c \cap J$ it follows that there exists $\epsilon \in \mathbb{R}^+$ such that $CE(Q, P) < CE(Q', P) - \epsilon$ for every $Q' \in H^c \cap J$. Thus $\mathfrak{R} \vDash \chi(\phi, \psi)$. By Theorem A.10 then $\mathfrak{F} \vDash \chi(\phi, \psi)$, which entails $\Pi_{\bar{J}}(P) \subset \bar{H}$. □

ACKNOWLEDGMENTS

The author thanks the anonymous reviewers of an earlier version of this paper for their perceptive and constructive comments. Specifically, comments and examples provided by one reviewer have been integrated into Sections 2.4 and 3.3 of the present paper.

NOTES

[1] Other names for this type of probability are "probability of the single case" (Reichenbach 1949), "probability" (Carnap 1950), "propositional probability" (Bacchus 1990b).
[2] Lewis (1976) proposes *imaging* as an alternative to conditioning, but imaging requires a similarity measure on the states of the probability space, which usually cannot be assumed as given.
[3] Jeffrey (1965) argues the same point for human reasoners with his "observation by candlelight"-example. That argument, however, is not directly transferable to an autonomous agent whose evidence – at least in principle – is always expressible by a single, well-defined, proposition.
[4] For $\mathbb{R}$ to be a formal model of LRCF one would have to define (arbitrary) values $Log(x) \in \mathbb{R}$ for $x \leqslant 0$. Note that in $\mathbb{R}$ the otherwise somewhat artificial identity (A.6) is given real meaning by the fact that $\lim_{x \to 0} x Log(x) = 0$.

[5] All the results in this section remain valid when we substitute "algebraic numbers" for "rational numbers" throughout.

REFERENCES

Abadi, M. and J. Y. Halpern: 1994, Decidability and expressiveness for first-order logics of probability, *Information and Computation* **112**, 1–36.

Bacchus, F.: 1990a, Lp, a logic for representing and reasoning with statistical knowledge, *Computational Intelligence* **6**, 209–231.

Bacchus, F.: 1990b, *Representing and Reasoning With Probabilistic Knowledge*, MIT Press, Cambridge.

Bacchus, F., A. Grove, J. Halpern, and D. Koller: 1992, From statistics to beliefs, In *Proceedings of National Conference on Artificial Intelligence (AAAI-92)*.

Bacchus, F., A. J. Grove, J. Y. Halpern, and D. Koller: 1997, From statistical knowledge bases to degrees of belief, *Artificial Intelligence* **87**, 75–143.

Boole, G.: 1854, *Investigations of Laws of Thought on which are Founded the Mathematical Theories of Logic and Probabilities*, London.

Carnap, R.: 1950, *Logical Foundations of Probability*, The University of Chicago Press, Chicago.

Carnap, R.: 1952, *The Continuum of Inductive Methods*, The University of Chicago Press, Chicago.

Dahn, B. I. and H. Wolter: 1983, On the theory of exponential fields, *Zeitschrift für mathematische Logik und Grundlagen der Mathematik* **29**, 465–480.

de Finetti, B.: 1937, La prévision: ses lois logiques, ses sources subjectives, *Annales de l'Institut Henri Poincaré*. English Translation in (Kyburg and Smokler 1964).

Dempster, A. P.: 1967, Upper and lower probabilities induced by a multivalued mapping, *Annals of Mathematical Statistics* **38**, 325–339.

Diaconis, P. and S. Zabell: 1982, Updating subjective probability, *Journal of the American Statistical Association* **77** (380), 822–830.

Dubois, D. and H. Prade: 1997, Focusing vs. belief revision: A fundamental distinction when dealing with generic knowledge, In *Proceedings of the First International Joint Conference on Qualitative and Quantitative Practical Reasoning*, Springer, pp. 96–107.

Fenstad, J. E.: 1967, Representations of probabilities defined on first order languages, In J. N. Crossley (ed.), *Sets, Models and Recursion Theory*, North Holland, Amsterdam, pp. 156–172.

Gaifman, H.: 1964, Concerning measures in first order calculi, *Israel Journal of Mathematics* 2.

Gaifman, H. and M. Snir: 1982, Probabilities over rich languages, testing and randomness, *Journal of Symbolic Logic* **47**(3), 495–548.

Gilboa, I. and D. Schmeidler: 1993, Updatin ambiguous beliefs, *Journal of Economic Theory* **59**, 33–49.

Grove, A. and J. Halpern: 1998, Updating sets of probabilities, In *Proceedings of the 14th Conference on Uncertainty in AI*, pp. 173–182.

Grove, A., J. Halpern, and D. Koller: 1992a, Asymptotic conditional probabilities for first-order logic, In *Proceedings of the 24th ACM Symposium on Theory of Computing*.

Grove, A., J. Halpern, and D. Koller: 1992b, Random worlds and maximum entropy, In *Proceedings of the 7th IEEE Symposium on Logic in Computer Science*.

Hailperin, T.: 1976, *Boole's Logic and Probability*, Vol. 85 of *Studies in Logic and the Foundations of Mathematics*, North Holland, Amsterdam.

Hailperin, T.: 1996, *Sentential Probability Logic*, Lehigh University Press, Bethlehem.

Halpern, J.: 1990, An analysis of first-order logics of probability, *Artificial Intelligence* **46**, 311–350.

Hoover, D. N.: 1978, Probability logic, *Annals of Mathematical Logic* **14**, 287–313.

Jaeger, M.: 1994a, A logic for default reasoning about probabilities, In R. Lopez de Mantaraz and D. Poole (eds.), *Proceedings of the 10th Conference on Uncertainty in Artificial Intelligence (UAI'94)*, Morgan Kaufmann, Seattle, USA, pp. 352–359.

Jaeger, M.: 1994b, Probabilistic reasoning in terminological logics, In J. Doyle, E. Sandewall, and P. Torasso (eds.), *Principles of Knowledge Representation an Reasoning: Proceedings of the 4th International Conference (KR94)*, Morgan Kaufmann, Bonn, Germany, pp. 305–316.

Jaeger, M.: 1995a, Default Reasoning about Probabilities, PhD thesis, Universität des Saarlandes.

Jaeger, M.: 1995b, Minimum cross-entropy reasoning: A statistical justification, In C. S. Mellish (ed.), *Proceedings of the 14th International Joint Conference on Artificial Intelligence (IJCAI-95)*, Morgan Kaufmann, Montréal, Canada, pp. 1847–1852.

Jaynes, E.: 1978, Where do we stand on maximum entropy?, In R. Levine and M. Tribus (eds.), *The Maximum Entropy Formalism*, MIT Press, Cambridge, pp. 15–118.

Jeffrey, R.: 1965, *The Logic of Decision*, McGraw-Hill, New York.

Jensen, F.: 2001, *Bayesian Networks and Decision Graphs*, Springer, Berlin.

Keisler, H.: 1985, Probability quantifiers, In J. Barwise and S. Feferman (eds.), *Model-Theoretic Logics*, Springer, pp. 509–556.

Kullback, S.: 1959, *Information Theory and Statistics*, Wiley, New York.

Kullback, S. and R. A. Leibler: 1951, On information and sufficiency, *Annals of mathematical statistics* **22**, 79–86.

Kyburg, H. E.: 1974, *The Logical Foundations of Statistical Inference*, D. Reidel Publishing Company.

Kyburg, H. E.: 1983, The reference class, *Philosophy of Science* **50**, 374–397.

Kyburg, H. E. and H. E. Smokler (eds.): 1964, *Studies in Subjective Probability*, Wiley, New York.

Lewis, D.: 1976, Probabilities of conditionals and conditional probabilities, *The Philosophical Review* **85**(3), 297–315.

McCarthy, J.: 1980, Circumscription – a form of non-monotonic reasoning, *Artificial Intelligence* **13**, 27–39.

Moral, S. and N. Wilson: 1995, Revision rules for convex sets of probabilities, In G. Coletti, D. Dubois, and R. Scozzafava (eds.), *Mathematical Models for Handling Partial Knowledge in Artificial Intelligence*, Kluwer, Dordrecht.

Nilsson, N.: 1986, Probabilistic logic, *Artificial Intelligence* **28**, 71–88.

Paris, J. and A. Vencovská: 1990, A note on the inevitability of maximum entropy, *International Journal of Approximate Reasoning* **4**, 183–223.

Paris, J. and A. Vencovská: 1992, A method for updating that justifies minimum cross entropy, *International Journal of Approximate Reasoning* **7**, 1–18.

Paris, J. B.: 1994, *The Uncertain Reasoner's Companion*, Cambridge University Press, Cambridge.

Pearl, J.: 1988, *Probabilistic Reasoning in Intelligent Systems : Networks of Plausible Inference*, The Morgan Kaufmann series in representation and reasoning, rev. 2nd pr. edn, Morgan Kaufmann, San Mateo, CA.

Pollock, J. L.: 1983, A theory of direct inference, *Theory and Decision* **15**, 29–95.

Rabin, M. O.: 1977, Decidable theories, In J. Barwise (ed.), *Handbook of mathematical logic*, Elsevier Science Publishers, Amsterdam.

Reichenbach, H.: 1949, *The Theory of Probability*, University of California Press, Berkely, CA.

Savage, L. J.: 1954, *The Foundations of Statistics*, Wiley, New York.

Scott, D. and P. Krauss: 1966, Assigning probabilities to logical formulas, In J. Hintikka and P. Suppes (eds.), *Aspects of Inductive Logic*, North Holland, Amsterdam, pp. 219–264.

Shafer, G.: 1976, *A Mathematical Theory of Evidence*, Princeton University Press, New Jersey.

Shoham, Y.: 1987, Nonmonotonic logics: Meaning and utility, In *Proceedings of IJCAI-87*.

Shore, J. and R. Johnson: 1980, Axiomatic derivation of the principle of maximum entropy and the principle of minimum cross-entropy, *IEEE Transactions on Information Theory* **IT-26**(1), 26–37.

Shore, J. and R. Johnson: 1983, Comments on and correction to "Axiomatic derivation of the principle of maximum entropy and the principle of minimum cross-entropy", *IEEE Transactions on Information Theory* **IT-29**(6), 942–943.

von Mises, R.: 1951, *Wahrscheinlichkeit Statistik und Wahrheit*, Springer.

von Mises, R.: 1957, *Probability, Statistics and Truth*, George Allen & Unwin, London.

Walley, P.: 1991, *Statistical Reasoning with Imprecise Probabilities*, Chapman & Hall, London.

Department for Computer Science
Aalborg University, Fredrik Bajers Vej 7E
DK-9220 Aalborg Ø, Denmark
E-mail: jaeger@cs.aau.dk

HYKEL HOSNI* and JEFF PARIS

RATIONALITY AS CONFORMITY

ABSTRACT. We argue in favour of identifying one aspect of rational choice with the tendency to conform to the choice you expect another like-minded, but non-communicating, agent to make and study this idea in the very basic case where the choice is from a non-empty subset K of 2^A and no further structure or knowledge of A is assumed.

1. INTRODUCTION

The investigation described in this paper has its origins in Paris and Vencovská (1990, 1997, 2001) (see also Paris (1999) for a general overview). In those papers it was shown that as far as probabilistic uncertain reasoning is concerned there are a small set of so called 'common sense' principles which, if adhered to, completely determine any further assignment of beliefs, i.e. probabilities. Interesting as these results may be this raises the question *why* we consider these principles to be 'common sense' (or, more exactly, why we consider transgressing them to be *contra* common sense).

It is a question we have spent some effort trying to resolve. The principles looked to us like common sense, and indeed the general consensus of colleagues was that, certainly, to flout them was to display a lack of common sense. Nevertheless we could find no more basic element to which they could be reduced (for example showing as in the Dutch Book argument that if you fail to obey them then you are certain to lose that most basic of all substances, money). From this apparent impasse one explanation did however suggest itself. Namely, that these principles appeared common sensical to us all *exactly because their observance forced us to assign similar probabilities*. It is this idea, of common sense, or rationality, as conformity, that we shall investigate in this paper.

Certainly in the real world some one not acting in the way that people expect would be described as having *no common sense*, for

Synthese (2005) 144: 249–285
Knowledge, Rationality & Action 79–115
DOI 10.1007/s11229-004-4684-1

example filling-up the home fridge with fresh food the day before leaving for a long holiday, or, in more serious situations, such as declaring war on your ally when already fully stretched, of acting *illogically or irrationally*. Despite the numerous meanings or even intuitions that have been attached to these terms, see for example the volume (Elio 2002), for the limited purposes of this work we shall use them synonymously.

To motivate the sort of problem we are interested in suppose that your wife is coming to your office to collect the car keys but unexpectedly you have to go out before she arrives. Your problem is where to leave the keys so that she can find them. In other words your problem is *choosing* a point in the room where you think your wife will also choose to look. Being a logical sort of person you ask yourself "where would I expect someone to leave the keys?". If there was a vanity table by the door that might seem an obvious choice, because people tend to leave 'outdoor things' at this point. On the other hand if you had only just moved into the office and it only contained packing cases scattered around the walls then you might feel the centre of the carpet was the best option available to you, it being the only place, as far as you could see, that *stood out*.

It would seem in this situation that there are two considerations you could be drawing on. One is *common knowledge*, you assume that your wife is also aware of the typical use that vanity tables by entrances are put to. The other is what one might call *common reasoning*, you assume that your wife will also reason that the centre of the room 'stands out', so given the common intent to locate the same spot in the room, you place the keys right there. In the first case, conformity would be characterized as a consequence of learned and possibly arbitrary conventions. A formalization of this is not, however, what we are pursuing here. Indeed part of what we aim at understanding is how certain conventions might arise in the first place: why certain choices look more rational than others given that both agents intend to conform. So it is the second aspect of common sense – common reasoning – that we wish to investigate in this paper.

To do this we shall take what might be described as a mathematician's approach to this problem. We shall strip away all the inessentials, all the additional considerations which one normally carries with one in problems such as the one described above,[1] and consider a highly idealized and abstract simplification of the problem. Our justification for this is that if one cannot resolve this problem satisfactorily how could one expect to be successful on the infinitely more complicated real world examples?

2. THE PROBLEM

The problem we wish to consider is that of trying to choose one from a number of options so that your choice *conforms* with that of another like-minded, but otherwise inaccessible, agent (the payoff for success, ditto failure, being the same in all cases).

What is arguably the simplest possible choice situation of this sort is the one in which we have some finite non-empty set K of otherwise entirely structureless options f. In other words options that whilst different are otherwise entirely indistinguishable. Then the very definition of 'indistinguishable' seems to suggest that in this case there is no better strategy available to us than to make a choice from K entirely at random (i.e. according to the uniform distribution).

The inevitable next step then is to consider the case when we *do* have some structure on the options, or as we may henceforth call them, *worlds*, $f \in K$. In this case, as logicians, the most obvious minimal structure on these worlds is that there are some finite number of unary predicates which each of them may or may not satisfy. To simplify matters for the present we shall further assume that each world is uniquely determined by the predicates it does or does not satisfy. In other words we are moving up from the language of equality to a finite unary language. What this amounts to then is that K is a non-empty subset of 2^A, the set of maps f from the finite non-empty set A into $2 = \{0, 1\}$.

To give a concrete example of what is involved here we might have $A = 4$ and K the set of functions (worlds) $\{f_1, f_2, f_3, f_4, f_5\}$ where

$$
\begin{array}{c|cccc}
 & 0 & 1 & 2 & 3 \\
\hline
f_1 & 0 & 0 & 0 & 1 \\
f_2 & 0 & 1 & 0 & 0 \\
f_3 & 0 & 1 & 1 & 0 \\
f_4 & 1 & 1 & 1 & 1 \\
f_5 & 0 & 0 & 1 & 0 \\
\end{array}
$$

and the problem, for an agent, is to pick one of these so as to agree with the choice made by another like-minded, but otherwise non-communicating and indeed, inaccessible, agent. However

[81]

in presenting the problem like this we should be aware that as far as the agents are concerned there is not supposed to be any structure on A or $\{0, 1\}$, nor even on K beyond the fact that it is the (unordered) set $\{f_1, f_2, f_3, f_4, f_5\}$. For practical examples this can be accomplished by informing the first agent that his or her counterpart may receive the matrix

$$0\ 0\ 0\ 1$$
$$0\ 1\ 0\ 0$$
$$0\ 1\ 1\ 0$$
$$1\ 1\ 1\ 1$$
$$0\ 0\ 1\ 0$$

with the columns permuted and the rows permuted.

We understand a non-empty subset K of 2^A as *knowledge*, indeed knowledge that among the elements of K only one of them corresponds to the world chosen by another like-minded agent facing the same choice. In this way we implicitly introduce a qualitative measure of uncertainty: the bigger the size of K, the greater the agent's uncertainty about which choice of worlds qualifies as rational. This corresponds to a very general and fundamental idea in the formalization of reasoning under uncertainty (see e.g. Halpern 2003) and plays a likewise important role here.

It is clear that in general there will be situations, as in the case where we assumed no structure at all, when the agent is reduced to making some purely random choices. We shall therefore assume that the agent acts by first applying some considerations to reduce the set of possible choices $K (\neq \emptyset)$ to a non-empty subset $R(K)$ of K and then picks at random from $R(K)$. A function

$$R : \wp^+(2^A) \longmapsto \wp^+(2^A),$$

where $\wp^+(2^A)$ is the set of non-empty subsets of 2^A (which for brevity we sometimes denote as $\mathbb{K}$), will be called a *Reason* if $R(K) \subseteq K$ for all $K \in \wp^+(2^A)$.

Clearly, then, an optimal reason R, is one that always returns a singleton $R(K)$ for all $K \in \wp^+(2^A)$, as this would amount to entail conformity with probability 1. We shall see, however, that this situation represents the exception rather than the rule in the formalization to follow.

One might question at this point whether a better model for the agent's actions might be to have him or her put a probability distribution over K and then pick according to that distribution.

[82]

In fact in such a case the agent would do at least as well by instead selecting the most probable elements of K according to this distribution and then randomly (i.e. according to the uniform distribution) selecting from them – which puts us back into the original situation.

In the next three sections we consider three different Reasons which are suggested by the context of this, and related, problems.

3. THE REGULATIVE REASON

As mentioned already the work in this paper was in part motivated by considering why the principles of probabilistic uncertain reasoning introduced in Paris and Vencovská (1990, 1997, 2001) warranted the description 'common sense'. The underlying problem in those papers was analogous to the one we are considering here, how to sensibly choose one probability function out of a set of probability functions. The solution we developed there was not to directly specify a choice but instead to require that the choice process should satisfy these principles and see where that landed us. In fact it turned out well in the linear cases considered in Paris and Vencovská (1990, 1997) since the imposed principles happily permitted only one possible choice.

Given that fortunate outcome there it would seem natural to attempt a similar procedure here, namely to specify certain 'common sense' principles we would wish the agent's Reasons to satisfy and see what comes out. Clearly, the present problem is much less structured then the one in which knowledge and belief are represented via subjective probability functions. Indeed the current setting is arguably one of the simplest ones in which we can make sense of rational choice concerning "knowledge" and "possibilities". It therefore follows that if choice processes analogous to the ones that characterize probabilistic common sense could be specified, those would have an undoubtedly high level of generality.

Our next step then is to introduce 'common sense principles' or rules that, arguably, Reasons *should* satisfy if they are to prevent agents from undertaking "unreasonable steps".[2] Hence, we call the resulting Reason, *Regulative*. The key result of this section is that their observance leads to a characterization of a set $R(K)$ of "naturally outstanding elements" of K, formulated in Theorem 1.

Renaming
Let $K \in \mathbb{K}$ and let σ be a permutation of A. R satisfies *Renaming* if whenever

$$K\sigma = \{f\sigma \mid f \in K\}$$

then $R(K\sigma) = R(K)\sigma$.

In this definition $K\sigma$ is, as usual, the set $\{f\sigma \mid f \in K\}$, and similarly for $R(K)\sigma$ etc. The justification for this seems evident given the discussion in the previous section. Since the elements of A have no further structure any permutation of these elements simply produces an exact replica of what we started with. More precisely if you feel that the most popular choices of worlds from K are the set of worlds $R(K)$ then you should feel the same for these replicas, i.e. that the most popular choices of worlds from $K\sigma$ should be $R(K)\sigma$.

Obstinacy
R satisfies Obstinacy if whenever $K_1, K_2 \in \mathbb{K}$ and $R(K_1) \cap K_2 \neq \emptyset$ then $R(K_1 \cap K_2) = R(K_1) \cap K_2$.

The justification for this principle is that if you feel the most popular choices in K_1 are $R(K_1)$ and some of these choices are in K_2 then such worlds will remain the most popular even when the choice is restricted to $K_1 \cap K_2$.

This 'justification' *in general* is more than a little suspect. For consider $f \in R(K_1) - K_2$. In that case one might imagine those agents who chose f from K_1 having to re-choose when K_1 was refined to $K_1 \cap K_2$. The assumption is that they went back to $R(K_1)$ and randomly chose from there an element which *was* in K_2. An argument against this is that by intersecting K_1 with K_2 some otherwise rather non-descript world from K_1 becomes, within $K_1 \cap K_2$, sufficiently distinguished to be a natural choice. Whilst this will become clearer later when we have other Reasons to hand it can nevertheless still be illustrated informally at this point.

Suppose that K is

```
1 1 0 0
0 1 1 0
0 0 1 1
1 0 0 1
0 0 0 0
1 1 1 1
```

In this case the two most obvious choices would appear (to most people at least) to be 0 0 0 0 and 1 1 1 1. However if we take instead the subset

$$1\ 1\ 0\ 0$$
$$0\ 0\ 0\ 0$$
$$1\ 1\ 1\ 1$$

of K then it would seem that now 1 1 0 0 has become the obvious choice, not 0 0 0 0 or 1 1 1 1.

Despite this shortcoming in certain cases we still feel is of some theoretical interest at least to persevere with this principle, and also because of the conclusions it leads to. We note that in so as far as nothing is known about the nature of the options, the property captured by Obstinacy is widely endorsed by the social choice community (see, e.g. Kalai et al., 2002). Indeed there are also a number of related principles in that discipline which may warrant consideration vis-a-vis our present intention, though our initial investigations to date along these lines have not yielded any worthwhile new insights.

In order to introduce our final principle we need a little notation. For $K \in \mathbb{K}$ we say that $X \subseteq A$ is a *support* of K if whenever $f, g \in 2^A$ and f restricted to X (i.e. $f \restriction X$) agrees with g restricted to X then $f \in K$ if and only if $g \in K$.

The set A itself is trivially a support for every $K \in \mathbb{K}$. More significantly it is straightforward to show that the intersection of two supports of K is also a support, and hence that every $K \in \mathbb{K}$ has a unique smallest support. Notice that if K has support X then $K\sigma$ has support $\sigma^{-1}X$.

If K has support X then it is useful to think of this knowledge as telling the agent (just) how elements of K act on X. Namely, for f to be in K it is necessary and sufficient that $f \restriction X = g$ for some

$$g \in \{h \restriction X | h \in K\}.$$

Irrelevance
Suppose $K_1, K_2 \in \mathbb{K}$ with supports X_1, X_2 respectively and for any $f_1 \in K_1$ and $f_2 \in K_2$ there exists $f_3 \in \mathbb{W}$ such that $f_3 \restriction X_1 = f_1 \restriction X_1$ and $f_3 \restriction X_2 = f_2 \restriction X_2$. Then

(Irr) $R(K_1) \restriction X_1 = R(K_1 \cap K_2) \restriction X_1$

where

$$R(K) \restriction X = \{f \restriction X | f \in R(K)\}.$$

[85]

The condition on K_1, K_2 amounts to saying that *as far as* K_1 *is concerned* K_2 *is irrelevant (and conversely)* because given that we know (only) that f satisfies the requirement for membership of K_1 (i.e. that $f \upharpoonright X_1$ is amongst some particular set of functions on X_1) the additional information that $f \in K_2$ tells us nothing we didn't already know about $f \upharpoonright X_1$.

The principle then amounts to saying that in these circumstances the choices from K_1 and K_2 should also reflect that irrelevance. That is, if $f_1 \in R(K_1)$, then there is an $f_3 \in R(K_1 \cap K_2)$ such that $f_3 \upharpoonright X_1 = f_1 \upharpoonright X_1$ and conversely given $f_3 \in R(K_1 \cap K_2)$ there exists such a f_1 (and similarly for K_2).

The justification for this is along the following lines. In choosing a most popular point from K_1 we are effectively choosing from $K_1 \upharpoonright X_1$ and then choosing from all possible extensions (in $\mathbb{W}$) of these maps to domain A, and similarly for K_2. The given conditions allow that in choosing from $K_1 \cap K_2$ we can first freely choose from $K_1 \upharpoonright X_1$ then from $K_2 \upharpoonright X_2$ and finally freely choose from all possible extensions to domain A. Viewed in this way it seems then that any function in $R(K_1) \upharpoonright X_1$ should also be represented in $R(K_1 \cap K_2) \upharpoonright X_1$.[3]

DEFINITION. We shall say that a reason R is a Regulative Reason if is satisfies Renaming, Obstinacy and Irrelevance.

3.1. *The Regulative Reason Characterized*

We start by noticing that there certainly is one Reason satisfying the common sense properties defined above, namely the *trivial Reason R* such that $R(K) = K$ for all $K \in \mathbb{K}$, though of course in practice this 'reason' amounts to nothing at all.[4]

THEOREM 1. Let R be a Regulative Reason. Then either R is trivial or $R = R_0$ or $R = R_1$ where for $i = 0, 1$ R_i is defined by

$$R_i(K) = \{f \in K \mid \forall g \in K, |f^{-1}(i)| \geq |g^{-1}(i)|\}.$$

Conversely each of these three Reasons are Regulative, i.e. satisfy Renaming, Obstinacy and Irrelevance.

We begin with the proof of the "if" part. As usual, $\vec{0} : A \to 2$ is defined by $\vec{0}(x) = 0$ for all $x \in A$ and similarly, $\vec{1} : A \to 2$ is defined by $\vec{1}(x) = 1$ for all $x \in A$.

The first step consists in showing that Regulative Reasons are indeed threefold.

LEMMA 2. Let R be Regulative. Then either $R(2^A) = 2^A$ or $R(2^A) = \{\vec{0}\}$ or $R(2^A) = \{\vec{1}\}$.

Proof. We first show the following claim:

If $f, g \in R(2^A)$ (possibly $f = g$) are such that 0, 1 are in the ranges of f, g respectively, then $R(2^A) = 2^A$.

To this end let $f, g \in R(2^A)$ and $f(x) = 0$ and $g(y) = 1$ for some $x, y \in A$. For σ a permutation on A transposing only x and y we have that $2^A \sigma = 2^A$. Hence, by Renaming, $R(2^A)\sigma = R(2^A\sigma)$. In particular:

(1) $f \in R(2^A) \implies f\sigma \in R(2^A)$.

Now let $K = \{h \in 2^A | h(y) = 0\}$. Since $f\sigma \in R(2^A) \cap K \neq \emptyset$ then:

$$R(2^A) \cap K = R(2^A \cap K)\text{(by Obstinacy)}$$
$$= R(K).$$
(2) $\therefore f\sigma \in R(K).$

Put $K_1 = 2^A$ and $K_2 = K$ with support $X_1 = A - \{y\}$ and $X_2 = \{y\}$, respectively. As $\emptyset = \{y\} \cap X_1$, we can, for any $f_1 \in 2^A$ and $f_2 \in K$, construct a function $f_3 \in 2^A$ such that $f_3 \upharpoonright X_1 = f_1 \upharpoonright X_1$ and $f_3 \upharpoonright X_2 = f_2 \upharpoonright X_2$. Thus

$$R(2^A) \upharpoonright X_1 = R(2^A \cap K) \upharpoonright X_1 \quad \text{(by Irrelevance)}$$
(3) $$= R(K) \upharpoonright X_1.$$

Therefore, $g \upharpoonright X_1 \in R(K) \upharpoonright X_1$. Furthermore for

(4) $g'(z) = \begin{cases} g(z) & \text{if } z \neq y \\ 0 & \text{if } z = y. \end{cases}$

we have that $g' \in R(K)$. Hence $g' \in R(2^A)$, by (2) above.

The claim now follows since we have shown that if we take any function $h \in R(2^A)$ and change its value on one argument the resulting function is also in $R(2^A)$.

The proof of Lemma 2 now follows by noticing that if $R(2^A) \neq 2^A$ then by the claim either 0 or 1 is not in the range of any $f \in R(2^A)$. Therefore, since $R(2^A) \neq \emptyset$ it must either be that $R(2^A) = \{\vec{0}\}$ or $R(2^A) = \{\vec{1}\}$.   □

Our next step is to prove the required result for trivial Reasons.

[87]

LEMMA 3. If $R(2^A) = 2^A$, then $R(K) = K$ for any $K \in \mathbb{K}$.

Proof. Notice that if $R(2^A) = 2^A$ then for $K \in \mathbb{K}$,

$$K \cap R(2^A) = K \neq \emptyset$$

so by Obstinacy,

$$R(K) = K \cap R(2^A) = K. \qquad \square$$

Hence, the final step in the proof of the "if" direction of Theorem 1 deals with the more interesting case of non-trivial Reasons.

It will be useful here to introduce a little notation. For the remainder of this section, let $\pi : dom(\pi) \to \{0, 1\}$, where the domain of π, $dom(\pi)$, $dom(\pi)$, is a subset of A. Similarly for $\pi_1, \ldots, \pi_k$. For such a π let

$$X_\pi = \{f \in 2^A | f \restriction dom(\pi) = \pi\}.$$

LEMMA 4. If $R(2^A) = \{\vec{1}\}$, then

$$R(X_\pi) = \{\pi \vee \vec{1}\},$$

where

$$(5) \qquad \pi \vee \vec{1}(x) = \begin{cases} \pi(x) & \text{if } x \in dom(\pi) \\ \vec{1}(x) & \text{otherwise.} \end{cases}$$

Proof. Suppose that $Z \in A - dom(\pi)$. To prove the result it is enough to show that $f(z) = 1$ for $f \in R(X_\pi)$. Let $K_1 = 2^A$ with support $\{z\}$ and $K_2 = X_\pi$ with support $dom(\pi)$. Notice that the conditions for the applications of Irrelevance are met since $\emptyset = \{z\} \cap dom(\pi)$. Hence

$$R(2^A) \restriction \{z\} = R(X_\pi) \restriction \{z\}.$$

Therefore, for $f \in R(X_\pi)$

$$(6) \qquad f(z) = \begin{cases} 1 & \text{if } z \in A - dom(\pi), \\ \pi(x) & \text{if } z \in dom(\pi), \end{cases}$$

making $f = \pi \vee \vec{1}$. $\qquad \square$

This can be immediately generalized as follows.

[88]

LEMMA 5. Suppose $R(2^A) = \{\vec{1}\}$ and let $Z = \{z_1, z_2, \ldots, z_n\} \subseteq A$ with $0 \leqslant r \leqslant n$. Let $\tau_1^r, \tau_2^r, \ldots, \tau_q^r$ be all the maps from a subset of size r of Z to $\{0\}$. Then

$$R\big(X_{\tau_1^r} \cup X_{\tau_2^r} \cup \cdots \cup X_{\tau_q^r}\big) = \{\tau_1^r \vee \vec{1}, \tau_2^r \vee \vec{1}, \ldots, \tau_q^r \vee \vec{1}\}.$$

Proof. We first recall that, by the definition of R,

$$(7) \qquad R\big(X_{\tau_1^r} \cup X_{\tau_2^r} \cup \cdots \cup X_{\tau_q^r}\big) \subseteq \big(X_{\tau_1^r} \cup X_{\tau_2^r} \cup \cdots \cup X_{\tau_q^r}\big)$$

Now let π be a permutation of A such that $Z\pi = Z$. Then

$$\big(X_{\tau_1^r} \cup X_{\tau_2^r} \cup \cdots \cup X_{\tau_q^r}\big)\pi = \big(X_{\tau_1^r} \cup X_{\tau_2^r} \cup \cdots \cup X_{\tau_q^r}\big).$$

Hence, by Renaming:

$$\begin{aligned} &f \in R\big(X_{\tau_1^r} \cup X_{\tau_2^r} \cup \cdots \cup X_{\tau_q^r}\big) \\ (8) \qquad &\Longleftrightarrow f\pi \in R\big(X_{\tau_1^r} \cup X_{\tau_2^r} \cup \cdots \cup X_{\tau_q^r}\big) \end{aligned}$$

By Equation (7), $R\big(X_{\tau_1^r} \cup X_{\tau_2^r} \cup \cdots \cup X_{\tau_q^r}\big) \cap X_{\tau_j^r} \neq 0$, for some $0 \leqslant j \leqslant q$. Thus, by Obstinacy,

$$\begin{aligned} &R\big(X_{\tau_1^r} \cup X_{\tau_2^r} \cup \cdots \cup X_{\tau_q^r}\big) \cap X_{\tau_j^r} \\ &= R\big((X_{\tau_1^r} \cup X_{\tau_2^r} \cup \cdots \cup X_{\tau_q^r}) \cap X_{\tau_j^r}\big) \\ (9) \qquad &= R(X_{\tau_j^r}) \quad \text{(for some } 0 \leqslant j \leqslant q). \end{aligned}$$

Recalling, from Lemma 4, that $R(X_{\tau_j^r}) = \{\tau_j^r \vee \vec{1}\}$ we have that $\tau_j^r \vee \vec{1} \in R\big(X_{\tau_1^r} \cup X_{\tau_2^r} \cup \cdots \cup X_{\tau_q^r}\big)$ for some $0 \leqslant j \leqslant q$. By equation (8), however, this can be generalized to any $0 \leqslant j \leqslant q$. Hence

$$(10) \qquad R\big(X_{\tau_1^r} \cup X_{\tau_2^r} \cup \cdots \cup X_{\tau_q^r}\big) \supseteq \{\tau_1^r \vee \vec{1}, \tau_2^r \vee \vec{1}, \ldots, \tau_q^r \vee \vec{1}\}.$$

To see that the converse is also true, suppose $h \in R(X_{\tau_1^r} \cup X_{\tau_2^r} \cup \cdots \cup X_{\tau_q^r})$. Then since

$$R\big(X_{\tau_1^r} \cup X_{\tau_2^r} \cup \cdots \cup X_{\tau_q^r}\big) \subseteq X_{\tau_1^r} \cup X_{\tau_2^r} \cup \cdots \cup X_{\tau_q^r},$$

$h \in X_{\tau_j^r}$, for some j. But as we have just observed,

$$R\big(X_{\tau_1^r} \cup X_{\tau_2^r} \cup \cdots \cup X_{\tau_q^r}\big) \cap X_{\tau_j^r} = R\big(X_{\tau_j^r}\big),$$

so $h = \{\tau_j^r \vee \vec{1}\}$ as required. $\qquad \square$

[89]

LEMMA 6. Suppose $Z = \{z_1, z_2, \ldots, z_n\} \subseteq A$ and let $\tau_1^r, \tau_2^r, \ldots, \tau_p^r$ be some maps from a subset of Z of size r to $\{0\}$. Then

$$R\left(X_{\tau_1^r} \cup X_{\tau_2^r} \cup \cdots \cup X_{\tau_p^r}\right) = \{\tau_1^r \vee \vec{1}, \tau_2^r \vee \vec{1}, \ldots, \tau_p^r \vee \vec{1}\}.$$

Proof. Let $\tau_1^r, \tau_1^r, \ldots, \tau_q^r$ be as in Lemma 5. Then by Obstinacy

$$R\left(X_{\tau_1^r} \cup X_{\tau_2^r} \cup \cdots \cup X_{\tau_q^r}\right) = R\left(X_{\tau_1^r} \cup X_{\tau_2^r} \cup \cdots \cup X_{\tau_q^r}\right)$$
$$\cap \left(X_{\tau_1^r} \cup X_{\tau_2^r} \cup \cdots \cup X_{\tau_p^r}\right)$$
$$= \{\tau_1^r \vee \vec{1}, \tau_2^r \vee \vec{1}, \ldots, \tau_p^r \vee \vec{1}\}. \qquad \square$$

We now have all the devices necessary to move on to the crucial step.

LEMMA 7. Let $\tau_1^{r_1}, \tau_2^{r_2}, \ldots, \tau_p^{r_p}$ be maps each from some subset of Z of cardinality $r_1, \ldots, r_p$ to $\{0\}$ respectively. If $R(2^A) = \{\vec{1}\}$, then for $r = \min \{r_i | i = 1, \ldots, p\}$

$$R\left(X_{\tau_1^{r_1}} \cup X_{\tau_2^{r_2}} \cup \cdots \cup X_{\tau_p^{r_p}}\right) = \left\{\tau_j^{r_j} \vee \vec{1} | r_j = r\right\}.$$

Proof. Let $\delta_1^r, \delta_2^r, \ldots, \delta_q^r$ be all the maps from a subset of size r of Z to $\{0\}$. Then

(11) $\{\tau_j^{r_j} \vee \vec{1} | r_j = r\} \subseteq \{\delta_i^r \vee \vec{1} | i = 1, \ldots, q\}.$

Now, since each $X_{\tau_i^{r_i}} \subseteq X_{\delta_k^r}$, for some k, by Lemma 6 above and (11)

$$R\left(X_{\tau_1^{r_1}} \cup X_{\tau_2^{r_2}} \cup \cdots \cup X_{\tau_p^{r_p}}\right) = R\left(X_{\delta_1^r} \cup X_{\delta_2^r} \cup \cdots \cup X_{\delta_q^r}\right)$$
$$\cap \left(X_{\tau_1^{r_1}} \cup X_{\tau_2^{r_2}} \cup \cdots \cup X_{\tau_p^{r_p}}\right)$$
$$= \{\tau_j^{r_j} \vee \vec{1} | r_j = r\}. \qquad \square$$

COROLLARY 8. For $X \in \mathbb{K}$, if $R(2^A) = \{\vec{1}\}$ then

$$R(X) = \left\{f \in X \mid |f^{-1}\{0\}| = r\right\},$$

where r is minimal such that $|f^{-1}\{0\}| = r$ for some $f \in X$.

Proof. The result follows as an immediate consequence of Obstinacy and Lemma 7. $\qquad \square$

[90]

Notice that by duality, Corollary 8 holds for $\vec{1}$ being replaced by $\vec{0}$.

This completes the proof of the "if" direction of Theorem 1. We now move on to show its converse, namely that if a Reason $R(\cdot)$ is defined in any of the above three ways, then Renaming, Irrelevance and Obstinacy are satisfied. This clearly characterizes completely Regulative Reasons for the special case in which worlds are maps from a finite set A.

Again, we start with the trivial Reasons, and then we move on to the case of the non-trivial ones.

LEMMA 9. Suppose $R(X) = X$, for all $X \in \mathbb{K}$. Then Renaming, Obstinacy and Irrelevance are satisfied.

Proof. (Renaming) Suppose $K \in \mathbb{K}$ with support $X \subseteq A$ and π is a permutation of A. Then

$$R(K)\pi = K\pi = R(K\pi)$$

as required.

(Obstinacy) For $K_1, K_2 \in \mathbb{K}$, with supports $X_1, X_2 \subseteq A$ respectively,

$$R(K_1) \cap K_2 = K_1 \cap K_2 = R(K_1 \cap K_2)$$

as required.

(Irrelevance) Suppose $K_1, K_2 \in \mathbb{K}$ (with supports X_1, X_2 respectively) are such that for any $f_1 \in K_1$, $f_2 \in K_2$, there exists $f_3 \in \mathbb{W}$ such that $f_3 \restriction X_1 = f_1 \restriction X_1$ and $f_3 \restriction X_2 = f_2 \restriction X_2$. We have to show that $R(K_1) \restriction X_1 = R(K_1 \cap K_2) \restriction X_1$. Let $g \in 2^{X_1}$. If $g \in R(K_1 \cap K_2) \restriction X_1$ then obviously $g \in R(K_1) \restriction X_1$. As to the other direction, suppose $g = f_1 \restriction X_1$ with $f_1 \in K_1$. Then we are given that for $f_2 \in K_2$ there is $f_3 \in \mathbb{W}$ such that $f_3 \restriction X_1 = f_1 \restriction X_1 = g$ and $f_3 \restriction X_2 = f_2 \restriction X_2$. Thus, $f_3 \in K_1 \cap K_2$ and $g = f_3 \restriction X_1 \in R(K_1 \cap K_2) \restriction X_1$, as required. $\square$

LEMMA 10. $R_1(K)$ satisfies Renaming, Obstinacy and Irrelevance.

Proof. (Renaming) Let σ be a permutation of A. Then

$$f \in R_1(K)\sigma \Longleftrightarrow f = g\sigma, \quad \text{for some } g \in R_1(K)$$

$$(12) \qquad \Longleftrightarrow f = g\sigma, \quad \text{for some } g \in \{h \in K \mid |h^{-1}\{1\}| = r\}$$

where $r = \max\{|h^{-1}\{1\}| \mid h \in K\}$. But since $|h^{-1}\{1\}| = |(h\sigma)^{-1}\{1\}|$, then

$$h \in K \quad \text{and} \quad |h^{-1}\{1\}| = r \Longleftrightarrow h\sigma \in K\sigma \quad \text{and} \quad |(h\sigma)^{-1}\{1\}| = r.$$

[91]

and $r = \max\{|(h\sigma)^{-1}\{1\}|\,|\,h\sigma \in K\sigma\}$. Hence

$$f \in R_1(X) \Longleftrightarrow f\sigma \in R_1(X\sigma),$$

as required.

(Obstinacy) Let $K_1, K_2 \in \mathbb{K}$ and let $R_1(K_1) \cap K_2 \neq \emptyset$ and set

$$r' = \max\{|g^{-1}\{1\}|\,|\,g \in K_1 \cap K_2\}$$

We claim that $r' = r$, where r is defined as above. To see that the result follows from this claim notice that if $r' = r$, then

$$\begin{aligned}
R(K_1 \cap K_2) &= \{f \in K_1 \cap K_2 \,|\, |f^{-1}\{1\}| = r\} \\
&= \{f \in K_1 \,|\, |f^{-1}\{1\}| = r\} \cap K_2 \\
&= R_1(K_1) \cap K_2.
\end{aligned}$$

We show the claim by contradiction. Since $K_1 \cap K_2 \subseteq K_1$, the case $r' > r$ is clearly not possible. To see that $r' < r$ is not possible either, and hence that $r' = r$, let $h \in R_1(K_1) \cap K_2$. Then r' would be the largest n for which there exists $h' \in K_1 \cap K_2$ such that $|h'^{-1}\{1\}| = n$. But since $h \in R_1(K_1)$, r would be such an n, giving $r' \geqslant r$ as required.

(Irrelevance) Suppose $K_1, K_2 \in \mathbb{K}$ (with supports X_1, X_2, respectively) and for any $f_1 \in K_1$, $f_2 \in K_2$, there exists $f_3 \in \mathbb{W}$ such that $f_3 \upharpoonright X_1 = f_1 \upharpoonright X_1$ and $f_3 \upharpoonright X_2 = f_2 \upharpoonright X_2$. We have to show that

$$R_1(K_1) \upharpoonright X_1 = R_1(K_1 \cap K_2) \upharpoonright X_1.$$

So assume that $g \in R_1(K_1) \upharpoonright X_1$. Then $\exists f_1 \in R_1(K_1)$ such that $f_1 \upharpoonright X_1 = g$. We now claim that

(13) $\forall x \notin X_1 \; f_1(x) = 1$.

Suppose otherwise and define

$$f'(x) = \begin{cases} f_1(x) & \text{if } x \in X_1 \\ 1 & \text{otherwise.} \end{cases}$$

Then $f' \in K_1$ but $|f'^{-1}\{1\}| > |f^{-1}\{1\}|$, which is impossible if $f_1 \in R_1(K_1)$. Hence $X_1 \supseteq \{x \,|\, f_1(x) = 0\}$ (and similarly, $X_2 \supseteq \{x \,|\, f_2(x) = 0\}$, for $f_2 \in R_1(K_2)$). Thus $\exists f \in K_1 \cap K_2$ such that $f \upharpoonright X_1 = f_1 \upharpoonright X_1$ and $f \upharpoonright X_2 = f_2 \upharpoonright X_2$. Moreover, since $X_1 \cup X_2$ is a support for $K_1 \cap K_2$, can also assume that

(14) $f(x) = 1$,   for all $x \notin X_1 \cup X_2$.

[92]

Claim now that there is no $h \in K_1 \cap K_2$ such that

(15) $|h^{-1}\{1\}| > |f^{-1}\{1\}|$.

Suppose on the contrary that such an h existed. By (14) we may assume $h(x) = 1$ for all $x \notin X_1 \cup X_2$. Notice first that

(16) $x \in X_1 \cap X_2 \Rightarrow f(x) = h(x)$.

To see this, notice that $f \in K_1$, $h \in K_2$. So $\exists g'$ such that $g' \restriction X_1 = f \restriction X_1$ and $g' \restriction X_2 = h \restriction X_2$. Hence $f(x) = g'(x) = h(x)$, as required. Now,

$$|h^{-1}\{1\}| = \overbrace{|\{y \in X_1 - X_2 | h(y) = 1\}|}^{\alpha^h} + |\{y \in X_2 - X_1 | h(y) = 1\}|$$
$$+ |\{y \in X_2 \cap X_1 | h(y) = 1\}|.$$

and

$$|f^{-1}\{1\}| = \overbrace{|\{y \in X_1 - X_2 | f(y) = 1\}|}^{\alpha^f} + |\{y \in X_2 - X_1 | f(y) = 1\}|$$
$$+ |\{y \in X_2 \cap X_1 | f(y) = 1\}|.$$

Without loss of generality then, if $|h^{-1}\{1\}| > |f^{-1}\{1\}|$ then $\alpha^h > \alpha^f$. But this leads to the required contradiction. To see that define

$$h'(z) = \begin{cases} h(z) & \text{if } z \in X_1 \\ 1 & \text{otherwise.} \end{cases}$$

Then $h' \in K_1$ but $|h'^{-1}\{1\}| = |h^{-1}\{1\} \cap X_1| > |f_1^{-1}\{1\}|$, and this is clearly inconsistent with $f_1 \in R(K_1)$. So $f \in R(K_1 \cap K_2)$ and hence $g \in R(K_1 \cap K_2) \restriction X_1$, as required for this direction of the proof.

As to the other direction for Irrelevance, assume that $g \in R(K_1 \cap K_2) \restriction X_1$ but $g \notin R(K_1) \restriction X_1$. Define

$$g'(x) = \begin{cases} g(x) & \text{if } x \in X_1 \\ 1 & \text{otherwise.} \end{cases}$$

Then, $g' \in K_1$ as it agrees on X_1 with $g \in K_1$. Indeed $g' \notin R(K_1) \restriction X_1$ too, since $g' \restriction X_1 = g \restriction X_1$. Hence $\exists f \in R(K_1)$ such that

(17) $|\{y \in X_1 | f(y) = 1\}| > |\{y \in X_1 | g(y) = 1\}|$.

Now pick $h \in R(K_1 \cap K_2)$ such that $h \restriction X_1 = g$ and define f' such that $f' \restriction X_1 = f \restriction X_1$ and $f' \restriction X_2 = h \restriction X_2$. As above we can assume that

(18) $f'(x) = 1$ for all $x \notin X_1 \cup X_2$

[93]

Then $f' \in K_1 \cap K_2$ and $|f'^{-1}\{1\} \cap X_1| > |h^{-1}\{1\} \cap X_1|$ (by (17) and the facts $f' \upharpoonright X_1 = f \upharpoonright X_1$ and $h \upharpoonright X_1 = g$). Thus, since $|f'^{-1}\{1\} \cap X_2| = |h^{-1}\{1\} \cap X_2|$ and $f' \upharpoonright X_1 \cap X_2 = h \upharpoonright X_1 \cap X_2$, we have that

$$|f'^{-1}\{1\} \cap (X_1 \cup X_2)| > |h^{-1} \upharpoonright \{1\} \cap (X_1 \cup X_2)|.$$

But this is inconsistent with the maximality of $|h^{-1}\{1\}|$, concluding the proof of the converse of Theorem 1.   □

A pleasing aspect of Theorem 1 is that it seems to us to point to precisely the answer(s) that people commonly do come up with when presented with this choice problem. For example in the case

0 0 0 1
0 1 0 0
0 1 1 0
1 1 1 1
0 0 1 0

it is our experience that the fourth row, 1 1 1 1, is the favoured choice. In other words the (unique) choice according to R_1. Of course that is not the only Regulative Reason, R_0 gives {0001, 0100, 0010} whilst the trivial reason gives us back the whole set. Clearly though those two Reasons could be seen as inferior to R_1 here because they ultimately require a random choice from a larger set, thus increasing the probability of non-agreement. (This idea will be explored further in the next chapter when we come to Reasons based on Ambiguity.) This seems to point to a further elaboration of our picture whereby the agent might for a particular K experiment with several Reasons and ultimately settle for a choice which depends on K itself.[5] We shall return to this point later.

Of course one might argue in this example that in making the choice of 1 1 1 1 one was not *consciously* aware of any obligation to satisfy Renaming, Obstinacy and Irrelevance. Be that as it may it is nevertheless interesting we feel that observance of these principles turns out to be both so restrictive and to rather frequently leads to 'the people's choice'. Notice too that if one does adopt a Regulative Reason then one automatically also observes Obstinacy. This *could* then be offered as a defense of Obstinacy against the earlier criticism, that it is no more unreasonable than adopting a Regulative Reason. Whether or not there are alternative sets of 'justified' principles which yield interesting families of reasons such as the one we have considered here remains a matter for further investigation.

4. THE MINIMUM AMBIGUITY REASON

4.1. *An Informal Procedure*

In the previous section we saw how an agent might arrive at a particular canonical Reason by adopting and adhering to certain principles, principles which (after some consideration) one might suppose any other like-minded agent might similarly come to. An alternative approach, which we shall investigate in this section, is to introduce a notion of 'distinguishability', or 'indistinguishability', between elements of K and chose as $R(K)$ those most distinguished, equivalently *least ambiguous*, elements. Instead of being based on principles this $R(K)$ will in the first instance be specified by a procedure, or algorithm, for constructing it.

The idea behind the construction of $R(K)$ is based on trying to fulfill two requirements. The first requirement is that if f and g are, as elements of K, *indistinguishable*, then $R(K)$ should not contain one of them, f say, without also containing the other, g. In other words an agent should not give positive probability to picking one of them but zero probability to picking the other. The argument for this is that if they are 'indistinguishable' on the basis of K then another agent could just as well be making a choice of $R(K)$ which included g but not f. Since agents are trying to make the same ultimate choice of element of K this surely looks like an undesirable situation (and indeed, as will later become clear, taking that route may be worse, and will never be better, than avoiding it).

According to this first requirement then $R(K)$ should be closed under the 'undistinguishability relation'.

The second requirement is that the agent's choice of $R(K)$ should be as small as possible (in order to maximize the probability of randomly picking the same element as another agent) subject to the additional restriction that this way of thinking should not equally permit another like-minded agent (so also, globally, satisfying the first requirement) to make a different choice, since in that case any advantage of picking from the small set is lost.

The first consequence of this is that initially the agent should be looking to choose from those minimal subsets of K closed under indistinguishability, 'minimal' here in the sense that they do not have any proper non-empty subset closed under indistinguishability. Clearly if this set has a unique smallest element then the elements of this set are the least ambiguous, most outstanding, in K and this would be a natural choice for $R(K)$. However, if there are two or more potential choices $X_1, X_2, \ldots, X_k$ at this stage with the same number of elements then the agent could do no worse than combine them into a single potential choice $X_1 \cup X_2 \cup \cdots \cup X_k$ since the choice of any one of them would be open to the obvious criticism that another 'like-minded agent' could make a different (in this case disjoint) choice, which would not improve the chances of a match (and may make them considerably worse if the first agent subsequently rejected $X_1 \cup X_2 \cup \cdots \cup X_k$ in favour of a better choice). Faced with this revelation our agent would realize that the 'smallest' way open to reconcile these alternatives is to now permit $X_1 \cup X_2 \cup \cdots \cup X_k$ as a potential choice whilst dropping $X_1, X_2, \ldots, X_k$.[6]

The agent now looks again for a smallest element from the current set of potential choices and carries on arguing and introspecting in this way until eventually at some stage a unique choice presents itself.

In what follows we shall give a formalization of this procedure.

4.2. Permutations and Ambiguity

The first step in the construction of the Minimum Ambiguity Reason consists in providing the agent with a notion of equivalence or indistinguishability among worlds in a given $K \subseteq 2^A$.

In fact with the minimal structure we have available here the notion we want is almost immediate: Elements g, h of K are *indistinguishable* (with respect to K) if there is a permutation σ of A such that

$$K = K\sigma (= \{f\sigma \mid f \in K\})$$

and $g\sigma = h$.

We shall say that a permutation σ of A is a permutation of K if $K = K\sigma$.

The idea here is that within the context of our choice problem a permutation σ of K maps $f \in K$ to an $f\sigma$ in $K\sigma$ which has essentially the standing within $K\sigma (= K)$ as f had within K. In other words as far as K is concerned f and $f\sigma$ are indistinguishable. The following Lemma is immediate.

LEMMA 11. If σ and τ are permutations of K then so are $\sigma\tau$ and σ^{-1}.

[96]

Having now disposed of what we mean by indistiguishability between elements of $K \subseteq 2^A$, we now recursively define for $f \in K$ *the ambiguity class of f within K* at *level m* by:

$$\mathbb{S}_0(K, f) = \{g \in K \mid \exists \text{ permutation } \sigma \text{ of } K \text{ such that } f\sigma = g\}$$

$$\mathbb{S}_{m+1}(K, f) = \begin{cases} \{g \in K \mid |\mathbb{S}_m(K, f)| = |\mathbb{S}_m(K, g)|\} & \text{if } |\mathbb{S}_m(K, f)| \leqslant m+1, \\ \mathbb{S}_m(K, f) & \text{otherwise.} \end{cases}$$

For $f, g \in K$ define the relation

$$g \sim_m f \Leftrightarrow g \in \mathbb{S}_m(K, f).$$

LEMMA 12. $\sim_m$ is an equivalence relation.

Proof. By induction on m. For the case $m = 0$ this is clear since if $f, g, h \in K$ and $f\sigma = g$, $g\tau = h$ with σ, τ permutations of K then $g\sigma^{-1} = f$, $f\sigma\tau = h$ and by Lemma 11 $\sigma^{-1}, \sigma\tau$ are also permutations of K.

Assume true for m. If $|\mathbb{S}_m(K, f)| > m+1$ then, by the definition of $\mathbb{S}_{m+1}(K, f)$, the result follows immediately from the inductive hypothesis. Otherwise, the reflexivity of $\sim_m$ is again immediate. For symmetry assume that $g \in \mathbb{S}_{m+1}(K, f)$. Then $g \in \{h \in K \mid |\mathbb{S}_m(K, h)| = |\mathbb{S}_m(K, f)|\}$, so $|\mathbb{S}_m(K, g)| = |\mathbb{S}_m(K, f)|$ and $f \in \{h \in K \mid |\mathbb{S}_m(K, h)| = |\mathbb{S}_m(K, g)|\}$. An analogous argument shows that $\sim_{m+1}$ is also transitive. □

Thus, as f ranges over K, $\sim_m$ induces a partition on K and the sets $\mathbb{S}_m(K, f)$ are its equivalence classes. Moreover, this m-th partition is a refinement of the $m+1$st partition. In other words, the sets $\mathbb{S}_m(K, f)$ are increasing and so eventually constant fixed at some set which we shall call $\mathbb{S}(K, f)$.

The ambiguity of f within K is then defined by:

$$\mathbb{A}(K, f) =_{def} |\mathbb{S}(K, f)|.$$

Finally, we can define the *Minimum Ambiguity Reason* $R_{\mathbb{A}}(K)$ by letting:

(19) $\quad R_{\mathbb{A}}(K) = \{f \in K \mid \forall g \in K, \mathbb{A}(K, f) \leqslant \mathbb{A}(K, g)\}.$

As a rather self evident consequence of the definition of $R_{\mathbb{A}}$ we have the following result.

PROPOSITION 13. $R_{\mathbb{A}}(K) = \mathbb{S}(K, f)$, for any $f \in R_{\mathbb{A}}(K)$

[97]

Proof. Let $f \in R_{\mathbb{A}}(K)$. To show that $\mathbb{S}(K, f) \subseteq R_{\mathbb{A}}(K)$ suppose $\mathbb{S}(K, f) = \mathbb{S}_m(K, f)$ and $g \in \mathbb{S}_m(K, f)$. Then $\mathbb{S}m(K, g) = \mathbb{S}_m(K, f)$ so $\mathbb{S}_m(K, g)$ must equal $\mathbb{S}(K, g)$ (since m could be taken arbitrarily large) and $|\mathbb{S}(K, g)| = |\mathbb{S}(K, f)|$, so $g \in R_{\mathbb{A}}(K)$. Conversely let $g \in R_{\mathbb{A}}(K)$ and fix some large m. If $g \notin \mathbb{S}(K, f)$, then $\mathbb{S}(K, f) \cap \mathbb{S}(K, g) = \emptyset$ and since both f and g are in $R_{\mathbb{A}}(K)$, then $|\mathbb{S}(K, f)| = |\mathbb{S}(K, g)|$. But this leads to the required contradiction since for m large enough, $|\mathbb{S}_m(K, f)| \leqslant m + 1$ so $\mathbb{S}_m(K, f)$ and $\mathbb{S}_m(K, g)$ would both be proper subsets of $\mathbb{S}_{m+1}(K, f)$. Thus g would eventually be in $\mathbb{S}_m(K, f)$, contradicting the hypothesis. $\square$

EXAMPLE. Let $K \in K$ and suppose that as f ranges over K the 0-ambiguity classes of f in K are given by the following partition of K

$$\{a_1, a_2\}, \{b_1, b_2\}, \{c_1, c_2\},$$
$$\{d_1, d_2, d_3\}, \{e_1, e_2, e_3\},$$
$$\{f_1, f_2, \ldots, f_6\}, \{g1, g2, \ldots, g_6\},$$
$$\{h_1, h_2, \ldots, h_{12}\},$$
$$\{i_1, i_2, \ldots, i_{24}\}.$$

For $m = 1$ the classes remain fixed. For $m = 2$ the first three classes get combined and the $\mathbb{S}_2(K, f)$ look like

$$\{a_1, a_2, b_1, b_2, c_1, c_2\},$$
$$\{d_1, d_2, d_3\}, \{e_1, e_2, e_3\},$$
$$\{f_1, f_2, \ldots, f_6\}, \{g_1, g_2, \ldots, g_6\},$$
$$\{h_1, h_2, \ldots, h_{12}\},$$
$$\{i_1, i_2, \ldots, i_{24}\}.$$

Similarly for $m = 3$ where the two classes of size 3 are combined so that the $\mathbb{S}_3(K, f)$ become

$$\{a_1, a_2, b_1, b_2, c_1, c_2\},$$
$$\{d_1, d_2, d_3, e_1, e_2, e_3\},$$
$$\{f_1, f_2, \ldots, f_6\}, \{g_1, g_2, \ldots, g_6\},$$
$$\{h_1, h_2, \ldots, h_{12}\},$$
$$\{i_1, i_2, \ldots, i_{24}\}.$$

The ambiguity classes do not change until step 6 when the four classes with six elements are combined making $\mathbb{S}_6(K, f)$ look like

$$\{a_1, a_2, b_1, b_2, c_1, c_2, d_1, d_2, d_3, e_1, e_2, e_3, f_1, f_2, \ldots, f_6, g_1, g_2, \ldots, g_6\},$$
$$\{h_1, h_2, \ldots, h_{12}\},$$
$$\{i_1, i_2, \ldots, i_{24}\}.$$

Finally, we combine the two classes with 24 elements and obtain $\mathbb{S}_{24}(K, f)$ with just two classes

$$\{a_1, a_2, b_1, b_2, c_1, c_2, d_1, d_2, d_3, e_1, e_2, e_3, f_1, f_2, \ldots,$$
$$f_6, g_1, g_2, \ldots, g_6, i_1, i_2, \ldots, i_{24}\},$$
$$\{h_1, h_2, \ldots, h_{12}\}.$$

Clearly the ambiguity classes stabilize at this 24-th step and hence the Minimum Ambiguity Reason for this K gives the 12-set $\{h_1, h_2, \ldots, h_{12}\}$.

Notice that, in the definition of the ambiguity classes of K, the splitting of the inductive step into two cases is indeed necessary to ensure that some sets closed under permutations of K are not dismissed unnecessarily early. This same example shows that if we allowed the inductive step in the definition to be replaced by the (somehow more intuitive) equation

(20) $\mathbb{S}_{m+1}(K, f) = \{g \in K \mid |\mathbb{S}_m(K, f)| = |\mathbb{S}_m(K, g)|\}$

we would fail to pick the "obvious" smallest such subset of K. To see this suppose again that K is as above but this time the alternative procedure based on (20) was used to construct $R_\mathbb{A}$. Then we would have all the classes of the same size all merged in one step so that the 1-ambiguity classes $\mathbb{S}_1(K, f)$ would look like:

$$\{a_1, a_2, b_1, b_2, c_1, c_2\},$$
$$\{d_1, d_2, d_3, e_1, e_2, e_3\},$$
$$\{f_1, f_2, \ldots, f_6, g_1, g_2, \ldots, g_6\},$$
$$\{h_1, h_2, \ldots, h_{12}\},$$
$$\{i_1, i_2, \ldots, i_{24}\}.$$

Then $\mathbb{S}_2(K, f)$ would look like this:

$$\{a_1, a_2, b_1, b_2, c_1, c_2, d_1, d_2, d_3, e_1, e_2, e_3\},$$
$$\{f_1, f_2, \ldots, f_6, g_1, g_2, \ldots, g_6, h_1, h_2, \ldots, h_{12}\},$$
$$\{i_1, i_2, \ldots, i_{24}\},$$

so that the procedure stabilizes at $m = 3$ with $\mathbb{S}(K, f)$ of the form:

$$\{a_1, a_2, b_1, b_2, c_1, c_2, d_1, d_2, d_3, e_1, e_2, e_3\},$$
$$\{f_1, f_2, \ldots, f_6, g_1, g_2, \ldots, g_6, h_1, h_2, \ldots, h_{12}, i_1, i_2, \ldots, i_{24}\},$$

Hence, the construction that follows the alternative definition of ambiguity classes, which imposes no restriction on appropriate stage for the combination of the classes, leads again to a 12-set. However, this alternative procedure appears to miss out what naturally seems to be a more distinguished subset of K.

4.3. *Justifying the Minimum Ambiguity Reason*

We now want to show that the Minimum Ambiguity Reason defined in (19) is an adequate formalization of the informal description given in Section 4.1. Recall that we put forward two informal desiderata for the resulting selection from K, firstly that it should be closed under indistinguishability and secondly that it should be the unique smallest possible such subset not eliminated by there being a like-minded agent who by similar reasoning could arrive at a different answer.

As far as the former is concerned notice that by Lemma 13 $R_{\mathbb{A}}(K)$ is closed under all the $\sim_m$, not just $\sim_0$. Thus this requirement of closure under indistinguishability is met, *assuming* of course that one accepts this interpretation of 'indistinguishability'. Indeed $R_{\mathbb{A}}$ satisfies Renaming as we now show.

THEOREM 14. $R_{\mathbb{A}}$ satisfies Renaming.

Proof. As usual let σ be a permutation of A. We need to prove that

$$R_{\mathbb{A}}(K)\sigma = R_{\mathbb{A}}(K\sigma).$$

We first show by induction on m that for all $f \in K, \mathbb{S}_m(K, f)\sigma = \mathbb{S}_m(K\sigma, f\sigma)$. To show the base case $m = 0$ for all $f \in K$, let

$$\mathbb{S}_0(K, f) = \{g_1, \ldots, g_q\}.$$

Choose a permutation τ of K such that $f\tau = g_i$. Then $\sigma^{-1}\tau\sigma$ is a permutation of $K\sigma$ and $(f\sigma)\sigma^{-1}\tau\sigma = g_i\sigma$. Hence, $\mathbb{S}_0(K, f)\sigma \subseteq \mathbb{S}_0(K\sigma, f\sigma)$. Similarly, $\mathbb{S}_0(K\sigma, f\sigma)\sigma^{-1} \subseteq \mathbb{S}_0(K, f)$, so equality must hold here.

[100]

Assume now the result for the $\mathbb{S}_m$-th ambiguity class, so we want to prove that

$$\mathbb{S}_{m+1}(K, f)\sigma = \mathbb{S}_{m+1}(K\sigma, f\sigma).$$

We distinguish between two cases, corresponding to the ones appearing in the construction of the ambiguity classes. Recall that $\mathbb{S}_{m+1}(K, f) = \mathbb{S}_m(K, f)$ if $m + 1 > |\mathbb{S}_m(K, f)|$. So, in this case, the result follows immediately by the inductive hypothesis. Otherwise, since σ (on 2^A) is 1-1, it is enough to see that

$$
\begin{aligned}
\mathbb{S}_{m+1}(K, f)\sigma &= \{g \in K \,|\, |\mathbb{S}_m(K, f)| = |\mathbb{S}_m(K, g)|\}\sigma \\
&= \{g\sigma \in K\sigma \,|\, |\mathbb{S}_m(K\sigma, f\sigma)| = |\mathbb{S}_m(K\sigma, f\sigma)|\} \quad \text{(i.h.)} \\
&= \mathbb{S}_{m+1}(K\sigma, f\sigma).
\end{aligned}
$$

Since, by Lemma 13, $R_\mathbb{A}(K)$ is the smallest $\mathbb{S}(K, f)$, this concludes the proof of the Lemma. $\qquad\square$

Before further considering how far our formal construction of $R_\mathbb{A}(K)$ matches the informal description in Section 3.1, it will be useful to have the next result to hand.

THEOREM 15. A non-empty $K' \subseteq K$ is closed under permutations of K into itself if and only if there exists a Reason R satisfying Renaming such that $R(K) = K'$.

Proof. The direction from right to left follows immediately from the Renaming principle. For the other direction define, for $K_1 \subseteq 2^A$, $K_1 \neq \emptyset$,

$$
(21) \quad R(K_1) = \begin{cases} K'\sigma & \text{if } K_1 = K\sigma \text{ for some permutation } \sigma \text{ of A;} \\ K_1 & \text{otherwise.} \end{cases}
$$

Note that in the first case $R(K_1)$ is defined unambiguously, that is to say, whenever we have two permutations σ_1, σ_2 of A such that $K_1 = K\sigma_1 = K\sigma_2$, then $K'\sigma_1 = K'\sigma_2$. This follows since in this case, $\sigma_2\sigma_1^{-1}$ is a permutation of A and $K\sigma_2\sigma_1^{-1} = K$ so $K'\sigma_2\sigma_1^{-1} = K'$, i.e. $K'\sigma_1 = K'\sigma_2$.

We now want to show that if σ is a permutation of A and $K_1\sigma = K_2$ then $R(K_2) = R(K_1)\sigma$. If K_1 is covered by the first case of (21), then so is K_2, for if τ is a permutation of A such that $K_1 = K\tau$, then $K_2 = K\tau\sigma$ and $R(K_1\sigma) = R(K_2) = K'\tau\sigma = R(K_1)\sigma$. If K_1 is covered by the second case of (21), so is K_2 since if $K_2 = K\tau$ for some permutation τ of A, then $K_1 = K\tau\sigma^{-1}$ so $R(K_1)$ would be defined by the first case. It follows then that here we must have $R(K_1\sigma) = R(K_2) = K_2 = K_1\sigma = R(K_1)\sigma$ as required. $\square$

The importance of this result is that in the construction of $R_{\mathbb{A}}(K)$ the choices $\mathbb{S}_m(K, f)$ which were eliminated (by coalescing) because of there currently being available an alternative choice of a $\mathbb{S}_m(K, g)$ of the same size are indeed equivalently being eliminated on the grounds that there is a likeminded agent, even one satisfying Renaming, who could pick $\mathbb{S}_m(K, g)$ in place of $\mathbb{S}_m(K, f)$. In other words it is not as if some of these choices are barred because no agent could make them whilst still satisfying Renaming. Once a level m is reached at which there is a unique smallest $\mathbb{S}_m(K, f)$ this will be the choice for the informal procedure. It is also easy to see that this set will remain the unique smallest set amongst all the subsequent $\mathbb{S}_n(K, g)$, and hence will qualify as $R_{\mathbb{A}}(K)$. In this sense then our formal procedure fulfills the intentions of the informal description of Section 3.1.

4.4. Comparing Regulative and Minimum Ambiguity Reasons

In this and the previous section we have put forward arguments for both the Regulative and Minimum Ambiguity Reasons being considered as 'rational' within the understanding of that term in this paper. Interestingly in practice neither seems to come out self evidently better in all cases. For example, in the case considered earlier of

$$0\,0\,1\,1$$
$$0\,1\,1\,0$$
$$1\,1\,0\,0$$
$$1\,1\,1\,1$$

R_1 gives the singleton $\{1111\}$ whilst $R_{\mathbb{A}}$ gives the somewhat unexceptional $\{0011, 1100\}$ and R_0 the rather useless $\{0011, 0110, 1100\}$. On the other hand if we take the subset

$$0\,0\,1\,1$$
$$0\,1\,1\,0$$
$$1\,1\,0\,0$$

of this set $R_{\mathbb{A}}$ gives $\{0110\}$ whilst both R_1 and R_0 give the whole set.

Concerning the defining principles of the Regulative Reasons, whilst as we have seen $R_{\mathbb{A}}$ does satisfy Renaming the above example shows that it fails to satisfy Obstinacy. Indeed with a little more work we can show that it does not even satisfy Idempotence, that is $R(R(K)) = R(K)$, a consequence of Obstinacy. Finally $R_\mathbb{A}$ does not satisfy Irrelevance either. For an example to show this let K_1 consist of

```
1 0 0 1 * * * * * * *
1 1 0 1 * * * * * * *
1 1 1 1 * * * * * * *
1 0 0 0 * * * * * * *
0 0 0 1 * * * * * * *
```

and let K_2 consist of

```
* * * * 1 0 0 0 0 0 0
* * * * 0 1 0 0 0 0 0
* * * * 0 0 1 0 0 0 0
* * * * 0 0 0 1 1 0 0
* * * * 0 0 0 1 0 1 0
* * * * 0 0 0 1 0 0 1
* * * * 0 0 0 0 1 1 0
* * * * 0 0 0 0 1 0 1
* * * * 0 0 0 0 0 1 1
```

where $*$ indicates a free choice of 0 or 1. Then K_1, K_2 satisfy the requirements of Irrelevance and $R_\mathbb{A}(K_1)$, $R_\mathbb{A}(K_2)$ are respectively

```
                1 1 1 1 1 0 0 0 0 0 0
1 0 0 0 1 1 1 1 1 1 1   0 0 0 0 1 0 0 0 0 0 0
1 0 0 0 0 0 0 0 0 0 0   1 1 1 1 0 1 0 0 0 0 0
0 0 0 1 1 1 1 1 1 1 1   0 0 0 0 0 1 0 0 0 0 0
0 0 0 1 0 0 0 0 0 0 0   1 1 1 1 0 0 1 0 0 0 0
                0 0 0 0 0 0 1 0 0 0 0
```

whereas $R_\mathbb{A}(K_1 \cap K_2)$ is

```
1 0 0 1 1 0 0 0 0 0 0
1 0 0 1 0 1 0 0 0 0 0
1 0 0 1 0 0 1 0 0 0 0
1 1 0 1 1 0 0 0 0 0 0
1 1 0 1 0 1 0 0 0 0 0
1 1 0 1 0 0 1 0 0 0 0
1 1 1 1 1 0 0 0 0 0 0
1 1 1 1 0 1 0 0 0 0 0
1 1 1 1 0 0 1 0 0 0 0
```

[103]

5. THE SMALLEST UNIQUELY DEFINABLE REASON

In this section we present another Reason which, at first sight, looks a serious challenger to the Regulative and Minimum Ambiguity Reasons so far introduced.

Consider again an agent who is given a non-empty subset K of 2^A from which to attempt to make a choice which is common to another like-minded agent. A natural approach here might be for the agent to consider all non-empty subsets of K that could be described, or to use a more formal term, defined, within the *structure available* to the agent. If some individual element was definable (meaning definable in this structure *without parameters*) then this would surely be a natural choice, unless of course there were other such elements. Similarly choosing a small definable set and then choosing randomly from within it would seem a good strategy, *provided there were no other definable sets of the same size*. Reasoning along these lines then suggests that our agent could reach the conclusion that s/he should choose the smallest definable set for which there was no other definable set of the same size.

Of course all this depends on what we take to be the *structure available* to the agent. In what follows we shall consider the case when the agent can recognize 0 and 1, elements of A, $\{0, 1\}$ and K, composition and equality.[7] Precisely, let $\mathcal{M}$ be the structure

$$\langle \{0, 1\} \cup A \cup K, \{0, 1\}, A, K, =, Comp, 0, 1 \rangle$$

where $=$ is equality for $\{0, 1\} \cup A \cup K$ (we assume of course that A, $\{0, 1\}$, 2^A are all disjoint) and $Comp$ is a binary function which on $f \in K$, $a \in A$ gives $f(a)$ (and, say, the first coordinate on arguments not of this form). As usual we shall write $f(a) = i$ in place of $Comp(f, a) = i$ etc.

We define the Uniquely Smallest Definable Reason, $R_{\mathbb{U}}$, by setting $R_{\mathbb{U}}(K)$ to be that smallest $\emptyset \neq K' \subseteq K$ first order definable in $\mathcal{M}$ for which there is no other definable subset of the same size.

The results that follow are directed towards understanding the structure of $R_{\mathbb{U}}(K)$ and its relationship to $R_{\mathbb{A}}(K)$.

LEMMA 16. Every permutation σ_0 of K determines an automorphism j_{σ_0} of $\mathcal{M}$ given by the identity on $\{0, 1\}$ and

(22) $a \in A \mapsto \sigma_0^{-1}(a)$,

and

(23) $f \in K \longrightarrow f\sigma_0.$

Conversely every automorphism j_0 of $\mathcal{M}$ determines a permutation σ_{j_0} of K given by

(24) $\sigma_{j0}(a) = j_0^{-1}(a)$

for $a \in A$.

Furthermore for $f \in K$, $f\sigma_{j_0} = j_0(f)$ and the corresponding automorphism determined by $j_{\sigma_{j_0}}$ is j_0 again.

Proof. For σ_0 a permutation of K it is clear that j_{σ_0} defined by (22) and (23) gives a 1-1 onto mapping from A and K into themselves. All that remains to show this first part is to notice that by direct substitution,

$$j_{\sigma_0}(Comp(f, a)) = Comp(f, a) = f(a) = f\sigma_0(\sigma_0^{-1}(a))$$
$$= Comp(j_{\sigma_0}(f), j_{\sigma_0}(a)).$$

In the other direction let j_0 be an automorphism of $\mathcal{M}$ and define σ_{j_0} by (24). Then since j_0 is an automorphism of $\mathcal{M}, \sigma_{j_0}$ is a permutation of A and for $f \in K, a \in A$,

$$f(a) = j_0(f(a)) = j_0(Comp(f, a)) = Comp(j_0(f), j_0(a)),$$

equivalently,

$$f(a) = j_0(f)(j_0(a)) = j_0(f\sigma_{j_0}^{-1})(a).$$

Hence

$$j_0^{-1}(f)(a) = f\sigma_{j_0}^{-1}(a)$$

so σ_{j0}^{-1} (and hence σ_{j_0} by Lemma 11) is a permutation of K since $j_0^{-1}(f) \in K$, as required.

The last part now follows immediately from the definitions (22), (23), (24).   □

We say that $K' \subseteq K$ satisfies Renaming within K if for all permutations σ of $K, K' = K'\sigma$. Thus 'standard Renaming' is just Renaming within 2^A.

THEOREM 17. A non-empty subset K' of K is definable (without parameters) in $\mathcal{M}$ if and only if K' satisfies Renaming within K.

Proof. Suppose that K' is definable in $\mathcal{M}$. Then clearly K' is fixed under all automorphisms of $\mathcal{M}$. In particular if σ is a permutation of K then by Lemma 16 j_σ is an automorphism of $\mathcal{M}$ so

$$K' = j_\sigma(K') = K'\sigma$$

Conversely suppose that K' satisfies Renaming within K. Then since every automorphism of $\mathcal{M}$ is of the form j_σ for some permutation σ of K and $j_\sigma(K') = K'\sigma = K'$ it follows that K' is fixed under all automorphisms of $\mathcal{M}$. Consider now the types $\theta_1^i(x), \theta_2^i(x), \theta_3^i(x), \ldots$ of the elements f_i of K in $\mathcal{M}$. If there were $f_i \in K'$ and $f_j \notin K'$ with the same type then by a back and forth argument (see for example Marker 2002) we could construct an automorphism of $\mathcal{M}$ sending f_i to f_j, contradicting the fact that K' is fixed under automorphisms. It follows that for some n the formulae $\theta_1^i(x) \wedge \theta_2^i(x) \wedge \cdots \wedge \theta_n^i(x)$ and $\theta_1^j(x) \wedge \theta_2^j(x) \wedge \cdots \wedge \theta_n^j(x)$ are mutually contradictory when $f_i \in K'$ and $f_j \notin K'$. From this it clearly follows that the formula

$$\bigvee_{f_i \in K'} \bigwedge_{m=1}^{n} \theta_m^i(x)$$

defines K' in $\mathcal{M}$ for suitably large n.   □

COROLLARY 18. The sets $\mathbb{S}_m(K, f)$ are definable in $\mathcal{M}$

Proof. These sets are clearly closed under permutations of K so the result follows from Theorem 17.   □

THEOREM 19. For all $K \in \mathbb{K}, |R_{\mathbb{A}}(K)| \leqslant |R_{\mathbb{U}}(K)|$, with equality just if $R_{\mathbb{A}}(K) = R_{\mathbb{U}}(K)$.

Proof. We shall show that for all m. If $f \in R_{\mathbb{U}}(K)$ then $\mathbb{S}_m(K, f) \subseteq R_{\mathbb{U}}(K)$. For $m = 0$ this is clear since $R_{\mathbb{U}}(K)$, being definable must be closed under permutations of K. Assume the result for m and let $f \in R_{\mathbb{U}}(K)$. If $\mathbb{S}_{m+1}(K, f)$ were not a subset of $R_{\mathbb{U}}(K)$ there would be $g \in K$ such that $|\mathbb{S}_m(K, f)| = |\mathbb{S}_m(K, g)|$ but $g \notin R_{\mathbb{U}}(K)$. Indeed $\mathbb{S}_m(K, g)$ would have to be entirely disjoint from $R_{\mathbb{U}}(K)$ by the inductive hypothesis. By Corollary 18 $\mathbb{S}_m(K, f)$ and $\mathbb{S}_m(K, g)$ are both definable, and hence so is

$$R_{\mathbb{U}}(K) \cup \mathbb{S}_m(K, g) - \mathbb{S}_m(K, f).$$

But this set is different from $R_{\mathbb{U}}(K)$ yet has the same size, contradiction.

Having established this fact we notice that for $f \in R_{\mathbb{U}}(K)$ we must have $\mathbb{S}(K, f) \subseteq R_{\mathbb{U}}(K)$ so since $R_{\mathbb{A}}(K)$ is the smallest of the $\mathbb{S}(K, g)$ the result follows.   □

In a way Theorem 19 is rather surprising in that one might initially have imagined that $R_{\mathbb{U}}(K)$, by its very definition, was about as specific a set as one could hope to describe. That $R_{\mathbb{A}}(K)$ can be strictly smaller than $R_{\mathbb{U}}(K)$ can be seen from the case when the $\sim_0$ equivalence classes look like

$$\{a_1, a_2\}, \{b_1, b_2\}, \{c_1, c_2\}, \{d_1, d_2, d_3, d_4\}.$$

In this case $R_{\mathbb{A}}$ gives $\{d_1, d_2, d_3, d_4\}$ whereas $R_{\mathbb{U}}$ just gives the union of all these sets.

We now briefly consider the relationship between the Regulative Reasons and $R_{\mathbb{U}}$. Since the set

$$R_i(K) = \{f \in K \,|\, \forall g \in K, |f^{-1}(i)| \geqslant |g^{-1}(i)|\}.$$

is definable in $\mathcal{M}$ $R_i(K)$ *is* a candidate for $R_{\mathbb{U}}(K)$. So if $|R_i(K)| < |R_{\mathbb{U}}(K)|$ it must be the case that there is another definable subset of $\mathcal{M}$ with the same size as $R_i(K)$. If $|R_i(K)| = |R_{\mathbb{U}}(K)|$ then in fact $R_i(K) = R_{\mathbb{U}}(K)$. From this point of view then $R_{\mathbb{U}}$ (and by Theorem 19 also $R_{\mathbb{A}}$) might be seen to be always at least as satisfactory as the R_i. On the other hand the R_i are in a practical sense computationally undemanding. [The computational complexity of the relation $f \sim_0 g$ between elements of K is currently unresolved, which strongly suggests that even if a polynominal time algorithm does exist it is far from transparent.]

We finally remark that, using the same examples as for $R_{\mathbb{A}}, R_{\mathbb{U}}$ also fails Obstinacy and Irrelevance.

6. AN ANALOGY WITH GAME THEORY

The situation we've been focussing on in this paper can be put quite naturally into game theoretic form. Although a full discussion of this, and the related reinterpretation of the results of the previous section in game theoretic terms are beyond the scope of this initial investigation on rationality-as-conformity, we nonetheless sketch here the main lines of this analysis.

The *conformity game*, as we might call it, is a two-person, non-cooperative game of complete yet imperfect information whose normal form goes like this. Each agent is to choose one strategy out of a set of possible choices, identical for both agents. Each strategy corresponds to one element of $K = \{f_1, \ldots, f_k\}$, say. Agents get a (unique) positive payoff p if they play the same strategy, and nothing otherwise, all this being common knowledge. But since agents are to play simultaneously, they are clearly inaccessible to each other. Since it is a game of multiple Nash-equilibria, the conformity game is therefore a typical example of a (pure) *coordination game*, and as such, it is generally considered to be unsolvable within the framework of traditional game theory.

Recall that in Section 2 we hinted at two general ways of solving the conformity game, corresponding to the following situations. Either worlds in K have no structure other than being distinct elements of a set, or worlds in K do have some structure, and in particular there are properties that might hold (be true) in (of) some worlds. In the former case we seem to be forced to accept that agents have no better way of playing the conformity game other than picking some world $f_i \in K$ at random (i.e. according to the uniform distribution). In the latter case, however, agents might use the information about the structure of the worlds in K to focus on some particularly distinguished possible world. On the assumption of like-mindedness, i.e. common reasoning, if one of those, say f_j should stand out as having some distinguished properties, agents will conclude that such properties are indeed intersubjectively accessible and hence select f_j. In the phraseology of coordination games those distinguished properties essentially contribute towards identifying a *salient* strategy, the corresponding equilibrium being called a *focal point*.[8]

Though the analysis of the relation between rationality-as-conformity on the one hand and the selection of multiple Nash equilibria in (pure) coordination games on the other, goes beyond the scope of this paper, we note here that the Reasons we have been investigating in this paper qualify as natural candidates for a formalization of choice processes leading to focal points.

7. THE RATIONALITY OF CONFORMITY

The general results of this paper can be seen as formalizing the intuition according to which it would be irrational for two commonsensical agents to disagree *systematically* on their world view provided that it can be assumed that they are like-minded and that they are facing essentially the same choice problem. But why is this intuition reasonable within a general understanding of "rationality"?

There are surely several philosophical accounts of rationality that not only seem to be consistent with this intuition but seem to offer it some support. Nozick's theory of practical rationality (Nozick 1993) is surely to be included among those. According to the latter it is a sound principle of rationality that agents "sometimes accept something because others in our society do" (Nozick 1993, p. 129). This clearly finds it underpinnings in the intrinsic fallibility of human-level intelligent agents, indeed in the fact that within a society of rational agents, the systematic error of the majority is somehow less likely that the individual's. Moreover, as noted by Keynes, "Worldly wisdom teaches that it is better for reputation to fail conventionally than to succeed unconventionally" (Keynes 1951 p.158).

The formalisation of rational choice behaviour we have pursued here is subtended by the assumption that reasons are devices agents apply to restrict their options, to go part, or sometimes even all, of the way to choosing a course of action or making a decision. Indeed, we have investigated choice behaviour as a two-stage process. In the first such stage agents apply reasons to *discard* those possible choices that are recognised as being unsuitable for the agent's purpose of conforming. If at the end of this process the agent is left with more than one *equally acceptable* option, then the actual choice is to be finalised by picking randomly (i.e. according to the uniform distribution) from this set.

Thus, it turns out that the general intuitions we have been following in the construction of the Regulative Reasons are remarkably close to those considered by Carnap (in the context of probabilistic confirmation theory) when developing his programme on Inductive Logic:

The person X wishes to assign rational credence values to unknown propositions on the basis of the observations he has made. It is the purpose of inductive logic to help him to do that in a rational way; or, more precisely, to give him some general rules, each of which warns him against certain unreasonable steps. The rules do not in general lead him to specific values; they leave some freedom of choice within certain limits. What he chooses is not a credence value for a given

proposition but rather certain features of a general policy for determining credence values. Hilpinen (1973)

Indeed, as noted in passing above, the principles that make up
the Regulative Reasons, are understood exactly as *policies* helping
agents to achieve their goal by forbidding them to undertake certain
unreasonable steps that could prevent them from conforming.

Hence, again in consonance with the Carnapian perspective, we
can go on and argue that our (idealised) modelling of Reasons does
(ideally) also provide a justified definition of "rational" within the
context, though what will be ultimately meant by this term is, like
the scent of a rose, more easily felt than described.

Notice that neither Carnap's view nor the present account imply
that whenever agents apply Reasons they will necessarily conform
with probability 1. As we have seen the "rational choice" can simply
be underdetermined with respect to the logical tools, the common
reasoning, available to the agents, so that the possibility of disagreement in the final choice of a unique element from K cannot simply
be ruled out in general (and indeed it would be rather exceptional
for $R(K)$ to have size 1).

As a last point concerning the rationality of conformity, some
illuminating suggestions can be found in the discussion of
radical interpretation mainly championed by Davidson in a number
of works (see the collections Davidson 1984, 2001). The situation is one in which two agents are trying to establish successful communication despite their lack of a common language
and without knowning anything about each other's view of the
world.

According to Davidson, who inherits this intuition from Quine's
analysis of radical translation (Quine 1960, ch. 2), it is simply not possible for an agent to intepret successfully a speaker
without assuming that she structures the world pretty much the
same way the interpreter does. In other words, radical interpretation can only take place under an assumption which *mutatis
mutandis* is entirely analogous to what we have been referring to
here as like-mindedness: agents must assume that they share *common reasoning*. According to Davidson, this *Principle of Charity*,
which ultimately provides fundamental clues about the others' cognitive make-up, is a necessary condition that agents must satisfy in
order to solve a problem of radical interpretation, hence for estab-
lishing communication. But this amounts to activating the kind of

structure that Davidson considers necessary in order for agents to be considered rational (Davidson 2001).

Again in connection with the Carnapian view, we note that according to the Principle of Charity agents should *discard* those possible interpretations that would make, to their eyes, the interpretee systematically wrong or (logically) inconsistent hence, yet without going into any of the subtleties of this topic, systematically irrational. In the formalization of Reasons this, as we have seen, amounts to discarding those possible worlds that are believed, on the fundamental assumption of their like-mindedness to prevent agents from converging or, in Davidson's felicitous terminology, *triangulating* on the same possible world.

It is also interesting to notice here that a more or less implicit feature of (the radical) interpretation problems, which is shared by our rationality-as-conformity, consists in the fact that agents must share a common intention. Both the interpreter and the interpretee must in fact aim at assigning similar meanings to similar linguistic behaviours, that is, must aim at conforming.[9]

As one might expect any ideas about the nature of rationality are likely to resonate with at least some of the multitude of viewpoints on the subject. The idea of rationality-as-conformity as we have presented it here is no exception and for this section we have just briefly noted some links with established positions on this matter. A fuller discussion may be found in the forthcoming (Hosni 2005).

8. TOO MANY REASONS?

In this paper we have focussed on characterizing the choice process that would lead one isolated, common sensical, agent to conform to the behaviour of another like-minded yet inaccessible agent facing (essentially) the same choice problem. To this effect we have introduced what amount to four working Reasons, $R_0, R_1, R_{\mathbb{A}}, R_{\mathbb{U}}$. These arose through very different considerations. In the case of the Regulative Reasons through an adherence to rules, for $R_{\mathbb{A}}$ through an algorithm based on repeatedly trying to fulfill two desiderata, and for $R_{\mathbb{U}}$ through picking the smallest uniquely definable set within the given structure of the problem. This plurality of approaches and answers raises a vexing question.

How can we feel any confidence that there are not other approaches which will lead to entirely different answers?

As we have noted above, ideas and concepts from Game Theory would seem to have very definite application in generating Reasons. Furthermore similar hopes might be extended to other areas of mathematics, for example Social Choice Theory, which we have already alluded to in passing, Group Theory (the construction of $R_{\mathbb{A}}(K)$ could be seen as simply talking about permutation subgroups), Model Theory with its interests in definable subsets and Kolmogorov Complexity, with its emphasis on minimum description length. In short the answer to the question which headed this paragraph is that we can have little such confidence beyond the modicum which comes from having failed to find any ourselves.

Moreover, even with the candidates we already do have we have seen, from the examples given, that both the Regulative and Minimum Ambiguity Reasons appear capable, on their day, of monopolizing the right answer, the 'common sense' answer. Does that mean that even in this very simple context (let alone in the real world) there can be multiple common sense arguments? Or does it mean that we should try them all and pick the 'best answer'? (though that might seem to land us right back with the sort of problems we set out to answer in the first place!)

One advantage however that we should mention that the Minimum Ambiguity Reason and the Smallest Uniquely Definable Reason would appear to have over the Regulative Reasons is that they are easily generalizable. In place of permutations of K, equivalently automorphisms of $\mathcal{M}$, we take all automorphisms of the structure given to the agents and then define the corresponding $R_{\mathbb{A}}$ and $R_{\mathbb{U}}$ exactly analogously to the way we have here. To take a particular example if at the very start we had said that agents might not only receive the matrix with the rows and columns permuted but also possibly with 0 and 1 transposed then the natural structure would have been $\mathcal{M}$ with the constants 0 and 1 removed, i.e.

$$\langle \{0, 1\} \cup A \cup K, \{0, 1\}, A, K, =, Comp \rangle.$$

In this case an automorphism j corresponds to a permutation σ of A and a 1-1 function $\delta : \{0, 1\} \mapsto \{0, 1\}$ such that

$$j \restriction \{0, 1\} = \delta, \ j \restriction A = \sigma^{-1}, \quad j(f) = \delta f \sigma \quad \text{for } f \in K.$$

Again the corresponding $R_{\mathbb{A}}$ and $R_{\mathbb{U}}$ can be defined and give in general practically worthwhile answers (i.e. non-trivial). However with this change the requirement of Renaming cannot be strengthened to what is expected here, i.e.

$$\delta R(K)\sigma = R(\delta K\sigma)$$

without reducing the possible Regulative Reasons to the trivial one alone – as can be seen by considering the initial step in the proof of Theorem 1.

ACKNOWLEDGEMENTS

We would like to thank the editor for his patience and encouragment and the referees whose unrestrained comments certainly resulted in this being a better paper.

NOTES

* Partially supported by EPSRC research studentship and a grant from the Alexander von Humboldt Foundation, the Federal Ministry of Education and Research and the Program for the Investment in the Future (ZIP) of the German Government.
[1] You have doubtless already though whyever doesn't he leave a message stuck to the door, or call her mobile, or leave the keys with his secretary, …!
[2] See Section 7 for more on this.
[3] There seems to be an implicit assumption in this argument that for $f \in K_1$, $f \upharpoonright X_1$ and $f \upharpoonright A - X_1$ are somehow independent of each other. In the current simple case of $\mathbb{w} = 2^A$ this is true but it fails in the case, not considered here, in which the worlds are probability functions.
[4] It can be shown that if we had taken A to be infinite then this would have been the only Regulative Reason.
[5] Alternatively one might hedge one's bets and adopt the "collected extremal choice", $R_\cup(K) = R_0(K) \cup R_1(K)$, in the sense of Aizermann and Malishevski (1981) (see also Rott (2001), p. 163) and by the Aizerman-Malishevski Theorem (Theorem 4 of Aizermann and Malishevski (1981)) $R_\cup$ is a *Plott function*, that is to say a function that satisfies the so-called Path Independence property introduced in (Plott 1973).
[6] It is noted in Rott (2001) that this strategy mirrors the "sceptical" as opposed to the "credulous" approach to non-monotonic inference.
[7] One might subsequently argue that the agent could then also recognize automorphisms of $\mathcal{M}$ so the set of these too should be added to our structure, and the whole process repeated, and repeated … In fact this does not change the definable subsets of K so it turns out there is no point in going down this path.
[8] The *loci classici* for pure coordination games and the related notions of salience and focal points are (Schelling 1960; Lewis 1969). Since then the literature on this topic developed enormously, yet particularly relevant to the present proposal are

the more recent (Mehta et al., 1994; Janssen 2001; Camerer 2003; Morris and Shin 2003).
[9] We touch upon some of these intriguing connection between radical interpretation, rationality-as-conformity and coordination games in Hosni (2004).

REFERENCES

Aizermann, M. and A. Malishevski: 1981, 'General Theory of Best Variants Choice: Some Aspects', *IEEE Transactions on Automatic Control* **26**, 1030–1040.

Camerer, C.: 2003, *Behavioral Game Theory: Experiments on Strategic Interaction*, Princeton.

Davidson, D.: 1984, *Inquiries into Truth and Interpretation*, Oxford University Press.

Davidson, D.: 2001, 'Rational Animals', In *Subjective, Intersubjective, Objective*, Oxford University Press. pp. 95–105.

Davidson, D.: 2001, *Subjective, Intersubjective, Objective*, Oxford University Press.

Elio, R.: (ed.) 2002, *Commonsense, Reasoning, and Rationality*, Oxford University Press, New York.

Halpern, J.: 2003, *Reasoning About Uncertainty*, MIT Press.

Hilpinen, R.: 1973, 'Carnap's New System of Inductive Logic', *Synthese* **25**, 307–333.

Hosni, H.: 2004, 'Conformity and Interpretation', In *Prague International Colloquium on Logic, Games and Philosophy: Foundational Perspectives,* Prague, Czech Republic, October 2004.

Hosni, H.: 2005, *Doctoral Thesis*. School of Mathematics, The University of Manchester, Manchester, UK, http: www.maths.man.ac.uk/~hykel/.

Janssen, M.: 2001, 'Rationalizing Focal Points', *Theory and Decision* **50**, 119–148.

Kalai, G., A. Rubinstein and R. Spiegler: 2002, 'Rationalizing Choice Functions by Multiple Rationales', *Econometrica* **70**(6), 2481–2488.

Keynes, J. M.: [1936] 1951, *The General Theory of Employment Interest and Money*, McMillan, London.

Kraus, S., J. S. Rosenschein and M. Fenster: 2000, 'Exploiting focal points among alternative solutions: Two approaches', *Annals of Mathematics and Artificial Intelligence* **28**(1–4), 187–258.

Lewis, D.: 1969, *Convention: A Philosophical Study*, Harvard University Press.

Marker, D.: 2002, *Model theory: An Introduction*, Graduate Texts in Mathematics 217. Springer.

Mehta, J., C. Strarmer and R. Sugden: 1994, 'The Nature of Salience: An Experimental Investigation of Pure Coordination', *The Americal Economic Review* **84**(3), 658–673.

Morris, S. and H. S. Shin: 2003, 'Global Games: Theory and Application', In M. Dewatripont, L. Hanson and S. Turnovsky (eds.), *Advances in Economics and Econometrics, Proceedings of the 8th World Congress of the Econometric Society*, Cambridge University Press.

Nozick, R.: 1993, *The Nature of Rationality,* Princeton University Press, Princeton.

Paris, J. B: 1999, 'Common Sense and Maximum Entropy', *Synthese* **117**(1), 73–93.

Paris, J. B. and A. Vencovská: 1990, 'A Note on the Inevitability of Maximum Entropy, *International Journal of Approximated Reasoning* **4**, 183–224.

Paris, J. B. and A. Vencovská: 1997, 'In Defence of the Maximum Entropy Inference Process', *International Journal of Approximate Reasoning* **17**, 77–103.

Paris, J. B. and A. Vencovská: 2001, 'Common Sense and Stochastic Independence', In D. Corfield and J. Williamson (eds.), *Foundations of Bayesianism*, Kluwer Academic Press, pp. 230–240.

Plott, C. R.: 1973, 'Path Independence, Rationality and Social Choice', *Econometrica* **41**(6), 1075–1091.

Quine, W. V.: 1960, *Word and Object,* MIT Press, Cambridge, Massachusetts.

Rott, H.: 2004, *Change, Choice and Inference: A Study of Belief Revision and Nonmonotonic Reasoning*, Oxford University Press.

Schelling, T.: 1960, *The Strategy of Conflict*, Harvard University Press.

Department of Mathematics,
Manchester University,
Manchester, U.K.
E-mail: hykel@maths.man.ac.uk

GREGORY R. WHEELER

ON THE STRUCTURE OF RATIONAL ACCEPTANCE: COMMENTS ON HAWTHORNE AND BOVENS

ABSTRACT. The structural view of rational acceptance is a commitment to developing a logical calculus to express rationally accepted propositions sufficient to represent valid argument forms constructed from rationally accepted formulas. This essay argues for this project by observing that a satisfactory solution to the lottery paradox and the paradox of the preface calls for a theory that both (i) offers the facilities to represent accepting less than certain propositions within an interpreted artificial language and (ii) provides a logical calculus of rationally accepted formulas that preserves rational acceptance under consequence. The essay explores the merit and scope of the structural view by observing that some limitations to a recent framework advanced James Hawthorne and Luc Bovens are traced to their framework satisfying the first of these two conditions but not the second.

1.

The *lottery paradox* (Kyburg 1961) arises from considering a fair 1000 ticket lottery that has exactly one winning ticket. If this much is known about the execution of the lottery it is therefore rational to accept that one ticket will win. Suppose that an event is very likely if the probability of its occurring is greater than 0.99. On these grounds it is rational to accept the proposition that ticket 1 of the lottery will not win. Since the lottery is fair, it is rational to accept that ticket 2 won't win either – indeed, it is rational to accept for any individual ticket i of the lottery that ticket i will not win. However, accepting that ticket 1 won't win, accepting that ticket 2 won't win,..., and accepting that ticket 1000 won't win entails that it is rational to accept that no ticket will win, which entails that it is rational to accept the contradictory proposition that one ticket will win and no ticket will win.

The *paradox of the preface* (Makinson 1965) arises from considering an earnest and careful author who writes a preface for a

Synthese (2005) 144: 287–304
Knowledge, Rationality & Action 117–134
DOI 10.1007/s11229-005-2722-2

book he has just completed. For each page of the book, the author believes that it is without error. Yet in writing the preface the author believes that there is surely a mistake in the book, somewhere, so offers an apology to his readers. Hence, the author appears to be committed to both the claim that every page of his book is without error and the claim that at least one page contains an error.

Abstracted from their particulars, the lottery paradox and the paradox of the preface are each designed to demonstrate that three attractive principles for governing rational acceptance lead to contradiction, namely that

1. It is rational to accept a proposition that is very likely true,
2. It is not rational to accept a proposition that you are aware is inconsistent,
3. If it is rational to accept a proposition A and it is rational to accept another proposition A', then it is rational to accept $A \wedge A'$

are jointly inconsistent. For this reason, these two paradoxes are sometimes referred to as *the paradoxes of rational acceptance*.

These paradoxes are interesting because of the apparent price exacted for giving up any of the three principles governing rational acceptance. Abandoning the first principle by restricting rational acceptance to only certainly true propositions severely restricts the range of topics to which we may apply logic to draw "sound" conclusions, thereby threatening to exclude the class of strongly supported but possibly false claims from use as non-vacuous premises in formally represented arguments. Giving up the last principle by abandoning logical closure operations for accepted propositions clouds our understanding of the logical form of arguments whose premises are rationally accepted but perhaps false, thereby threatening our ability to distinguish good argument forms from bad. Finally, adopting a strategy that denies the second principle offers little advantage on its own, since even a paraconsistent approach that offers a consequence operation that does not trivialize when applied to a set containing a contradictory proposition must still specify a closure operation for rationally accepted propositions that reconciles the general conflict between the first and third legislative principles.

When considering a strategy to resolve a paradox it is worth remarking that simply avoiding inconsistency is not necessarily sufficient to yield a satisfactory solution since consistency may be achieved merely by dropping one of the conditions necessary to generate the antinomy. Besides restoring consistency, a satisfactory resolution must also address the motivations behind the principles that generate the paradox in the first place. There are two ways one can do this. The first type of response is to reject one or more of the principles and then explain how to get along without principles of this kind. The thrust of this approach is to claim that a purported paradox is really no paradox at all but rather a mistake arising from a commitment to a dubious principle. The second type of reply regards the constituent principles of a paradox as all well-motivated, if ill-formulated, so regards the paradox as genuine. The aim of this type of reply is to offer a substantive solution to the paradox, which is a solution that revises one or more of the original principles so that they consistently capture the key features that motivated adopting the original principles.

In the case of the lottery paradox and the paradox of the preface a solution of the second type is required. Namely, a satisfactory solution to these paradoxes should provide a sufficiently expressive language for representing accepting less than certain propositions and also provide a sufficiently powerful logic to model entailments made in cogent arguments involving uncertain but rationally accepted premises.[1]

2.

This description of the paradoxes of rational acceptance and what should be expected from a solution is fairly standard. However, in this essay I propose refining the standard view by adding a requirement that every proposed solution should satisfy. The requirement concerns minimal syntactic capabilities that a formal system's language should possess. More specifically, the proposal is to require that a system's formal language be expressive enough to construct compound rationally accepted formulas. This language requirement may be thought of as a *structural constraint* on the formal system underlying any proposed solution to the paradoxes. For this reason, I refer to this proposal as the *structural view of rational acceptance.*

The structural view is motivated by observing that the problem raised by the paradoxes of rational acceptance is a general one of how to reconcile the first and third legislative principles. But to study the general relationship between rational acceptance and logical consequence, we need to understand valid *forms* of arguments whose premises are rationally accepted propositions. This point suggests three conditions for us to observe. First, it is important to define the notion of rational acceptance independently of any particular interpreted structure, since this notion is serving as a semantic property that is thought to be preserved (in a restricted sense) under entailment. Second, to formally represent an argument composed of rationally accepted propositions we must have facilities for formally representing their combination within an object language. Finally, of formal languages that satisfy the first two properties, preference should be given to those within systems that make the relationship between rational acceptance and logical consequence transparent.

Notice that these conditions correspond to general properties that well-designed logical calculi enjoy. However, the most familiar logical calculus – the propositional calculus defined on the primitive Boolean connectives ¬ and ∨ – is precisely the calculus that generates the paradoxes of rational acceptance. The structural view of rational acceptance then sees the problem raised by the paradoxes to be one of selecting the right calculus for rationally accepted formulas.

It is worth mentioning that the structural constraints are not jointly sufficient conditions for resolving the paradoxes since there are several ways a formal language could meet the first two constraints, and the notion of transparency that figures in the third condition is imprecise. No doubt other concerns will need to be brought to bear to select the correct class of logics for rational acceptance. However, the point of this essay is to argue that a logic of rational accepted formulas should at least satisfy these conditions. Viewing these paradoxes within frameworks that attempt to satisfy these minimal constraints will allow us to focus more precisely on the key open questions surrounding rational acceptance.

It is also important to note that there isn't anything necessarily mistaken about an unstructured logic. For instance, the operator Γ, defined over a set of accepted sentences X such that a proposition A is in the image set of $\Gamma(X)$ if and only if $X \cup \{\neg A\}$ is inconsistent, is an unstructured operator; yet, Γ is also sound. If Γ and $\neg$ were the only operators a logic featured, that logic also would be unstructured and sound. The problem with an unstructured consequence operator like Γ is that if we have a question whose answer turns on the syntactic details of how the manipulation of elements in X affects the appearance of A in the image set of $\Gamma(X)$, then Γ is the wrong theoretical tool to expect an answer to that question.

[120]

Thus the difference between structured and unstructured systems is not logical in the sense that each type of framework necessarily identifies different classes of rationally accepted propositions. Rather, the fundamental disagreement rests in what analytical resources are necessary to study arguments composed of rationally accepted propositions. The structural view holds that it is necessary for the object language to include connectives in order to express compound rationally accepted formulas and to define restricted logical consequence for rationally accepted formulas in terms of these connectives. An unstructured view does not.

Finally, note that the structural view has an important methodological consequence for rational acceptance studies. For if one accepts that what is needed is a formal language for rational accepted formulas, then research should move away from purely semantic approaches and toward the study of probabilistic logical calculi.

3.

To motivate the structural view it will be useful to consider an important framework that does not satisfy the structural constraints just discussed, the *logic of belief* developed by James Hawthorne and Luc Bovens in (Hawthorne and Bovens 1999). The point behind criticizing this particular framework is to show that certain limitations of that theory's solution to the paradoxes is traced to the unstructured logic underpinning the account. The reason that Hawthorne and Bovens's system is an excellent one for making my general point is that their system is very well developed: it is doubtful that their theory can be improved without adopting the structural constraints that are the focus of this essay.

Hawthorne and Bovens view the lottery paradox and the paradox of the preface to be problems involving how to identify rational beliefs resulting from composing probabilistic events (e.g., how many of the tickets I each judge as losing tickets may I conjoin and still rationally regard as losing tickets?) and how to identify rational beliefs resulting from the decomposition of compound probabilistic events (e.g., how short can my book be before my apology for mistakes in the preface becomes incoherent?). They frame their discussion of the paradoxes in terms of belief states for ideal agents who satisfy certain rational coherence constraints.

Hawthorne and Bovens's approach follows a proposal made by Richard Foley (1992) about how to construe the first legislative principle for rational acceptance. In Foley's paper he advances the *Lockean thesis*, which states that rational acceptance should be viewed as rational belief and that a rational belief is just a rational degree of confidence above some threshold level that an agent deems sufficient for belief. We'll say more about this principle shortly. Hawthorne and Bovens's project is to use the rationality constraints that come with probabilistic models of doxastic states – and which are built into the Lockean thesis by virtue of framing rational acceptance in terms of rational belief – in order to establish a correspondence between quantitative degrees of confidence and a qualitative notion of full belief. This correspondence then allows them to reconstruct a probabilistic model of belief for an agent given only that he is in a suitable context in which he has full beliefs that satisfy the Lockean thesis, and *vice versa*. The Hawthorne and Bovens proposal, then, is that from the constraints imposed by the Lockean thesis, an ideal agent's report of his (full) beliefs provides us with rational lottery-states and rational preface-states, which in effect yields a solution to the paradoxes since these states will include rational beliefs that are combinations of individual rational beliefs and rational beliefs that are detached from compound rational beliefs.

Hawthorne and Bovens regard this approach as a powerful frame-work for resolving the paradoxes of rational acceptance, stating that there is "a precise relationship between ... qualitative and quantitative doxastic notions" that "may be exploited to provide a completely satisfactory treatment of the *preface* and the *lottery*" paradoxes (Hawthorne and Bovens 1999, 244). The gist of their proposal is that representing the lottery paradox and the preface paradox in terms of the logic of belief yields enough insight into the relationship between qualitative and quantitative notions of rational acceptance to provide "a foundation for a very plausible account of the logic of rationally coherent belief" (Hawthorne and Bovens 1999, 244).

After summarizing their proposal in Section 4, I will advance reasons for resisting both of these claims in Sections 5 and 6. Specifically, I will argue that the relationship between qualitative and quantitative notions of belief does not afford us results sufficient to construct a satisfactory solution to these paradoxes but, on the contrary, introduces obstacles to constructing such an account. Furthermore, I suggest that there is reason to doubt that Hawthorne and

Bovens's framework provides a suitable foundation for the logic of rationally coherent belief.

4.

Hawthorne and Bovens's logic of belief is a theory that yields consistency constraints for an ideally rational agent α who grasps all logical truths. Two types of doxastic states for α are considered, a quantitative doxastic notion, called a *degree of confidence function*, and the qualitative notions of full belief and its complement.[2] The degree of confidence function, defined over a countable set $\mathcal{F}$ of propositions, is isomorphic to the classical probability measure, whereas belief and its complement are defined relative to a threshold point in the unit interval. The relationship between these two notions is given by the *Lockean thesis*: α is said to believe a proposition A in $\mathcal{F}$ if and only if α's degree of confidence measure of A is greater than or equal to a threshold value q in the closed unit interval [0,1]. It is from this equivalence relation and the logical omniscience assumption for α that Hawthorne and Bovens derive the central results underpinning their proposal.

Their idea is to consider descriptions of α entertaining beliefs sufficient to generate an instance of the preface paradox and also of α entertaining beliefs sufficient to generate an instance of the lottery paradox, yielding belief states that are called *preface states* and *lottery states*, respectively. Within a sub-class of belief states the consistency constraints imposed by the degree of confidence measure allow one to derive a precise estimate of α's threshold value q, in cases where q is unknown. This is achieved by using α as an oracle to determine whether a proposed belief state (in an appropriately constrained context) satisfies both the Lockean thesis and the consistency constraints imposed by the degree of confidence measure over $\mathcal{F}$. The idea is that an ideally rational agent satisfying the Lockean thesis may be used as a semantic reference for determining the class of full beliefs, from which a quantitative probability model may be constructed. One may pass in the other direction as well – from a quantitative probabilistic doxastic notion to a qualitative model of the ideal agent's set of full beliefs – so long as the agent satisfies the Lockean thesis.

In the case of the preface, suppose there is a particular book with n pages and an agent α who believes that each page in this book is without an error yet also believes that there is an error on at least one of the pages. Hawthorne and Bovens refer to this as an *n-page preface state*. If q is a threshold value for belief, then α can consistently be in an n-page preface state only if $n \geq q/(1-q)$. Hawthorne and Bovens propose exploiting this inequality to fix a least upper bound on q when its value is unknown by solving the inequality for q rather than n, that is $n/(n+1) \geq q$. The idea is that if one doesn't know the value of an agent's threshold point for acceptance, q, one can provide a least upper bound for this value by placing α in preface states of varying size n and record the least value n where α satisfies the rationality constraints of the Lockean thesis.

In the case of the lottery, matters are slightly more complicated. Whereas placing α in various sized preface states is designed to fix the least upper bound on α's threshold value q, an analogous method for placing α in a restricted class of lottery states is intended to fix the greatest lower bound on q. The restricted class of lottery states, called *weak lottery contexts*, are just those that have at most one winning ticket – that is, for all $W_{i,j} \in \mathcal{F}$, α is certain that $\neg(W_i \wedge W_j)$ where $i \neq j$. Adopting the phrase 'deems it possible that W' to express that α does not believe $\neg W$, Hawthorne and Bovens define an *m-ticket optimistic state* in a weak lottery context to be one in which an agent deems it possible that each m tickets of a lottery may win but that at most one ticket will win. When the threshold point q is defined, the agent may be in an m-ticket optimistic state only if $m < 1/(1-q)$. When q is unknown, the greatest lower bound may be calculated by solving for q rather than n, that is $(m-1)/m < q$. The idea here is that if one doesn't know the value of an agent's threshold point for acceptance, then place α in various sized optimistic states and record the greatest value m where α satisfies the rationality constraints of the Lockean thesis.

The idea then is to combine these two results to fix an upper and lower bound on α's *quantitative* threshold of belief, q, by determining what α *qualitatively* believes in preface states that satisfy the weak lottery context restriction for optimistic states – namely, those contexts in which not more than one ticket wins and the set $\mathcal{F}$ of tickets (propositions) is finite – while satisfying the Lockean thesis. Suppose that α is in a context for belief that is an n-page preface state and also an n-optimistic state such that α believes he will not win with only $n-1$ tickets but deems it genuinely possible that he may win with n tickets. Hawthorne and Bovens's first result then is that α's threshold value for belief is some q such that $(n-1)/n < q \leq n/(n+1)$.

This estimate for q may be improved if additional restrictions are introduced. For instance, Hawthorne and Bovens introduce the notion of a *strong equiplausible lottery context*, which holds when α is certain that exactly one ticket wins and that the outcomes are equiprobable. Then a more precise estimate for q may be derived.

In general, if in a *strong equiplausible lottery context* for an n ticket lottery an agent believes she will not win with only $m - 1$ tickets [for $m \leq n$], but deems it *genuinely possible* that she may win with m tickets, then the agent's threshold value for belief is some number q such that $1 - (m/n) < q \leq 1 - (m-1)/n$ (Hawthorne and Bovens 1999, 254).

Their claim then is that the two kinds of propositional attitudes, qualitative belief and quantitative degree of confidence, show that ideal agents that satisfy the Lockean thesis may rationally entertain lottery beliefs and preface beliefs without contradiction.

If *preface* and *lottery* beliefs are re-described in quantitative doxastic terms, their paradoxical features evaporate. In the *lottery* we realize that the likelihood that any given ticket will win is extremely low, yet this in no way contradicts our certainty that some ticket will win. In the *preface* we judge that the likelihood that any given page still contains an error is extremely low, yet this is perfectly consistent with our high degree of confidence that at least one error has been missed in a lengthy book (Hawthorne and Bovens 1999, 243).

However, there is reason to resist the claim that each paradox evaporates, if by 'evaporate' it is intended that the proposal provides a satisfactory solution to the preface and lottery paradoxes. For, as we've observed, one may dissolve these paradoxes by denying any one of the three legislative principles with which we began. In Hawthorne and Bovens's proposal the Lockean thesis satisfies the first principle, while the second is satisfied by consequence of adopting the classical probability measure that underpins modeling α as an agent who satisfies the Lockean thesis, irrespective of whether α's doxastic states are determined by a quantitative degree of confidence function or qualitative full belief. Hence, it is the third legislative principle that is rejected. Hawthorne and Bovens's strategy is to extract closure conditions for *particular* collections of beliefs from the semantics of the theory. The question remaining is whether this strategy resolves the general conflict between the first and third legislative principles.

[125]

5.

Hawthorne and Bovens proposal is built on an important insight, namely that a rationally acceptable n-element conjunction must itself be above threshold for acceptance rather than just assuring that each of the n conjuncts is above threshold. Thus, a closure condition for sets of rationally accepted propositions must account for the possible depletion of probability mass of conjoined probabilistic events. We might then think that the logic of belief does provide an account that generates structural rules. Such an account would propose using α to determine conjunctions of propositions, if any, that are above threshold when each conjunct is – that is, when $\mathrm{Pr}_\alpha(\bigwedge_{1 \leq i \leq n} A_i) \geq q$ where $\{A_i : A_i \in \mathcal{F} \wedge \mathrm{Pr}_\alpha(A_i) \geq q\}$. The proposal would be to accept only those conjunctions that α does. Notice, however, that this isn't a structured closure operation since we do not have facilities within the object language for combining or decomposing accepted formulas to yield accepted formulas. Instead, what the theory provides is a description of a decision procedure in the metalanguage built around a semantic reference, α, who delivers a Yes or No reply to whether a candidate belief is rational to accept.

To illustrate this point, consider two rules that Hawthorne and Bovens discuss, labeled here as HB1 and HB2:

HB1. For all $n < q/(1-q)$, if α believes $\neg E_1, \alpha$ believes $\neg E_2, \ldots, \alpha$ believes $\neg E_n$, then α does not believe $(E_1 \vee E_2 \vee \cdots \vee E_n)$ (Hawthorne and Bovens 1999, p. 246).

HB2. For all $m \geq 1/(1-q)$ and each $i \neq j$, if α is certain that $\neg(W_i \wedge W_j)$, and if α does not believe $\neg W_1, \alpha$ does not believe $\neg W_2, \ldots, \alpha$ does not believe $\neg W_{m-1}$, then for each $k \geq m$, α believes $\neg W_k$ (Hawthorne and Bovens 1999, p. 251).

Notice that HB1 and HB2, although sound, are unstructured. The point to notice is that HB1 and HB2 do not operate upon formulas but rather on states of belief. A consequence of this observation of particular importance is that there are no logical operators within Hawthorne and Bovens's logic of belief corresponding to the coordinating conjunctions appearing on the left-hand side of each rule. Hence, HB1 and HB2 are essentially meta-linguistic descriptions of decision procedures rather than inference rule schemata, since there are no formulas within Hawthorne and Bovens's framework to stand in as substitution instances for either rule. A consequence of this is that one cannot construct a proof within the logic of belief since there are no formulas from which to construct one. HB1 and HB2 are thus not logical rules of inference.[3]

This omission marks an important limitation to unstructured accounts. For it is clear that there is a logical distinction between a conjunction (disjunction) of rationally accepted propositions and a rationally accepted conjunction (disjunction); indeed, the paradoxes of rational acceptance are examples of arguments that invite us to ignore this distinction. But to formally evaluate arguments, the structure of formulas should reflect their meaning in a manner that clearly demonstrates the difference between conjunctions and disjunctions of propositions versus conjunctions and disjunctions of rationally accepted propositions. As we observed, Hawthorne and Bovens's logic of belief does not provide these resources.

To summarize, the first conclusion to draw about Hawthorne and Bovens's logic of belief is that it is not a logical calculus but instead is a specification for decision procedures that work by determining whether a belief state satisfies the semantic constraints of the theory, precisely as the unstructured operator Γ behaves with respect to propositions.

6.

Hawthorne and Bovens's analysis of the paradoxes of rational acceptance holds that the Lockean conception of belief, based on a probabilistic semantics with an acceptance level, offers a suitable foundation for resolving the paradoxes because it explains the relationship between qualitative belief, quantitative belief, and a quantitative threshold level for rational acceptance. In considering whether Hawthorne and Bovens's proposal provides a suitable foundation for a logic of rationally coherent belief we'll need to discuss the role that *qualitative belief* plays in the theory.

The first point to observe is that α's qualitative notion of belief does not play an essential role in the logic. By this I mean that their core theory starts with a known value for q and then classifies propositions by virtue of their probability measure with respect to q; full belief and its complement thus serve as derived notions. The criticism that Hawthorne and Bovens's account is unstructured applies to this core theory. The addition of qualitative belief to the logic plays no constructive part in resolving the issues raised in the previous sections, which is to say that the addition of qualitative belief does not in itself provide a means to construct rationally accepted formulas.

The main role that qualitative belief plays in the account is to provide an estimate of q when q is unknown, provided the agent satisfies the Lockean thesis – which, recall, includes the rationality constraints for probabilistic doxastic states – and an important restriction that I will return to shortly. It is precisely these rationality constraints that are built-in to the Lockean thesis that allows the theory to pass back and forth between qualitative and quantitative notions of belief. The main point to note here is that this estimation problem of the threshold parameter q is distinct from the issues stemming from not having a structured closure operation for sets of rationally accepted formulas.

Before pressing on, a remark on the restrictions necessary for full qualitative belief to estimate a value for q. By building their account around rational belief states, Hawthorne and Bovens need to restrict the scope of their theory to agents who are working with a qualitative notion of belief in order to pass from this notion to a quantitative estimate for q. They do this by specifying the kind of belief states in which the theory operates, generating a particular model of rationally accepted beliefs for that particular collection of beliefs. Hawthorne and Bovens propose approximating q from full qualitative belief within what they call weak lottery contexts – that is, in cases where we know there is no more than one mistakenly-accepted (believed and false) proposition out of a set of otherwise correctly accepted (believed and true) propositions. Note that rule HB2 incorporates the weak lottery context restriction.

However, in many cases involving rational acceptance the conditions for weak lottery contexts are not satisfied. We mustn't be tricked into thinking that an accidental feature of the lottery paradox thought experiment – such as that it is known for certain that no more than one ticket wins – picks out essential features that a logic for rational acceptance may always rely upon. In very many cases involving rationally accepted propositions errors of acceptance are independent (Kyburg 1997). For instance, in most government sponsored lottery drawings we do not know that there will be at most one ticket that will win, nor do we know that at most one plane will crash in a given year, nor do we typically know that no more than one sample will be biased among a collection of measurements. But the assumption that there is at most one mistakenly accepted proposition is necessary to define an m-element optimistic state, which in turn is used to approximate the greatest lower bound on q given α's belief states.

Admittedly, there is a degree of idealization we accept in modeling rational acceptance as belief states. What is important to notice here is that the restrictions used to approximate q from α's full belief states are more demanding than logical omniscience and should not pass without a note accounting for their cost.

These remarks also apply to Hawthorne and Bovens's proposal to reformulate their logic of belief in terms of *qualitative probability* (Hawthorne and Bovens 1999, Appendix B). Considering why qualitative probability does not offer an improvement on their position with respect to providing a structured closure operation will put us in position to advance a reason to doubt their claim that the theory provides a good foundation for a logic of rational belief.

Qualitative probability theory stems from an observation of Frank Ramsey's (Ramsey 1931) that beliefs of ideally rational agents form a total order: it is the violation of this condition that underpins his Dutch book argument. The idea behind qualitative probability is that if we could provide qualitative axioms for belief that, when satisfied, were sufficient to yield a total order, then we would have grounds to consider the axioms rational – all without assigning numerical degrees of belief.

Consider a relation $\succeq$ on $\mathcal{F}$, where A, B, C, D are propositions in $\mathcal{F}$, and '$A \succeq B$' is interpreted to say that an ideal agent α deems A to be at least as plausible as B. Any relation $\succeq$ satisfying the following six axioms is a *qualitative probability relation*.

1. If $(A \equiv B)$ and $(C \equiv D)$ are logically true and $(A \succeq C)$, then $(B \succeq D)$;
2. It is not the case that $(A \wedge \neg A) \succeq (A \vee \neg A)$;
3. $B \succeq (A \wedge \neg A)$;
4. $A \succeq B$ or $B \succeq A$;
5. If $A \succeq B$ and $B \succeq C$, then $A \succeq C$;
6. If $\neg(A \wedge C)$ and $\neg(B \wedge C)$ are logically true, then $A \succeq B$ if and only if $(A \vee C) \succeq (B \vee C)$.

Given a qualitative probability relation $\succeq$ with respect to α, we may define an *equivalence relation*, $\simeq$, and also a strict plausibility relation, $\succ$, as follows. First, *equivalence*: $A \succeq B$ and $B \succeq A$ if and only if $A \simeq B$. Next, *strict plausibility*: $A \succ B$ if and only if $A \succeq B$ and not $B \succeq A$. Now let us consider a new axiom, Axiom 7.

7. If $A \succ B$, then, for some n, there are n propositions $S_1, \ldots, S_n$ where for all $S_{1 \leq i \leq j \leq n}$ and $i \neq j$, $\neg(S_i \wedge S_j)$ is logically true, and $(S_1 \vee \cdots \vee S_n)$ is logically true, and such that for each S_i, $A \succ (B \vee S_i)$.

A key result of Savage's (1972) is that if α exercises a qualitative probability relation $\succeq$ over $\mathcal{F}$ satisfying these seven axioms, then there is a unique quantitative probability measure such that $\Pr(A) \geq \Pr(B)$ if and only if $A \succeq B$.

Finally, Hawthorne and Bovens's proposal is to add to quantitative probability an axiom for full belief, namely

8. (Hawthorne and Bovens 1999, 262) if $A \succeq B$ (i.e., if α deems A to be at least as plausible as B) and α believes B, then α believes A.

An important point to notice about qualitative probability is that it demands more of ideal agents than just logical omniscience. While transitivity of the strict plausibility relation $\succ$ is uncontroversial, axioms for a strict plausibility relation are not sufficient to yield a total ordering of $\mathcal{F}$ — hence Axiom 5, the requirement that *weak* preference $\succeq$ satisfy transitivity. However, Axiom 4 demands that for any two beliefs A, B, a rational agent either finds A more plausible than B, B more plausible than A or will be indifferent between A and B, where indifference amounts to the agent judging each belief of equal epistemic bearing. But judging two beliefs A and B equally plausible is a stronger disposition than having *no* comparative judgment for one *vis a vis* another. It is perfectly consistent for α to be logically omniscient yet not be in a position to either stake one belief more or less plausible than another or judge them to be equally plausible. Indeed, to exclude indecision as a rationally possible state we must state that α is in a context in which all outcomes are comparable (cf., Hawthorne and Bovens 1999, 246, note 7),[4] has available to him a qualitative notion of confidence precise enough to put him in a position to satisfy the completeness axiom (Axiom 4) and the cognitive ability to apply this notion to his doxastic states. Logical omniscience alone is insufficient.

This said, notice that qualitative probability doesn't offer an improvement to Hawthorne and Bovens's original account with respect to providing a structured formal system. The crucial concept in modeling their logic of belief within qualitative probability is still α's notion of confidence and this notion must be precise enough for α to effect comparisons that satisfy the completeness axiom. We might pursue a strategy to maintain that confidence is a qualitative notion by introducing quantitative benefits that agents wish to maximize (e.g., money). But notice that this move takes us no closer to articulating a logical calculus for rationally accepted formulas. Hawthorne and Bovens's proposal, whether based directly on the classical probability measure or whether passing through qualitative probability as an intermediary theory, yields the same output: a metalinguistic description of a decision procedure that relies upon a table of α's rational beliefs for us to consult.

7.

To summarize, it was observed that Hawthorne and Bovens's logic of belief is not a logical calculus since it neither includes connectives in the object language for combining rationally accepted sentences nor does it provide logical inference rules that preserve rational acceptability under (restricted) entailment. Rather, what Hawthorne and Bovens do provide is an *unstructured* closure operation extracted from the semantic features of a Lockean conception of belief. It was observed that the rules HB1 and HB2 that the theory generates are sound decision procedures rather than sound rules of logical inference. The reason for this assessment is that the theory provides no formal language capable of expressing formulas that are substitution instances for either HB1 or HB2. So, HB1 and HB2 are necessarily meta-linguistic expressions which are better understood as describing how to calculate consistent rational belief states. For this reason it was concluded that the logic of belief is unstructured.

It was also observed that the notion of qualitative belief does little work in resolving the paradoxes and that there is reason to regard its use to estimate quantitative threshold parameters a handicap. First, the constructive role that full belief plays in the theory is to solve for the threshold parameter q and does not address how to extend the logic to include rationally accepted formulas. Second, even when considering the theory's capabilities for estimating threshold levels, it turns out that more is required to effect estimates than the assumption that α satisfy the Lockean thesis and logical omniscience. In the original theory the class of contexts in which we may precisely estimate the threshold value for q is constrained by the weak lottery context assumption, which is more restrictive than general circumstances involving rationally accepted propositions. The theory of qualitative probability does nothing to relax these constraints but rather adds additional restrictions by requiring that α only be in contexts in which all his beliefs are comparable and that he not be indecisive on pain of failing to satisfy the completeness axiom. With respect to this last point, it was remarked (Note 4) that this feature may present a problem for Hawthorne and Bovens's development of their contextualist interpretation of confidence if a set of propositions varies by being comparable in one context but fails to be comparable in another context. This point presents another type of limitation to applying the logic of belief, but I won't pursue this line here.

In short, a theory purporting to resolve the paradoxes of rational acceptance should address the general conflict between the first and third legislative principles for rational acceptance. The structural view holds that a logical calculus for rational accepted formulas should be a requirement for every formal framework designed to resolve the paradoxes of rational acceptance. The minimum expressive capabilities of an object language should be to express the difference between the probability of a conjunctive (disjunctive) event and the conjunction (disjunction) of probabilistic events. Furthermore, restricted consequence should be defined with respect to a formal language for expressing rationally accepted propositions. A calculus with at least these capabilities would allow us to evaluate the formal features of arguments composed from rationally accepted propositions, giving us a more precise understanding of the general conflict presented by the first and third legislative principles.[5]

NOTES

[1] It should be noted that there is disagreement in the literature over whether the lottery paradox and the paradox of the preface both require a solution of the second type. To take one example, John Pollock has argued that the lottery paradox should be resolved by a solution of the first type whereas the paradox of the preface should be resolved by a solution of the second type. Pollock considers the lottery paradox an invitation to commit a mistake in reasoning. He argues that since the lottery paradox is an instance of collective defeat, the correct position should be to deny that it is rational to accept that any ticket of the lottery loses. Pollock regards the paradox of the preface, in turn, to be generated from principles we should accept and, hence, thinks that the paradox of the preface requires

a solution of the second form (e.g., Pollock 1993). Although I think that both paradoxes call for a substantive (type two) solution and will assume this point in this essay, the general view I advance (i.e., the structural view of rational acceptance) does not turn on how these paradoxes are classified with respect to the appropriate type of solution. What is necessary is a considerably weaker claim, namely that there is at least one paradox generated from reasoning governed by the three displayed legislative principles along the lines that I've described that requires a solution of the second type.

[2] There is a slight deviation in my notation that warrants mentioning. Hawthorne and Bovens mark the distinction between quantitative doxastic states and qualitative doxastic states by discussing two different agents, α and β, who differ precisely with respect to the kind of doxastic states each may entertain: α is an agent whose doxastic states are exclusively quantitative, whereas β is an agent whose doxastic states are exclusively qualitative. In my presentation of their account, I simply use α to denote an ideal agent and then discuss the restrictions we may place on α, including the two distinct types of doxastic notions mentioned above. The main reason for my choosing not to follow their notation is because doing so would obscure a critical point I wish to make. Hawthorne and Bovens hold that there are subtle tensions when moving between quantitative and qualitative notions of belief that the paradoxes of rational acceptance make stand out (1999, 241). Indeed, they think that "the *preface* and the *lottery* illuminate complementary facets of the relationship between qualitative and quantitative belief" (1999, 244). I reject this analysis, for reasons that will become apparent. The short of it is that I maintain that the apparent tensions between qualitative and quantitative belief that they study are artifacts of their framework that have little to do with either the paradox of the preface or the lottery paradox.

[3] There are two points to mention. First, it is important to stress again that the dispute is not over semantics *per se*, but rather the logic of belief's use of semantics in place of a syntax for constructing rationally accepted formulas. For a general discussion of formalized languages and inference rules, see (Church 1944, §07.). Second, it is worth mentioning again the three conditions on a formal language observed in Section 2. The objection discussed here is that one cannot begin to evaluate the logic of belief with respect to these conditions because there isn't a formal language for the logic of belief to even construct formal proof objects.

[4] This point may present a problem for Hawthorne and Bovens's contextualism, for they intend their degree of confidence measure to be contextually determined. On their view, an agent who maintains the very same degree of belief across contexts may nevertheless assign different threshold points in different contexts to yield different sets of fully accepted beliefs. For instance, in one context (i.e., by one confidence measure) an agent may have a common sense belief that a train will arrive on time but fail to believe that two events will occur at the same moment in a controlled experiment, where the different doxastic attitudes is due to a different threshold point for full belief rather than a different degree of confidence assigned to each proposition. Hawthorne and Bovens's remark that their "analysis applies to any single belief standard, and may be applied to each of a number of standards, one by one" (1999, 246, note 7). However, in light of this condition of comparability of belief, notice that this picture of accounting

for contextually sensitive thresholds for belief threatens to break down: for qualitative belief in a set of propositions may vary among contexts in the sense that all beliefs are pairwise comparable in one class of contexts but fail to be comparable in another class of contexts. But contexts of the latter type fail to yield a single qualitative probability space, and so a meaningful comparison of thresholds could not be made.

[5] This research was supported in part by FCT grant SFRH/BPD-13688-2003 and by a DAAD supported visit to the Technical University of Dresden's Artificial Intelligence Institute in November 2003. I would like to thank Jim Hawthorne, Gabriel Uzquiano and an anonymous referee for their very helpful comments.

REFERENCES

Church, A.: 1944, *Introduction to Mathematical Logic*, NJ: Princeton University, Press, Princeton.

Foley, R.: 1992, 'The Epistemology of Belief and the Epistemology of Degrees of Belief', *American Philosophical Quarterly* **29**, 111–121.

Hawthorne, J. and L. Bovens: 1999, 'The Preface, the Lottery, and the Logic of Belief', *Mind* **108**, 241–264.

Kyburg, H.E.: 1961, *Probability and the Logic of Rational Belief*, Wesleyan University Press, Middletown.

Kyburg, H.E., Jr.: 1997, 'The Rule of Adjunction and Rational Inference', *Journal of Philosophy* **94**, 109–125.

Makinson, D.C.: 1965, 'The Paradox of the Preface', *Analysis*, **25**, 205–207.

Pollock, J.L.: 1993, 'Justification and Defeat', *Artificial Intelligence*, **67**, 377–407.

Ramsey, F.P.: 1931, *The Foundations of Mathematics and Other Essays, volume 1.* Humanities Press, New York.

Savage, L.: 1972, *Foundations of Statistics*, Dover, New York.

CENTRIA
Artificial Intelligence Center
Universidade Nova de Lisboa
2829-516 Caparica
Portugal
E-mail: greg@di.fct.unl.pt

GIACOMO BONANNO

LOGIC AND THE FOUNDATIONS OF THE THEORY OF GAMES AND DECISIONS: INTRODUCTION*

This special issue of *Knowledge, Rationality and Action* contains a selection of papers presented at the sixth conference on "Logic and the Foundations of the Theory of Games and Decisions" (LOFT6), which took place in Leipzig, in July 2004.

The LOFT conferences have been a regular biannual event since 1994.[1] The first conference was hosted by the Centre International de Recherches Mathematiques in Marseille (France), the next four took place at the International Center for Economic Research in Torino (Italy) and the most recent one was hosted by the Leipzig Graduate School of Management in Leipzig (Germany).

The LOFT conferences are interdisciplinary events that bring together researchers from a variety of fields: computer science, economics, game theory, linguistics, logic, mathematical psychology, philosophy and statistics. In its original conception, LOFT had as its central theme the application of logic, in particular modal epistemic logic, to foundational issues in the theory of games and individual decision-making. Epistemic considerations have been central to game theory for a long time. The expression interactive epistemology has been used in the game-theory literature to refer to the analysis of what individuals involved in a strategic interaction know about facts concerning the external world as well as facts concerning each other's knowledge and beliefs. What is relatively new is the realization that the tools and methodology that were used in game theory are closely related to those already used in other fields, notably computer science and philosophy. Modal logic turned out to be the common language that made it possible to bring together different professional communities. The insights gained and the methodology employed in one field can benefit researchers in a different field. Indeed, new and active areas of research have sprung from the interdisciplinary exposure provided by the LOFT conferences.[2]

Over time the scope of the LOFT conferences has broadened to encompass other tools, besides modal logic, that can be used to shed light on the general issues of rationality and agency. Topics that have fallen within the LOFT umbrella include epistemic and

Synthese (2005) 147: 189–192
Knowledge, Rationality & Action 135–138
DOI 10.1007/s11229-005-1346-x

temporal logic, theories of information processing and belief revision, models of bounded rationality, non-monotonic reasoning, theories of learning and evolution, mental models, etc.

The papers collected in this issue of *Knowledge, Rationality and Action* reflect the interdisciplinary composition of the participants in the LOFT conferences and the cross-fertilization that has taken place among different fields.

Giacomo Bonanno proposes a modal logic for belief revision based on three operators, representing initial beliefs, information and revised beliefs. Three simple axioms are shown to characterize the qualitative version of Bayes' rule, which can be stated as follows: if the event representing the information has a non-empty intersection with the support of the initial probability distribution, then the support of the new probability distribution coincides with that intersection. The three axioms capture three aspects of a policy of minimal revision: in the absence of surprises, (1) the information is believed, (2) everything that was believed before continues to be believed and (3) any new belief must be deducible from the initial beliefs and the information received. Soundness and completeness are proved and theorems are derived concerning the interaction between initial and revised beliefs.

Hans van Ditmarsch proposes a framework for studying iterated belief revision based on dynamic epistemic logic. Information states are modeled as Kripke models with several accessibility relations, representing knowledge and degrees of belief. The revision of an information state by a formula ϕ is described by a dynamic modal operator that is interpreted as a binary relation between information states. The author describes five different types of dynamic belief revision and discusses their properties, in particular whether they lead to successful revision, that is, whether revision by ϕ leads to a new information state where ϕ is believed.

The paper by Noël Laverny and Jérôme Lang is related to belief revision as well as to knowledge-based programs. The latter combine knowledge and action in programs by prescribing different actions depending on the current state of knowledge. The authors propose an extension of this approach by introducing graded beliefs, so that a program can prescribe an action conditional on the degree of belief being above a given threshold. They propose the notion of belief-based program built on ordinal conditional functions. The authors distinguish between pure sensing actions (that leave the state of the world unchanged and act only on the agent's mental state by giving her some feedback about the actual world) and purely ontic action (that change the state of the world without giving any feedback to the agent). The focus of the analysis is on the execution of belief-based programs.

The paper by Martin Peterson and Sven Ove Hansson deals with decision theory. The authors characterize order-independent transformative decision rules. A transformative decision rule alters the representation of a decision problem by changing one of its components (the set of acts, the set of states, the probability distribution or the value assignments). Example of such a rule is the principle of insufficient reason, which prescribes that when there is no reason to believe that one state of the world is more probable than another, the decision maker should transform the initial representation of the decision problem into one in which every state is assigned equal probability. A set of transformative decision rules is order-independent in case the order in which the rules are applied is irrelevant. The main result of the paper is an axiomatic characterization of order-independent transformative decision rules, based on a single axiom.

The paper by Katrin Schulz deals with the paradox of free choice permission in linguistics. An example of a free choice sentence is 'You may go to the beach or go to the cinema', which intuitively seems to convey that the addressee may go to the beach and he may go to the cinema. However, such inference is at odds with standard assumptions about logical analysis. There are two strategies for resolving the tension between intuition and logic: one may either give up standard principles of logic, or one may try to explain the inference in pragmatic terms. The author explores the second strategy and develops it in enough detail to judge its feasibility and adequacy.

The paper by Giacomo Sillari deals with the philosophical problem of how to interpret and understand social conventions, in particular his aim is to cast David Lewis's account of convention in a formal framework. The author proposes a multi-agent modal logic containing, for each agent, a reason-to-believe operator and an indication operator. Sillari discusses the distinction between epistemic and practical rationality and its relationship to the notion of indication and inference. He also argues that a modal logic formalization of belief, supplemented with awareness structures, can be a natural interpretation of the epistemic concepts involved in Lewis's analysis of convention.

[137]

I would like to thank Wiebe van der Hoek, managing editor of *Knowledge, Rationality and Action*, for devoting a special issue to LOFT6, Arnis Vilks, Dean of the Leipzig Graduate School of Management, for funding and hosting LOFT6 and Pierfrancesco La Mura and Beate Kanheissner for the excellent local organization of the conference. The papers went through a thorough refereeing and editorial process. I would like to thank the many referees who provided invaluable help and the authors for their cooperation during the revision stage.

NOTES

* The conference was organized by Giacomo Bonanno, Wiebe van der Hoek, Pierfrancesco La Mura and Arnis Vilks with the assistance of a program committee consisting of Johan van Benthem, Vincent Hendricks, Wlodek Rabinowicz, Hans Rott and Marciano Siniscalchi.
¹ Collections of papers from previous LOFT conferences can be found in a special issue of *Theory and Decision* (Vol. 37, 1994, edited by M. Bacharach and P. Mongin), the volume *Epistemic logic and the theory of games and decisions* (edited by M. Bacharach, L.-A. GérardVaret, P. Mongin and H. Shin and published by Kluwer Academic, 1997), two special issues of *Mathematical Social Sciences* (Vols. 36 and 38, 1998, edited by G. Bonanno, M. Kaneko and P. Mongin), two special issues of *Bulletin of Economic Research* (Vol. 53, October 2001 and Vol. 54, January 2002, edited by G. Bonanno and W. van der Hoek) and a special issue of *Research in Economics*, (Vol. 57, 2003, edited by G. Bonanno and W. van der Hoek).
² There is substantial overlap between the LOFT community and the community of researchers who are active in another regular, biannual event, namely the conferences on Theoretical Aspects of Rationality and Knowledge (see www.tark.org).

Department of Economics
University of California
Davis, CA 95616|8578
U.S.A.
E-mail: gfbonanno@ucdavis.edu

GIACOMO BONANNO*

A SIMPLE MODAL LOGIC FOR BELIEF REVISION

ABSTRACT. We propose a modal logic based on three operators, representing intial beliefs, information and revised beliefs. Three simple axioms are used to provide a sound and complete axiomatization of the qualitative part of Bayes' rule. Some theorems of this logic are derived concerning the interaction between current beliefs and future beliefs. Information flows and iterated revision are also discussed.

1. INTRODUCTION

The notions of static belief and of belief revision have been extensively studied in the literature. However, there is a surprising lack of uniformity in the two approaches. In the philosophy and logic literature, starting with Hintikka's (1962) seminal contribution, the notion of static belief has been studied mainly within the context of modal logic. On the syntactic side a belief operator B is introduced, with the intended interpretation of $B\phi$ as "the individual believes that ϕ". Various properties of beliefs are then expressed by means of axioms, such as the positive introspection axiom $B\phi \rightarrow BB\phi$, which says that if the individual believes ϕ then she believes that she believes ϕ. On the semantic side Kripke structures (Kripke 1963) are used, consisting of a set of states (or possible worlds) Ω together with a binary relation $\mathcal{B}$ on Ω, with the interpretation of $\alpha\mathcal{B}\beta$ as "at state α the individual considers state β possible". The connection between syntax and semantics is then obtained by means of a valuation V which associates with every atomic sentence p the set of states where p is true. The pair $\langle\Omega, \mathcal{B}\rangle$ is called a frame and the addition of a valuation V to a frame yields a model. Rules are given for determining the truth of an arbitrary formula at every state of a model; in particular, the formula $B\phi$ is true at state α if and only if ϕ is true at every β such that $\alpha\mathcal{B}\beta$, that is, if ϕ is true at every state that the individual considers possible at α. A property of the accessibility relation $\mathcal{B}$ is said to correspond to an axiom if every instance

Synthese (2005) 147: 193–228
Knowledge, Rationality & Action 139–174
DOI 10.1007/s11229-005-1348-8

© Springer 2005

of the axiom is true at every state of every model based on a frame that satisfies the property and *vice versa*. For example, the positive introspection axiom $B\phi \rightarrow BB\phi$ corresponds to transitivity of the relation B. This combined syntactic-semantic approach has turned out to be very useful. The syntax allows one to state properties of beliefs in a clear and transparent way, while the semantic approach is particularly useful in reasoning about complex issues, such as the implications of rationality in interactive situations.[1]

The theory of belief revision (known as the AGM theory due to the seminal work of Alchourron et al. (1985)), on the other hand, has followed a different path.[2] In this literature beliefs are modeled as sets of formulas in a given syntactic language and the problem that has been studied is how a belief set ought to be modified when new information, represented by a formula ϕ, becomes available. With a few exceptions (see Section 4), the tools of modal logic have not been explicitly employed in the analysis of belief revision.

In the economics and game theory literature, it is standard to represent beliefs by means of a probability measure over a set of states Ω and belief revision is modeled using Bayes' rule. Let P_0 be the prior probability measure representing the initial beliefs, $E \subseteq \Omega$ an event representing new information and P_1 the posterior probability measure representing the revised beliefs. Bayes' rule says that, if $P_0(E) > 0$, then, for every event A, $P_1(A) = \frac{P_0(A \cap E)}{P_0(E)}$. Bayes' rule thus implies the following, which we call the *Qualitative Bayes Rule*:

$$\text{if } \text{supp}(P_0) \cap E \neq \varnothing, \text{ then } \text{supp}(P_1) = \text{supp}(P_0) \cap E.$$

where $\text{supp}(P)$ denotes the support of the probability measure P.[3]

In this paper we propose a unifying framework for static beliefs and belief revision by bringing belief revision under the umbrella of modal logic and by providing an axiomatization of the Qualitative Bayes Rule in a simple logic based on three modal operators: B_0, B_1 and I, whose intended interpretation is as follows:

$B_0\phi$ initially (at time 0) the individual believes that ϕ

$I\phi$ (between time 0 and time 1) the individual is informed that ϕ

$B_1\phi$ at time 1 (after revising his beliefs in light of the information received) the individual believes that ϕ.

Semantically, it is clear that the Qualitative Bayes Rule embodies the *conservativity principle* for belief revision, according to which "When

changing beliefs in response to new evidence, you should continue
to believe as many of the old beliefs as possible" (Harman (1986),
p. 46). The set of all the propositions that the individual believes
corresponds to the set of states that she considers possible (in a
probabilistic setting a state is considered possible if it is assigned
positive probability). The conservativity principle requires that, if
the individual considers a state possible and her new information
does not exclude this state, then she continue to consider it possible.
Furthermore, if the individual regards a particular state as impossible, then she should continue to regard it as impossible, unless her
new information excludes *all* the states that she previously regarded
as possible. The axiomatization we propose gives a transparent syntactic expression to the conservativity principle.

The paper is organized as follows. In Section 2 we provide a
characterization of the Qualitative Bayes Rule in terms of three simple axioms. In Section 3 we provide a logic which is sound and complete with respect to the class of frames that satisfy the Qualitative
Bayes Rule and prove some theorems of this logic concerning the
interaction between current beliefs and future beliefs. In section 4
we discuss the relationship between our analysis and that of closely
related papers in the literature. Section 5 examines the relationship
between our approach and the AGM approach. In Section 6 we deal
with the issue of iterated revision and Section 7 concludes.

2. AXIOMATIC CHARACTERIZATION OF THE QUALITATIVE BAYES RULE

We begin with the semantics. A *frame* is a quadruple $\langle \Omega, \mathcal{B}_0, \mathcal{B}_1, \mathcal{I} \rangle$
where Ω is a set of *states* and $\mathcal{B}_0$, $\mathcal{B}_1$, and $\mathcal{I}$ are binary relations on
Ω, whose interpretation is as follows:

$\alpha \mathcal{B}_0 \beta$ at state α the individual initially (at time 0) considers state
         β possible
$\alpha \mathcal{I} \beta$ at state α, state β is compatible with the information
         received
$\alpha \mathcal{B}_1 \beta$ at state α the individual at time 1 (in light of the informa-
         tion received) considers state β possible.

Let $\mathcal{B}_0(\omega) = \{\omega' \in \Omega : \omega \mathcal{B}_0 \omega'\}$ denote the set of states that, initially,
the individual considers possible at state ω. Define $\mathcal{I}(\omega)$ and $\mathcal{B}_1(\omega)$
similarly.[4] By *Qualitative Bayes Rule* (QBR) we mean the following
property:

$$\forall \omega \in \Omega, \text{ if } \mathcal{B}_0(\omega) \cap \mathcal{I}(\omega) \neq \varnothing \text{ then } \mathcal{B}_1(\omega) = \mathcal{B}_0(\omega) \cap \mathcal{I}(\omega).$$
$$\text{(QBR)}$$

Thus QBR says that if at a state the information received is consistent with the initial beliefs – in the sense that there are states that were considered possible initially and are compatible with the information – then the states that are considered possible according to the revised beliefs are precisely those states.

On the syntactic side we consider a modal propositional logic based on three operators: B_0, B_1 and I whose intended interpretation is as explained in Section 1. The formal language is built in the usual way from a countable set S of atomic propositions, the connectives $\neg$ (for "not") and $\lor$ (for "or") and the modal operators.[5] Thus the set Φ of formulas is defined inductively as follows: $q \in \Phi$ for every atomic proposition $q \in S$, and if $\phi, \psi \in \Phi$ then all of the following belong to Φ: $\neg\phi, \phi \lor \psi, B_0\phi, B_1\phi$ and $I\phi$.

Remark 1. We have allowed $I\phi$ to be a well-formed formula for every formula ϕ. As pointed out by Friedman and Halpern (1999), this may be problematic. For example, it is not clear how one could be informed of a contradiction. Furthermore, one might want to restrict information to facts by not allowing $I\phi$ be a well-formed formula if ϕ contains any of the modal operators B_0, B_1 and I.[6] Without that restriction, in principle we admit situations like the following: the individual initially believes that ϕ and is later informed that he did not believe that ϕ: $B_0\phi \land I\neg B_0\phi$. It is not clear how such a situation could arise.[7] However, since our results remain true – whether or not we impose the restriction – we have chosen to follow the more general approach. The undesirable situations can then be eliminated by imposing suitable axioms, for example the axiom $B_0\phi \to \neg I\neg B_0\phi$, which says that if the individual initially believes that ϕ then it cannot be the case that he is informed that he did not believe that ϕ (see Section 7 for further discussion).

The connection between syntax and semantics is given by the notion of model. Given a frame $\langle \Omega, \mathcal{B}_0, \mathcal{B}_1, \mathcal{I} \rangle$, a *model* is obtained by adding a *valuation* $V: S \to 2^{\Omega}$ (where 2^{Ω} denotes the set of subsets of Ω, usually called *events*) which associates with every atomic proposition $p \in S$ the set of states at which p is true. The truth of an arbitrary formula at a state is then defined inductively as follows

($\omega \models \phi$ denotes that formula ϕ is true at state ω; $\| \phi \|$ is the truth set of ϕ, that is, $\| \phi \| = \{ \omega \in \Omega : \omega \models \phi \}$):

if q is an atomic proposition, $\omega \models q$ if and only if $\omega \in V(q)$,

$\omega \models \neg \phi$ if and only if $\omega \nvDash \phi$

$\omega \models \phi \vee \psi$ if and only if either $\omega \models \phi$ or $\omega \models \psi$ (or both),

$\omega \models B_0 \phi$ if and only if $\mathcal{B}_0(\omega) \subseteq \| \phi \|$,[8]

$\omega \models B_1 \phi$ if and only if $\mathcal{B}_1(\omega) \subseteq \| \phi \|$,

$\omega \models I \phi$ if and only if $\mathcal{I}(\omega) = \| \phi \|$.

Remark 2. Note that, while the truth conditions for $B_0 \phi$ and $B_1 \phi$ are the standard ones, the truth condition of $I \phi$ is unusual in that the requirement is $\mathcal{I}(\omega) = \| \phi \|$ rather than merely $\mathcal{I}(\omega) \subseteq \| \phi \|$.[9]

We say that a formula ϕ is *valid in a model* if $\omega \models \phi$ for all $\omega \in \Omega$, that is, if ϕ is true at every state. A formula ϕ is *valid in a frame* if it is valid in every model based on that frame. Finally, we say that a property of frames is *characterized* by (or characterizes) an axiom if (1) the axiom is valid in any frame that satisfies the property and, conversely, (2) whenever the axiom is valid in a frame, then the frame satisfies the property.

We now introduce three axioms that, together, provide a characterization of the Qualitative Bayes Rule.

QUALIFIED ACCEPTANCE: $(I \phi \wedge \neg B_0 \neg \phi) \to B_1 \phi$.

This axiom says that if the individual is informed that ϕ ($I \phi$) and he initially considered ϕ possible (that is, it is not the case that he believed its negation: $\neg B_0 \neg \phi$) then he accepts ϕ in his revised beliefs. That is, information that is not surprising is believed.

The next axiom says that if the individual receives non-surprising information (i.e. information that does not contradict his initial beliefs) then he continues to believe everything that he believed before:

PERSISTENCE: $(I \phi \wedge \neg B_0 \neg \phi) \to (B_0 \psi \to B_1 \psi)$.

The third axiom says that beliefs should be revised in a minimal way, in the sense that no new beliefs should be added unless they are implied by the old beliefs and the information received:

MINIMALITY: $(I \phi \wedge B_1 \psi) \to B_0(\phi \to \psi)$.

The Minimality axiom is not binding (that is, it is trivially satisfied) if the information is surprising; suppose that at a state, say α, the individual is informed that ϕ ($\alpha \models I\phi$) although he initially believed that ϕ was *not* the case ($\alpha \models B_0\neg\phi$). Then, for every formula ψ, the formula ($\phi \rightarrow \psi$) is trivially true at every state that the individual initially considered possible ($\mathcal{B}_0(\alpha) \subseteq \parallel \phi \rightarrow \psi \parallel$) and therefore he initially believed it ($\alpha \models B_0(\phi \rightarrow \psi)$). Thus the axiom restricts the new beliefs only when the information received is not surprising, that is, only if ($I\phi \wedge \neg B_0\neg\phi$) happens to be the case.

The above axioms are further discussed below. The following proposition gives the main result of this section.

PROPOSITION 3. The Qualitative Bayes Rule (QBR) is characterized by the conjunction of the three axioms Qualified Acceptance, Persistence and Minimality (that is, if a frame satisfies QBR then the three axioms are valid in it and – conversely – if the three axioms are valid in a frame then the frame satisfies QBR).

The proof of Proposition 3 is a corollary of the following three lemmas, which characterize the three axioms individually.

LEMMA 4. The Qualified Acceptance axiom (($I\phi \wedge \neg B_0\neg\phi$) $\rightarrow$ $B_1\phi$) is characterized by the property: $\forall \omega \in \Omega$, if $\mathcal{B}_0(\omega) \cap \mathcal{I}(\omega) \neq \varnothing$ then $\mathcal{B}_1(\omega) \subseteq \mathcal{I}(\omega)$.

Proof. Fix a frame where the property holds, an arbitrary model based on it, a state ω and a formula ϕ such that $\omega \models I\phi \wedge \neg B_0\neg\phi$. Then $\mathcal{I}(\omega) = \parallel \phi \parallel$. Since $\omega \models \neg B_0\neg\phi$ there exists a $\beta \in \mathcal{B}_0(\omega)$ such that $\beta \models \phi$. Thus $\mathcal{B}_0(\omega) \cap \mathcal{I}(\omega) \neq \varnothing$ and, by the property, $\mathcal{B}_1(\omega) \subseteq \mathcal{I}(\omega)$. Hence $\omega \models B_1\phi$. Conversely, fix a frame that does not satisfy the property. Then there exists a state α such that $\mathcal{B}_0(\alpha) \cap \mathcal{I}(\alpha) \neq \varnothing$ and $\mathcal{B}_1(\alpha) \not\subseteq \mathcal{I}(\alpha)$, that is, there is a $\beta \in \mathcal{B}_1(\alpha)$ such that $\beta \notin \mathcal{I}(\alpha)$. Let p be an atomic proposition and construct a model where $\parallel p \parallel = \mathcal{I}(\alpha)$. Then $\alpha \models Ip$ and, since $\mathcal{B}_0(\alpha) \cap \mathcal{I}(\alpha) \neq \varnothing$, $\alpha \models \neg B_0\neg p$. Furthermore, $\beta \not\models p$ (because $\beta \notin \mathcal{I}(\alpha)$). Thus, since $\beta \in \mathcal{B}_1(\alpha)$, $\alpha \not\models B_1 p$ and the axiom is falsified at α. $\square$

Note that if the truth condition for $I\phi$ were "$\omega \models I\phi$ if and only if $\mathcal{I}(\omega) \subseteq \parallel \phi \parallel$" (rather than $\mathcal{I}(\omega) = \parallel \phi \parallel$), then Lemma 4 would not be true. The implication "property violated $\Rightarrow$ axiom not valid" would still be true (identical proof). However, the implication "property holds $\Rightarrow$ axiom valid" would no longer be true, because it could happen that $\mathcal{I}(\omega)$ is a **proper** subset of $\|\phi\|$. For example, let $\Omega = \{\alpha, \beta, \gamma\}$, $\mathcal{B}_0(\alpha) = \{\alpha\}$, $\mathcal{B}_0(\beta) = \mathcal{B}_0(\gamma) = \{\gamma\}$, $\mathcal{I}(\alpha) = \{\alpha\}$, $\mathcal{I}(\beta) = \{\beta\}$, $\mathcal{I}(\gamma) = \{\gamma\}$, $\mathcal{B}_1(\alpha) = \mathcal{B}_1(\beta) = \{\alpha\}$ and $\mathcal{B}_1(\gamma) = \{\gamma\}$. Then the property $\forall \omega$, if $\mathcal{B}_0(\omega) \cap \mathcal{I}(\omega) \neq \varnothing$ then $\mathcal{B}_1(\omega) \subseteq \mathcal{I}(\omega)$ is satisfied (note, in particular, that $\mathcal{B}_0(\beta) \cap \mathcal{I}(\beta) = \varnothing$). Construct a model where, for some atomic proposition p, $\| p \| = \{\beta, \gamma\}$. Then, under the rule $\mathcal{I}(\beta) \subseteq \| p \|$, we would have $\beta \models Ip$ and $\beta \models \neg B_0 \neg p \wedge \neg B_1 p$, so that the Qualified Acceptance axiom would be falsified at β. This frame is illustrated in Figure 1. In all the figures we represent a binary relation $R \subseteq \Omega \times \Omega$ as follows: (1) if there is an arrow from ω to ω' then $\omega' \in R(\omega)$ (i.e., $\omega R \omega'$), (2) if a rounded rectangle encloses a set of states then, for any two states ω and ω' in that rectangle, $\omega' \in R(\omega)$ and (3) if there is an arrow from a state ω to a rounded rectangle, then for any state ω' in that rectangle, $\omega' \in R(\omega)$.

LEMMA 5. The Persistence axiom $((\neg B_0 \neg \phi \wedge I\phi) \rightarrow (B_0 \psi \rightarrow B_1 \psi))$ is characterized by the property: $\forall \omega \in \Omega$, if $\mathcal{B}_0(\omega) \cap \mathcal{I}(\omega) \neq \varnothing$ then $\mathcal{B}_1(\omega) \subseteq \mathcal{B}_0(\omega)$.

Proof. Fix a frame where the property holds, an arbitrary model based on it, a state ω and formulas ϕ and ψ such that $\omega \models B_0 \psi \wedge \neg B_0 \neg \phi \wedge I\phi$. Then $\mathcal{I}(\omega) = \|\phi\|$. Since $\omega \models \neg B_0 \neg \phi$, $\mathcal{B}_0(\omega) \cap \mathcal{I}(\omega) \neq \varnothing$. Then, by the property, $\mathcal{B}_1(\omega) \subseteq \mathcal{B}_0(\omega)$. Since $\omega \models B_0 \psi$, $\mathcal{B}_0(\omega) \subseteq \|\psi\|$. Thus $\omega \models B_1 \psi$. Conversely, fix a frame that does not satisfy the property. Then there exists a state α such that $\mathcal{B}_0(\alpha) \cap \mathcal{I}(\alpha) \neq \varnothing$ and $\mathcal{B}_1(\alpha) \not\subseteq \mathcal{B}_0(\alpha)$, that is, there exists a $\beta \in \mathcal{B}_1(\alpha)$ such that $\beta \notin$

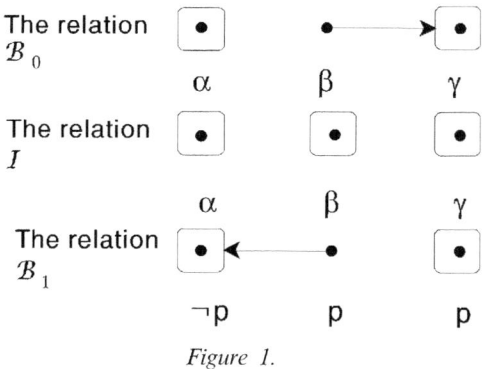

Figure 1.

$\mathcal{B}_0(\alpha)$. Let p and q be atomic propositions and construct a model where $\|p\| = \mathcal{B}_0(\alpha)$ and $\|q\| = \mathcal{I}(\alpha)$. Then $\alpha \models \mathcal{B}_0 p \wedge Iq$ and, since $\mathcal{B}_0(\alpha) \cap \mathcal{I}(\alpha) \neq \varnothing$, $\alpha \models \neg \mathcal{B}_0 \neg q$. Since $\beta \notin \mathcal{B}_0(\alpha)$, $\beta \not\models p$. Thus, since $\beta \in \mathcal{B}_1(\alpha), \alpha \not\models \mathcal{B}_1 p$. Thus the instance of the axiom with $\psi = p$ and $\phi = q$ is falsified at α.   □

Note again that with the standard validation rule for the operator I, the above lemma would not be true. The implication "property violated $\implies$ axiom not valid" would still be true (identical proof). However, the implication "property holds $\implies$ axiom valid" would no longer be true. This can be seen in the example of Figure 1 at state β with $\phi = \psi = p$. In fact, under the rule $\beta \models Ip$ if and only if $\mathcal{I}(\beta) \subseteq \|p\|$ (rather than $\mathcal{I}(\beta) = \|p\|$) we would have $\beta \models Ip$ and $\beta \models \mathcal{B}_0 p \wedge \neg \mathcal{B}_0 \neg p \wedge \neg \mathcal{B}_1 p$, so that the Persistence axiom would be falsified at β, despite the fact that the frame of Figure 1 satisfies the property that, $\forall \omega \in \Omega$, if $\mathcal{B}_0(\omega) \cap \mathcal{I}(\omega) \neq \varnothing$ then $\mathcal{B}_1(\omega) \subseteq \mathcal{B}_0(\omega)$ (notice, in particular, that $\mathcal{B}_0(\beta) \cap \mathcal{I}(\beta) = \varnothing$).

LEMMA 6. The Minimality axiom $((I\phi \wedge \mathcal{B}_1 \psi) \to \mathcal{B}_0(\phi \to \psi))$ is characterized by the following property: $\forall \omega \in \Omega, \mathcal{B}_0(\omega) \cap \mathcal{I}(\omega) \subseteq \mathcal{B}_1(\omega)$.

Proof. Fix a frame that satisfies the property and an arbitrary model based on it. Let α be a state and ϕ and ψ formulas such that $\alpha \models I\phi \wedge \mathcal{B}_1 \psi$. Then $\mathcal{I}(\alpha) = \|\phi\|$. By the property, $\mathcal{B}_0(\alpha) \cap \mathcal{I}(\alpha) \subseteq \mathcal{B}_1(\alpha)$. Since $\alpha \models \mathcal{B}_1 \psi$, $\mathcal{B}_1(\alpha) \subseteq \|\psi\|$. Thus, for every $\omega \in \mathcal{B}_0(\alpha) \cap \mathcal{I}(\alpha), \omega \models \psi$ and therefore $\omega \models \phi \to \psi$. On the other hand, for every $\omega \in \mathcal{B}_0(\alpha)$, if $\omega \notin \mathcal{I}(\alpha)$, then $\omega \models \neg \phi$ and therefore $\omega \models \phi \to \psi$. Thus $\mathcal{B}_0(\alpha) \subseteq \|\phi \to \psi\|$, i.e. $\alpha \models \mathcal{B}_0(\phi \to \psi)$.

Conversely, suppose the property is violated. Then there exists a state α such that $\mathcal{B}_0(\alpha) \cap \mathcal{I}(\alpha) \not\subseteq \mathcal{B}_1(\alpha)$, that is, there exists a $\beta \in \mathcal{B}_0(\alpha) \cap \mathcal{I}(\alpha)$ such that $\beta \notin \mathcal{B}_1(\alpha)$. Let p and q be atomic propositions and construct a model where $\|p\| = \mathcal{I}(\alpha)$ and $\|q\| = \mathcal{B}_1(\alpha)$. Then $\alpha \models Ip \wedge \mathcal{B}_1 q$. Since $\beta \in \mathcal{I}(\alpha)$ and $\beta \notin \mathcal{B}_1(\alpha)$, $\beta \models p \wedge \neg q$, i.e. $\beta \models \neg(p \to q)$. Thus, since $\beta \in \mathcal{B}_0(\alpha)$, $\alpha \not\models \mathcal{B}_0(p \to q)$. Hence the axiom is falsified at α.   □

Once again, it can be seen from Figure 1 that under the standard validation rule for I ($\omega \models I\phi$ if and only if $\mathcal{I}(\omega) \subseteq \|\phi\|$, rather than $\mathcal{I}(\omega) = \|\phi\|$) it is not true that satisfaction of the property $\forall \omega \in \Omega$, $\mathcal{B}_0(\omega) \cap \mathcal{I}(\omega) \subseteq \mathcal{B}_1(\omega)$ guarantees validity of the Minimality axiom. In fact, under the standard validation rule, Minimality would be falsified at state β with $\phi = p$ and $\psi = \neg p$.

The Qualitative Bayes Rule captures the following conservativity principle for belief revision: if the information received involves no surprises, then beliefs should be changed in a minimal way, in the sense that all the previous beliefs ought to be maintained and any new belief should be deducible from the old beliefs and the information. The extreme case of "no surprise" is the case where the individual is informed of something which he already believes, In this case the notion of minimal change would require that there be no change at all. This requirement is expressed by the following axiom:

NO CHANGE: $(B_0\phi \wedge I\phi) \to (B_1\psi \leftrightarrow B_0\psi)$.

PROPOSITION 7. Assume that initial beliefs satisfy axiom $K(B_0\phi \wedge B_0(\phi \to \psi) \to B_0\psi)$ and the consistency axiom $D(B_0\phi \to \neg B_0\neg\phi)$. Then the conjunction of Persistence and Minimality implies No Change.

Proof. We give a syntactic proof (PL stands for 'Propositional Logic'):

1. $B_0\phi \to \neg B_0\neg\phi$		Consistency of B_0
2. $B_0\phi \wedge I\phi \to \neg B_0\neg\phi \wedge I\phi$		1, PL
3. $\neg B_0\neg\phi \wedge I\phi \to (B_0\psi \to B_1\psi)$		Persistence
4. $B_0\phi \wedge I\phi \to (B_0\psi \to B_1\psi)$		2, 3, PL
5. $I\phi \wedge B_1\psi \to B_0(\phi \to \psi)$		Minimality
6. $I\phi \wedge B_1\psi \wedge B_0\phi \to B_0(\phi \to \psi) \wedge B_0\phi$		5, PL
7. $B_0(\phi \to \psi) \wedge B_0\phi \to B_0\psi$		Axiom K for B_0
8. $I\phi \wedge B_0\phi \wedge B_1\psi \to B_0\psi$		6, 7, PL
9. $I\phi \wedge B_0\phi \to (B_1\psi \to B_0\psi)$		8, PL
10. $I\phi \wedge B_0\phi \to (B_0\psi \to B_1\psi) \wedge (B_1\psi \to B_0\psi)$		4, 9, PL

□

Note that without consistency of initial beliefs Proposition 7 is not true,[10]

Note also that the converse of Proposition 7 does not hold: neither Persistence nor Minimality can be derived from No Change.[11]

We conclude this section with further discussion of the axioms studied above.

The relatively recent literature on dynamic epistemic logic studies how actions such as public announcements lead to revision of the interactive knowledge of a group of individuals (for a survey see van der Hoek and Pauly (forthcoming) and van Ditmarsch and

van der Hoek 2004). One of the issues studied in this literature is
what kind of public announcements can be successful in the sense
that they produce common knowledge of the announced fact. Some
public announcements, although truthful, cannot be successful. For
example if individual a does not know that p $(\neg K_a p)$, the public announcement '$p \wedge \neg K_a p$', although truthful, "leaves a with a
difficult, if not impossible task to update his knowledge; it is hard
to see how to simultaneously incorporate p and $\neg K_a p$ into his
knowledge" (van der Hoek and Pauly (forthcoming) p. 23). In our
approach this difficulty does not arise, since we distinguish between
initial beliefs (B_0) and revised beliefs (B_1). It is therefore not problematic to be told "p is true and you did not believe it before this
announcement" ($p \wedge \neg B_0 p$) since this fact can be truthfully incorpo-
rated into the revised beliefs. That is, the formula $p \wedge \neg B_0 p \wedge B_1(p \wedge \neg B_0 p)$ is consistent.

If the revised beliefs satisfy positive introspection, that is, if the
operator B_1 satisfies the axiom $B_1 \phi \rightarrow B_1 B_1 \phi$, then the following
axiom can be derived from Minimality: $I\phi \wedge B_1 \phi \rightarrow B_0(\phi \rightarrow B_1 \phi)$.[12]
This may seem counterintuitive. However, one cannot consistently
reject this principle and at the same time embrace Bayes' rule for
belief revision, since the former is an implication of the latter. In
fact, letting P_0 be the probability measure that represents the initial
beliefs, and denoting its support by $supp(P_0)$, for every event F it
is trivially true that

(1) $\text{supp}(P_0) = (\text{supp}(P_0) \cap F) \cup (\text{supp}(P_0) \cap \neg F),$

(where $\neg F$ denotes the complement of F). Now, let E be an
event representing new information such that $P_0(E) > 0$, that is,
$\text{supp}(P_0) \cap E \neq \varnothing$. Let P_1 be the probability measure representing
the revised beliefs obtained by applying Bayes' rule, so that, for
every event A, $P_1(A) = \frac{P_0(A \cap E)}{P_0(E)}$. Then, as noted in Section 1,

(2) $\text{supp}(P_1) = \text{supp}(P_0) \cap E.$

It follows from (1) and (2) that

(3) $\text{supp}(P_0) \subseteq \neg E \cup \text{supp}(P_1),$

which says that for any state ω that the individual initially consid-
ers possible ($\omega \in \text{supp}(P_0)$) if event E is true at ω ($\omega \in E$) then he
will later assign positive probability to ω ($\omega \in \text{supp}(P_1)$). Since, by
(2), $\text{supp}(P_1) \subseteq E$, assigning prior probability 1 to the event $\neg E \cup$

$supp(P_1)$ corresponds to the syntactic formula $B_0(\phi \to B_1\phi)$, where $\|\phi\| = E$.

3. A SOUND AND COMPLETE LOGIC FOR BELIEF REVISION

We now provide a sound and complete logic for belief revision. Because of the non-standard validation rule for the information operator I, we need to add the universal or global modality A (see Goranko and Passy 1992; Blackburn et al. 2001, p. 415). The interpretation of $A\phi$ is "it is globally true that ϕ". As before, a frame is a quadruple $\langle \Omega, \mathcal{B}_0, \mathcal{B}_1, \mathcal{I} \rangle$. To the validation rules discussed in Section 2 we add the following;

$$\omega \models A\phi \text{ if and only if } \|\phi\| = \Omega.$$

We denote by $\mathfrak{L}$ the logic determined by the following axioms and rules of inference.

AXIOMS:

1. All propositional tautologies.
2. Axiom K for B_0, B_1 and A (note the absence of an analogous axiom for I):

$$B_0\phi \wedge B_0(\phi \to \psi) \to B_0\psi \quad (\text{K}_0)$$
$$B_1\phi \wedge B_1(\phi \to \psi) \to B_1\psi \quad (\text{K}_1)$$
$$A\phi \wedge A(\phi \to \psi) \to A\psi \quad (\text{K}_A)$$

3. S5 axioms for A:

$$A\phi \to \phi \quad (\text{T}_A)$$
$$\neg A\phi \to A\neg A\phi \quad (5_A)$$

4. inclusion axioms for B_0 and B_1. (note the absence of an analogous axiom for I);

$$A\phi \to B_0\phi \quad (\text{Incl}_0)$$
$$A\phi \to B_1\phi \quad (\text{Incl}_1)$$

5. Axioms to capture the non-standard semantics for I:

$$(I\phi \wedge I\psi) \to A(\phi \leftrightarrow \psi) \quad (\text{I}_1)$$
$$A(\phi \leftrightarrow \psi) \to (I\phi \leftrightarrow I\psi) \quad (\text{I}_2)$$

RULES OF INFERENCE:

1. Modus Ponens: $\frac{\phi, \phi \to \psi}{\psi}$ (MP)
2. Necessitation for A: $\frac{\phi}{A\phi}$ (Nec$_A$)

Remark 8. Note that from (Nec$_A$) and (Incl$_0$) one obtains necessitation for B_0 as a derived rule of inference: $\frac{\phi}{B_0\phi}$. The same is true for B_1. On the other hand, the necessitation rule for I is **not** a rule of inference of logic $\mathfrak{L}$. Indeed necessitation for I is not validity preserving.[13] Neither is the following rule for I (normally referred to as rule RK): $\frac{\phi \to \psi}{I\phi \to I\psi}$.[14] On the other hand, by Nec$_A$ and I_2, the following rule for I (normally referred to as rule RE): $\frac{\phi \leftrightarrow \psi}{I\phi \leftrightarrow I\psi}$ is a derived rule of inference of $\mathfrak{L}$.

Note that, despite the non-standard validation rule, axiom K for I, namely $I\phi \wedge I(\phi \to \psi) \to I\psi$, is trivially valid in every frame.[15] It follows from the completeness theorem proved below that axiom K for I is provable in $\mathfrak{L}$. The following proposition, however, provides a direct proof.

PROPOSITION 9. $I\phi \wedge I(\phi \to \psi) \to I\psi$ is a theorem of logic $\mathfrak{L}$.

Proof. We give a syntactic proof ('PL' stands for 'Propositional Logic'):

1.	$(I\phi \wedge I(\phi \to \psi)) \to A(\phi \leftrightarrow (\phi \to \psi))$	Axiom I_1
2.	$(\phi \leftrightarrow (\phi \to \psi)) \to (\phi \leftrightarrow \psi)$	Tautology
3.	$A(\phi \leftrightarrow (\phi \to \psi)) \to A(\phi \leftrightarrow \psi)$	2, necessitation for A, axiom K_A and Modus Ponens
4.	$A(\phi \leftrightarrow \psi) \to (I\phi \leftrightarrow I\psi)$	Axiom I_2
5.	$(I\phi \wedge I(\phi \to \psi)) \to (I\phi \leftrightarrow I\psi)$	1, 3, 4 PL
6.	$(I\phi \wedge I(\phi \to \psi)) \to I\psi$	5, PL □

Recall that a logic is *complete* with respect to a class of frames if every formula which is valid in every frame in that class is provable in the logic (that is, it is a theorem). The logic is *sound* with respect to a class of frames if every theorem of the logic is valid in every frame in that class. The following proposition is a straightforward adaptation of a result due to Goranko and Passy (1992) (Theorem 6.2, p. 24). Its proof is relegated to the appendix.

PROPOSITION 10. Logic $\mathcal{L}$ is sound and complete with respect to the class of all frames $\langle \Omega, \mathcal{B}_0, \mathcal{B}_1, \mathcal{I} \rangle$.

We are interested in extensions of $\mathcal{L}$ obtained by adding various axioms. Let $\mathfrak{R}$ ('R' stands for 'Revision') be the logic obtained by adding to $\mathcal{L}$ the axioms discussed in the previous section:

$$\mathfrak{R} = \mathcal{L} + \text{Qualified Acceptance} + \text{Persistence} + \text{Minimality}.$$

The following proposition is proved in the appendix (in light of Propositions 3 and 10 it suffices to show that the axioms Qualified Acceptance, Persistence and Minimality are canonical).

PROPOSITION 11. Logic $\mathfrak{R}$ is sound and complete with respect to the class of frames $\langle \Omega, \mathcal{B}_0, \mathcal{B}_1, \mathcal{I} \rangle$ that satisfy the Qualitative Bayes Rule.

So far we have not postulated any properties of beliefs, in particular, in the interest of generality, we have not required beliefs to satisfy the KD45 logic. In order to further explore the implications of the Qualitative Bayes Rule, we shall now consider additional axioms:

Consistency of initial beliefs	$B_0 \phi \to \neg B_0 \neg \phi$	(D$_0$)
Positive Introspection of initial beliefs	$B_0 \phi \to B_0 B_0 \phi$	(4$_0$)
Self Trust	$B_0(B_0 \phi \to \phi)$	(ST)
Information Trust	$B_0(I\phi \to \phi)$	(IT).

Self Trust says that the individual at time 0 believes that his beliefs are correct (he believes that if he believes ϕ then ϕ is true), while Information Trust says that the individual at time 0 believes that any information he will receive will be correct (he believes that if he is informed that ϕ then ϕ is true).

Remark 12. It is well-known that Consistency of initial beliefs corresponds to seriality of $\mathcal{B}_0$ ($\mathcal{B}_0(\omega) \neq \varnothing$, for all $\omega \in \Omega$) and Positive Introspection to transitivity of $\mathcal{B}_0$ (if $\beta \in \mathcal{B}_0(\alpha)$ then $\mathcal{B}_0(\beta) \subseteq \mathcal{B}_0(\alpha)$). It is also well-known that Self Trust is characterized by secondary reflexivity of $\mathcal{B}_0$ (if $\beta \in \mathcal{B}_0(\alpha)$ then $\beta \in \mathcal{B}_0(\beta)$).[16]

LEMMA 13. Information Trust $(B_0(I\phi \to \phi))$ is characterized by reflexivity of $\mathcal{I}$ over $\mathcal{B}_0$: $\forall \alpha, \beta \in \Omega$, if $\beta \in \mathcal{B}_0(\alpha)$ then $\beta \in \mathcal{I}(\beta)$.

Proof. Suppose the property is satisfied. Fix arbitrary α and ϕ. If $\mathcal{B}_0(\alpha) = \varnothing$ then $\alpha \models B_0\psi$ for every formula ψ, in particular for $\psi = I\phi \to \phi$. Suppose therefore that $\mathcal{B}_0(\alpha) \neq \varnothing$ and fix an arbitrary $\beta \in \mathcal{B}_0(\alpha)$. If $\mathcal{I}(\beta) \neq \|\phi\|$ then $\beta \nvDash I\phi$ and therefore $\beta \models I\phi \to \phi$. If $\mathcal{I}(\beta) = \|\phi\|$ then $\beta \models I\phi$. By the property, $\beta \in \mathcal{I}(\beta)$. Thus $\beta \models \phi$ and, therefore, $\beta \models I\phi \to \phi$. Conversely, suppose the property is violated; Then there exist α and β such that $\beta \in \mathcal{B}_0(\alpha)$ and $\beta \notin \mathcal{I}(\beta)$. Let p be an atomic proposition and construct a model where $\|p\| = \mathcal{I}(\beta)$. Then $\beta \models Ip$. Since $\beta \notin \mathcal{I}(\beta)$, $\beta \models \neg p$. Thus $\beta \nvDash Ip \to p$ and, therefore, $\alpha \nvDash B_0(Ip \to p)$.   □

Remark 14. Since the additional axioms listed above are canonical, it follows from Proposition 11 that if Σ is a set of axioms from the above list, then the logic $\mathfrak{R} + \Sigma$ obtained by adding to $\mathfrak{R}$ the axioms in Σ is sound and complete with respect to the class of frames that satisfy the Qualitative Bayes Rule and the properties corresponding to the axioms in Σ. For example, the logic $\mathfrak{R} + \{D_0, 4_0, ST\}$ is sound and complete with respect to the class of frames that satisfy the Qualitative Bayes Rule as well as seriality, transitivity and secondary reflexivity of $\mathcal{B}_0$.

By Proposition 7, No Change $(B_0\phi \wedge I\phi \to (B_1\psi \leftrightarrow B_0\psi))$ is a theorem of $\mathfrak{R} + D_0$. We now discuss some further theorems of extensions of $\mathfrak{R}$. Consider the following axiom:

$$B_0\phi \to B_0 B_1 \phi,$$

which says that if the individual initially believes that ϕ then she initially believes that she will continue to believe ϕ later.

PROPOSITION 15. $B_0\phi \to B_0 B_1 \phi$ is a theorem of $\mathfrak{R} + 4_0 + ST + IT$.

Proof. It is shown in van der Hoek (1993) (p. 183, Theorem 4.3(c)) that axiom $B_0\phi \to B_0 B_1 \phi$ is characterized by the following property:

$$\forall \alpha, \beta \in \Omega, \text{ if } \beta \in \mathcal{B}_0(\alpha) \text{ then } \mathcal{B}_1(\beta) \subseteq \mathcal{B}_0(\alpha). \qquad (P_0)$$

By Remark 14, the logic $\mathfrak{R} + 4_0 + ST + IT$ is sound and complete with respect to the class of frames that satisfy the Qualitative Bayes Rule as well as transitivity and secondary reflexivity of $\mathcal{B}_0$ and reflexivity of $\mathcal{I}$ over $\mathcal{B}_0$. Thus it is enough to show that this class of frames satisfies property (P_0). Fix an arbitrary frame in this class and arbitrary states α and β such that $\beta \in \mathcal{B}_0(\alpha)$. By Secondary Reflexivity of $\mathcal{B}_0, \beta \in \mathcal{B}_0(\beta)$. By Reflexivity of $\mathcal{I}$ over $\mathcal{B}_0, \beta \in \mathcal{I}(\beta)$. Thus $\mathcal{B}_0(\beta) \cap \mathcal{I}(\beta) \neq \varnothing$ and, by the Qualitative Bayes Rule, $\mathcal{B}_1(\beta) = \mathcal{B}_0(\beta) \cap \mathcal{I}(\beta)$, so that $\mathcal{B}_1(\beta) \subseteq \mathcal{B}_0(\beta)$. By transitivity of $\mathcal{B}_0, \mathcal{B}_0(\beta) \subseteq \mathcal{B}_0(\alpha)$. Thus $\mathcal{B}_1(\beta) \subseteq \mathcal{B}_0(\alpha)$.

Remark 16. Close inspection of the proof of Proposition 15 reveals that Qualified Acceptance and Minimality play no role (since we only used the fact that $\mathcal{B}_0(\beta) \cap \mathcal{I}(\beta) \neq \varnothing$ implies that $\mathcal{B}_1(\beta) \subseteq \mathcal{B}_0(\beta)$), that is, $B_0\phi \to B_0B_1\phi$ is in fact a theorem of the logic $\mathfrak{L}+$ *Persistence* $+4_0 + ST + IT$.

The following frame, illustrated in Figure 2, shows that Positive Introspection of initial beliefs is crucial for Proposition 15: $\Omega = \{\alpha, \beta, \gamma\}$, $\mathcal{B}_0 = \mathcal{B}_1, \mathcal{B}_0(\alpha) = \{\beta\}, \mathcal{B}_0(\beta) = \{\beta, \gamma\}, \mathcal{B}_0(\gamma) = \{\gamma\}, \mathcal{I}(\alpha) = \mathcal{I}(\beta) = \mathcal{I}(\gamma) = \{\alpha, \beta, \gamma\}$. This frame does not validate the axiom $B_0\phi \to B_0B_1\phi$. In fact, let $\|p\| = \{\alpha, \beta\}$. Then $\alpha \models B_0p$ but $\alpha \not\models B_0B_1p$. However, the frame satisfies the Qualitative Bayes Rule ($\forall\omega$, if $\mathcal{B}_0(\omega) \cap \mathcal{I}(\omega) \neq \varnothing$ then $\mathcal{B}_1(\omega) = \mathcal{B}_0(\omega) \cap \mathcal{I}(\omega)$) and validates Self Trust (since $\mathcal{B}_0$ is secondary reflexive) and Information Trust (since $\mathcal{I}$ is reflexive). On the other hand, Positive Introspection of Initial Beliefs does not hold, since $\mathcal{B}_0$ is not transitive (in fact, $\alpha \models B_0p$ but $\alpha \not\models B_0B_0p$).

The next example, illustrated in Figure 3, shows that also Self Trust is crucial for Proposition 15: $\Omega = \{\alpha, \beta, \gamma, \delta, \varepsilon\}, \mathcal{B}_0(\alpha) = \{\beta, \gamma\}$, $\mathcal{B}_0(\beta) = \mathcal{B}_0(\gamma) = \{\gamma\}, \mathcal{B}_0(\delta) = \mathcal{B}_0(\varepsilon) = \{\varepsilon\}, \mathcal{I}(\alpha) = \{\alpha\}, \mathcal{I}(\beta) = \mathcal{I}(\delta) = \{\beta, \delta\}, \mathcal{I}(\gamma) = \{\gamma\}, \mathcal{I}(\varepsilon) = \{\varepsilon\}, \mathcal{B}_1 = \mathcal{I}, \|p\| = \{\beta, \gamma\}$. This frame does not validate axiom $B_0\phi \to B_0B_1\phi$ since $\alpha \models B_0p$ but $\alpha \not\models B_0B_1p$ (since $\beta \in \mathcal{B}_0(\alpha)$ and $\beta \not\models B_1p$ because $\delta \in \mathcal{B}_1(\beta)$ and $\delta \not\models p$). This frame

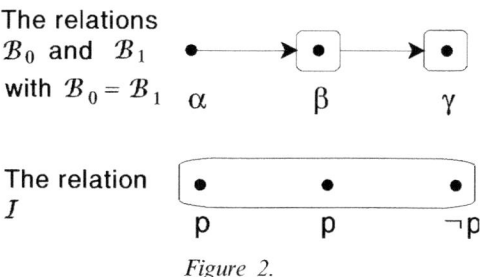

The relations
$\mathcal{B}_0$ and $\mathcal{B}_1$
with $\mathcal{B}_0 = \mathcal{B}_1$ α β γ

The relation
$\mathcal{I}$
    p   p   ¬p

Figure 2.

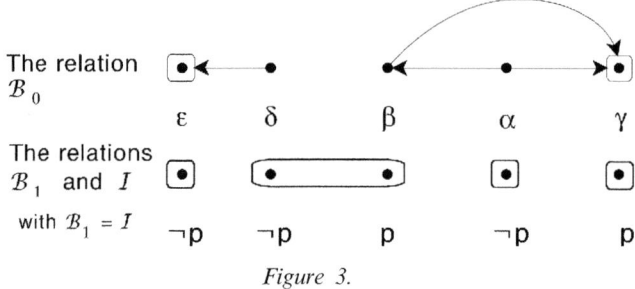

The relation $\mathcal{B}_0$

The relations $\mathcal{B}_1$ and $\mathcal{I}$
with $\mathcal{B}_1 = \mathcal{I}$

ε δ β α γ

$\neg p$ $\neg p$ p $\neg p$ p

Figure 3.

satisfies the Qualitative Bayes Rule and validates Information Trust (since $\mathcal{I}$ is reflexive) and Positive Introspection of Initial Beliefs (since $\mathcal{B}_0$ is transitive). However Self Trust $B_0(B_0\phi \to \phi)$ is not valid, since $\mathcal{B}_0$ is not secondary reflexive (for example, let q be such that $\|q\| = \{\gamma\}$, then $\alpha \nvDash B_0(B_0 q \to q)$, since $\beta \in \mathcal{B}_0(\alpha)$ and $\beta \models B_0 q \wedge \neg q$).

Similarly, it can be shown that Information Trust is necessary for Proposition 15 to be true.

Consider now the following axiom which is the converse of the previous one:

$$B_0 B_1 \phi \to B_0 \phi.$$

This axiom says that if the individual initially believes that later on she will believe ϕ then she must believe ϕ initially.

PROPOSITION 17. $B_0 B_1 \phi \to B_0 \phi$ is a theorem of $\mathfrak{R} + ST + IT$.

Proof. It is shown in van der Hoek (1993) (p. 183, Theorem 4.3 (e)) that axiom $B_0 B_1 \phi \to B_0 \phi$ is characterized by the following property: $\forall \alpha, \gamma \in \Omega,$

if $\gamma \in \mathcal{B}_0(\alpha)$ then there exists a $\beta \in \mathcal{B}_0(\alpha)$ such that
$\gamma \in \mathcal{B}_1(\beta).$ (P$_1$)

By Remark 14, the logic $\mathfrak{R} + ST + IT$ is sound and complete with respect to the class of frames that satisfy the Qualitative Bayes Rule as well as secondary reflexivity of $\mathcal{B}_0$ and reflexivity of $\mathcal{I}$ over $\mathcal{B}_0$. Thus it is enough to show that this class of frames satisfies property (P$_1$). Fix an arbitrary frame in this class and arbitrary states α and γ such that $\gamma \in \mathcal{B}_0(\alpha)$. By Secondary Reflexivity of $\mathcal{B}_0$, $\gamma \in \mathcal{B}_0(\gamma)$. By Reflexivity of $\mathcal{I}$ over $\mathcal{B}_0$, $\gamma \in \mathcal{I}(\gamma)$. Thus $\gamma \in \mathcal{B}_0(\gamma) \cap \mathcal{I}(\gamma)$ and, by the Qualitative Bayes Rule, $\mathcal{B}_0(\gamma) \cap \mathcal{I}(\gamma) = \mathcal{B}_1(\gamma)$, so that $\gamma \in \mathcal{B}_1(\gamma)$. Hence Property ($P_1$) is satisfied with $\beta = \gamma$.   □

Remark 18. Close inspection of the proof of Proposition 17 reveals that Qualified Acceptance and Persistence play no role (since we only used the fact that $\mathcal{B}_0(\gamma) \cap \mathcal{I}(\gamma) \subseteq \mathcal{B}_1(\gamma)$), that is, $\mathcal{B}_0\mathcal{B}_1\phi \rightarrow \mathcal{B}_0\phi$ is in fact a theorem of the logic $\mathfrak{L}+$ *Minimality* $+ST+IT$.

To see that Minimality is crucial for Proposition 17, consider the following frame: $\Omega = \{\alpha, \beta\}$ and, for every $\omega \in \Omega$, $\mathcal{B}_0(\omega) = \{\beta\}$, $\mathcal{I}(\omega) = \Omega$ and $\mathcal{B}_1(\omega) = \{\alpha\}$. This frame validates Self Trust (since $\mathcal{B}_0$ is secondary reflexive) and Information Trust (since $\mathcal{I}$ is reflexive). However, it does not validate Minimality, since $\mathcal{B}_0(\alpha) \cap \mathcal{I}(\alpha) = \{\beta\} \nsubseteq \mathcal{B}_1(\alpha) = \{\alpha\}$. Let p be such that $\|p\| = \{\alpha\}$. Then $\alpha \models \mathcal{B}_0\mathcal{B}_1 p \wedge \neg\mathcal{B}_0 p$.

The following example, illustrated in Figure 4, shows that also Self Trust is crucial for Proposition 17: $\Omega = \{\alpha, \beta, \gamma\}$, $\mathcal{B}_0(\alpha) = \{\beta, \gamma\}$, $\mathcal{B}_0(\beta) = \mathcal{B}_0(\gamma) = \{\gamma\}$, $\mathcal{I}(\alpha) = \{\alpha\}$, $\mathcal{I}(\beta) = \mathcal{I}(\gamma) = \{\beta, \gamma\}$, $\mathcal{B}_1(\alpha) = \{\alpha\}$, $\mathcal{B}_1(\beta) = \mathcal{B}_1(\gamma) = \{\gamma\}$, $\|p\| = \{\gamma\}$. Then $\alpha \models \mathcal{B}_0\mathcal{B}_1 p$ but $\alpha \nvDash \mathcal{B}_0 p$. This frame satisfies the Qualitative Bayes Rule ($\forall\omega$, if $\mathcal{B}_0(\omega) \cap \mathcal{I}(\omega) \neq \varnothing$ then $\mathcal{B}_1(\omega) = \mathcal{B}_0(\omega) \cap \mathcal{I}(\omega)$) as well as Information Trust (since $\mathcal{I}$ is reflexive).[17]

Putting together Propositions 15 and 17 we obtain the following corollary.

COROLLARY 19. $\mathcal{B}_0\phi \leftrightarrow \mathcal{B}_0\mathcal{B}_1\phi$ is a theorem of $\mathfrak{R}+4_0+ST+IT$.

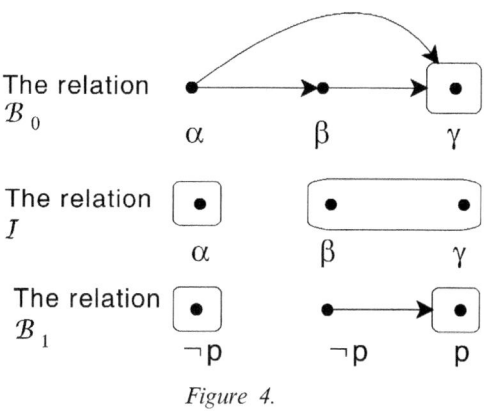

Figure 4.

Remark 20. In the proof of Propositions 15 and 17 it was shown that axiom $B_0\phi \leftrightarrow B_0B_1\phi$ is valid in every frame that satisfies the Qualitative Bayes Rule as well as the properties that characterize axioms 4_0, ST and IT (so that Corollary 19 follows from the completeness theorem: see Remark 14). On the other hand, if a frame validates axioms 4_0, ST, IT and $B_0\phi \leftrightarrow B_0B_1\phi$ then it does not necessarily satisfy the Qualitative Bayes Rule, as the example illustrated in Figure 5 shows.

The frame illustrated in Figure 5 is as follows; $\Omega = \{\alpha, \beta, \gamma\}$, $\mathcal{B}_0(\alpha) = \mathcal{B}_0(\beta) = \mathcal{B}_0(\gamma) = \{\gamma\}$, $\mathcal{I}(\alpha) = \mathcal{I}(\beta) = \mathcal{I}(\gamma) = \{\alpha, \beta, \gamma\}$, $\mathcal{B}_1(\alpha) = \mathcal{B}_1(\beta) = \{\beta\}$, $\mathcal{B}_1(\gamma) = \{\gamma\}$. This frame validates Self Trust (since $\mathcal{B}_0$ is secondary reflexive) and Information Trust (since $\mathcal{I}$ is reflexive). It also validates Positive Introspection of initial beliefs (since $\mathcal{B}_0$ is transitive). Furthermore, the frame satisfies properties P_0 and P_1 (see the proofs of Propositions 15 and 17) and thus validates axiom $B_0\phi \leftrightarrow B_0B_1\phi$. However, it does not validate Persistence.[18] In fact, let $\|p\| = \Omega$ and $\|q\| = \{\gamma\}$; then $\alpha \models Ip \wedge \neg B_0\neg p \wedge B_0q$ but $\alpha \nvDash B_1q$. Because of this, the Qualitative Bayes Rule is not satisfied: $\mathcal{B}_0(\alpha) \cap \mathcal{I}(\alpha) = \{\gamma\} \neq \varnothing$ and yet $\mathcal{B}_1(\alpha) \neq \{\gamma\}$.

4. CLOSELY RELATED LITERATURE

In this section we discuss the relationship between our approach and papers on belief revision that are closest to our analysis in that they make explicit use of modal logic. The relationship with the AGM literature will be discussed in Section 5.

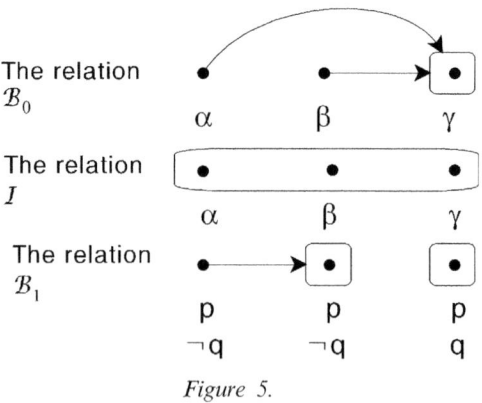

Figure 5.

Fuhrmann (1991) uses a simplified version of dynamic logic, which he calls update logic, to model belief contraction and belief revision. For every formula ϕ he considers a modal operator $[-\phi]$ with the interpretation of $[-\phi]\psi$ as "ψ holds after contracting by ϕ". Alternatively, he considers a modal operator $[*\phi]$, for every formula ϕ, with the intended interpretation of $[*\phi]\psi$ as "ψ holds after updating by ϕ". He provides soundness and completeness results with respect to the class of frames consisting of a set of states Ω and a collection $\{C_X\}$ of binary relations on Ω, one for every subset X of Ω (or for every X in an appropriate collection of subsets of Ω). In a similar vein, Segerberg (1999) notes the coexistence of two traditions in the literature on doxastic logic (the logic of belief), the one initiated by Hintikka (1962) and the AGM approach (Alchourron et al. 1985), and proposes a unifying framework for belief revision. His proposal is to use dynamic logic by thinking of expansion, revision and contraction as actions. Besides the belief operator B, he introduces three operators for every (purely Boolean) formula $\phi : [+\phi]$ for expansion, $[*\phi]$ for revision and $[-\phi]$ for contraction. Thus, for example, the intended interpretation of $[+\phi]B\chi$ is "after performing the action of expanding by ϕ the individual believes that χ". Fuhrmann's and Segerberg's logics are therefore considerably more complex than ours: besides requiring the extra apparatus of dynamic logic, they involves an *infinite* number of modal operators, while our logic uses only three.

A different axiomatization of the Qualitative Bayes Rule was provided by Battigalli and Bonanno (1997) within a framework where information is *not* modeled explicitly. The logic they consider is based on four modal operators: B_0 and B_1, representing – as in this paper – initial and revised beliefs, and two knowledge operators, K_0 and K_1. Knowledge at time 1 is thought of as implicitly based on information received by the individual between time 0 and time 1 and is the basis on which beliefs are revised. The knowledge operators satisfy the S5 logic (the Truth axiom, $K_t\phi \rightarrow \phi$, and negative introspection, $\neg K_t\phi \rightarrow K_t\neg K_t\phi$), while the belief operators satisfy the KD45 logic (consistency and positive and negative introspection). Furthermore, knowledge and belief are linked by two axioms: everything that is known is believed ($K_t\phi \rightarrow B_t\phi$) and the individual knows what he believes ($B_t\phi \rightarrow K_tB_t\phi$). Within this framework Battigalli and Bonanno express the Qualitative Bayes Rule as follows: $\forall\omega \in \Omega$, if $\mathcal{K}_1(\omega) \cap \mathcal{B}_0(\omega) \neq \varnothing$ then $\mathcal{B}_1(\omega) = \mathcal{K}_1(\omega) \cap \mathcal{B}_0(\omega)$, that is, if there are states that are compatible with what the individual knows at time 1 and what he believed at time 0, then the states that he considers possible at time 1 (according to his revised beliefs) are precisely those states. The authors show that, within this knowledge-belief framework the formula $B_0\phi \leftrightarrow B_0B_1\phi$ (which says that the individual believes something at time 0 if and only if he believes that he will continue to believe it at time 1) provides an axiomatization of the Qualitative Bayes Rule. We showed in Corollary 19 that this axiom is a theorem of our logic $\Re$ augmented with axioms 4_0 (one of the axioms postulated by Battigalli and Bonanno), ST (implied by the negative introspection axiom for B_0, which they assume) and IT (whose counterpart in their framework, since they do not model information explicitly, is $B_0(K_1\phi \to \phi)$, which is implied by the Truth axiom of K_1). However, as pointed out above (Remark 20) in a framework where information is modeled explicitly, it is no longer true that the Qualitative Bayes Rule is characterized by axiom $B_0\phi \leftrightarrow B_0B_1\phi$. Thus moving away from the knowledge-belief framework of Battigalli and Bonanno (1997) axiom $B_0\phi \leftrightarrow B_0B_1\phi$ becomes merely an implication of the Qualitative Bayes Rule under additional hypotheses.

In a recent paper, Board (2004) offers a syntactic analysis of belief revision. Like Segerberg, Board makes use of an infinite number of modal operators: for every formula ϕ, an operator B^ϕ is introduced representing the hypothetical beliefs of the individual in the case where she learns that ϕ. Thus the interpretation of $B^\phi\psi$ is "upon learning that ϕ, the individual believes that ψ". The initial beliefs are represented by an operator B. On the semantic side Board considers a set of states and a collection of binary relations, one for each state, representing the plausibility ordering of the individual at that state. The truth condition for the formula $B^\phi\psi$ at a state expresses the idea that the individual believes that ψ on learning that ϕ if and only if ψ is true in all the most plausible worlds in which ϕ is true. The author gives a list of axioms which is sound and complete with respect to the semantics. There are important differences between our framework and his. We model information explicitly by means of a single modal operator I, while Board models it through an infinite collection of hypothetical belief operators. While we model, at any state, only the information *actually* received by the individual, Board considers all possible hypothetical pieces of information: every formula represents a possible item of information, including contradictory formulas and modal formulas. Although, in principle, we also allowed information to be about an

arbitrary formula, in our approach it is possible to rule out problematic situations by imposing suitable axioms (see Remark 1 and
further discussion in Section 7).

Liau (2003) considers a multi-agent framework and is interested
in modeling the issue of trust. He introduces modal operators B_i, I_{ij}
and T_{ij} with the following intended meaning:

$B_i \psi$ Agent i believes that ψ
$I_{ij} \psi$ Agent i acquires information ψ from agent j
$T_{ij} \psi$ Agent i trusts the judgement of agent j on the truth of ψ.

On the semantic side Liau considers a set of states Ω and a col-
lection of binary relations $\mathcal{B}_i$ and $\mathcal{I}_{ij}$ on Ω, corresponding to the
operators B_i and I_{ij}. The truth conditions are the standard ones
for Kripke structures ($\omega \models B_i \psi$ if and only if $\mathcal{B}_i(\omega) \subseteq \|\psi\|$ and $\omega \models$
$I_{ij} \psi$ if and only if $\mathcal{I}_{ij}(\omega) \subseteq \|\psi\|$). Intuitively, $\mathcal{B}_i(\omega)$ is the set of
states that agent i considers possible at ω according to his belief,
whereas $\mathcal{I}_{ij}(\omega)$ is what agent i considers possible according to the
information acquired from j. The author also introduces a relation
$\mathcal{T}_{ij}$ that associates with every state $\omega \in \Omega$ a set of subsets of Ω. For
any $S \subseteq \Omega, S \in \mathcal{T}_{ij}(\omega)$ means that agent i trusts j's judgement on
the truth of the proposition corresponding to event S. Liau con-
siders various axioms and proves that the corresponding logics are
sound and complete with respect to the semantics. One of the axi-
oms the author discusses is $I_{ij} \psi \rightarrow B_i I_{ij} \psi$, which says that if agent
i is informed that ψ by agent j then she believes that this is the
case. Liau notes that, in general, this axiom does not hold, since
when i receives a message from j, she may not be able to exclude
the possibility that someone pretending to be j has sent the message; however, in a secure communication environment this would
not happen and the axiom would hold. There are important differences between our analysis and Liau's. We don't discuss the issue of
trust (although introducing an axiom such as $I\phi \rightarrow B_1\phi$ would capture the notion that information is trusted and therefore believed).
On the other hand, we explicitly distinguish between beliefs held
before the information is received and revised beliefs. Liau has only
one belief operator and therefore does not make this distinction. Yet
this distinction is very important. Suppose first that we take B to be
the *initial* belief (of some agent). Then an axiom like $I\psi \rightarrow BI\psi$
would not be acceptable on conceptual grounds, even if communication is secure. For example, consider a doctor who initially in uncer-

tain whether the patient has an infection (represented by the atomic proposition p) or not ($\neg p$). Let α be a state where p is true (the patient has an infection) and β a state where it is not. Thus the initial uncertainty can be expressed by setting $\mathcal{B}(\alpha) = \mathcal{B}(\beta) = \{\alpha, \beta\}$. The doctor orders a blood test, which, if positive, reveals that there is an infection and, if negative, reveals that there is no infection. Thus $\mathcal{I}(\alpha) = \{\alpha\}$ and $\mathcal{I}(\beta) = \{\beta\}$, so that $\alpha \models Ip$ and $\beta \models I\neg p$. Then $\alpha \models Ip$ but $\alpha \not\models BIp$. On the other hand, if we take B to be the *revised* belief (after the information is received) then postulating the axiom $I\phi \rightarrow BI\phi$ would imply in this example that $\mathcal{B}(\alpha) = \mathcal{I}(\alpha) = \{\alpha\}$ and $\mathcal{B}(\beta) = \mathcal{I}(\beta) = \{\beta\}$, that is, that the information is necessarily believed, thus making it impossible to separate the issues of information and trust. For example, we would not be able to model a situation where the doctor receives the result of the blood test but does not trust the report because of mistakes made in the past by the same lab technician.

The above discussion focussed on contributions that tried to explicitly cast belief revision in a modal logic. There are also discussions of belief revision which follow the AGM approach of considering belief sets where in addition the underlying logic is assumed to contain one or more modal operators (see for example Levi (1988) and Fuhrmann (1989)). Hansson (1994) contains a brief discussion of a restricted modal language for belief change, based on two operators, B (for belief) and L (for necessity).[19] Thus, for example, $LB\phi$ means that ϕ is necessarily believed. The author provides some results on the irreducible modalities of this logic and proposes a semantics for this logic.

5. RELATIONSHIP TO THE AGM FRAMEWORK

The AGM theory of belief revision has been developed within the framework of belief sets. Let Φ be the set of formulas in a propositional language.[20] Given a subset $S \subseteq \Phi$, its PL-deductive closure $[S]^{PL}$ (where 'PL' stands for 'Propositional Logic') is defined as follows: $\psi \in [S]^{PL}$ if and only if there exist $\phi_1, \ldots, \phi_n \in S$ such that $(\phi_1 \wedge \cdots \wedge \phi_n) \rightarrow \psi$ is a truth-functional tautology (that is, a theorem of Propositional Logic). A *belief set* is a set $K \subseteq \Phi$ such that $K = [K]^{PL}$. A belief set K is consistent if $K \neq \Phi$ (equivalently, if there is no formula ϕ such that both ϕ and $\neg\phi$ belong to K). Given a belief set K (thought of as the initial beliefs of the individual) and a formula ϕ (thought of as a new piece of information), the *revision of K by* ϕ, denoted by K_ϕ^* is a subset of Φ that satisfies the following conditions, known as the AGM postulates:

(K*1) K_ϕ^* is a belief set

(K*2) $\phi \in K_\phi^*$

(K*3) $K_\phi^* \subseteq [K \cup \{\phi\}]^{\text{PL}}$

(K*4)   if $\neg\phi \notin K$, then $[K \cup \{\phi\}]^{\text{PL}} \subseteq K_\phi^*$

(K*5) $K_\phi^* = \Phi$ if and only if ϕ is a contradiction

(K*6)   if $\phi \leftrightarrow \psi$ is a tautology then $K_\phi^* = K_\psi^*$

(K*7) $K_{\phi \wedge \psi}^* \subseteq [K_\phi^* \cup \{\psi\}]^{\text{PL}}$

(K*8)   if $\neg\psi \notin K_\phi^*$, then $[K_\phi^* \cup \{\psi\}]^{\text{PL}} \subseteq K_{\phi \wedge \psi}^*$

(K*1) requires the revised belief set to be deductively closed. In our framework this corresponds to requiring the B_1 operator to be a normal operator, that is, to satisfy axiom K ($B_1(\phi \to \psi) \wedge B_1\phi \to B_1\psi$) and the inference rule of necessitation (from ϕ to infer $B_1\phi$).

(K*2) requires that the information be believed. In our framework, this corresponds to imposing axiom $I\phi \to B_1\phi$, which is a strengthening of Qualified Acceptance, in that it requires that if the individual is informed that ϕ then he believes that ϕ even if he previously believed that $\neg\phi$. It is straightforward to prove that this axiom is characterized by the following property: $\forall\omega \in \Omega, \mathcal{B}_1(\omega) \subseteq \mathcal{I}(\omega)$.

(K*3) says that beliefs should be revised minimally, in the sense that no new belief should be added unless it can be deduced from the information received and the initial beliefs. As we will show later, this requirement corresponds to our Minimality axiom ($I\phi \wedge B_1\psi) \to B_0(\phi \to \psi)$.

(K*4) says that if the information received is compatible with the initial beliefs, then any formula that can be deduced from the information and the initial beliefs should be part of the revised beliefs. As shown below, this requirement corresponds to our Persistence axiom ($I\phi \wedge \neg B_0\neg\phi) \to (B_0\psi \to B_1\psi)$.

(K*5) requires the revised beliefs to be consistent, unless the information is contradictory. As pointed out by Friedman and Halpern (1999), it is not clear how information could consist of a contradiction. In our framework we can eliminate this possibility by imposing the axiom $\neg I(\phi \wedge \neg\phi)$, which is characterized by seriality of $\mathcal{I}$ ($\forall\omega \in \Omega, \mathcal{I}(\omega) \neq \varnothing$) (see Section 7). Furthermore, the requirement that revised beliefs be consistent can be captured by the consistency axiom (axiom D): $B_1\phi \rightarrow \neg B_1\neg\phi$, which is characterized by seriality of $\mathcal{B}_1$ ($\forall \omega \in \Omega, \mathcal{B}_1(\omega) \neq \varnothing$). Together with the axiom corresponding to (K*2), consistency of revised beliefs guarantees that information itself is consistent, that is, the conjunction of $B_1\phi \rightarrow \neg B_1\neg\phi$ and $I\phi \rightarrow B_1\phi$ implies $\neg I(\phi \wedge \neg\phi)$ (since $\mathcal{B}_1(\omega) \neq \varnothing$ and $\mathcal{B}_1(\omega) \subseteq \mathcal{I}(\omega)$ implies that $\mathcal{I}(\omega) \neq \varnothing$).

(K*6) is automatically satisfied in our framework, since if $\phi \leftrightarrow \psi$ is a tautology then $\|\phi\| = \|\psi\|$ in every model and therefore the formula $I\phi \leftrightarrow I\psi$ is valid in every frame. Hence revision based on $I\phi$ must coincide with revision based on $I\psi$.

(K*7) and (K*8) are a generalization of (K*3) and (K*4) that

"applies to *iterated* changes of belief. The idea is that if K_ϕ^* is a revision of K and K_ϕ^* is to be changed by adding further sentences, such a change should be made by using expansions of K_ϕ^* whenever possible. More generally, the minimal change of K to include both ϕ and ψ (that is, $K_{\phi\wedge\psi}^*$) ought to be the same as the expansion of K_ϕ^* by ψ, so long as ψ does not contradict the beliefs in K_ϕ^*" (Gärdenfors 1988, p. 55).[21]

We postpone a discussion of iterated revision to the next section, where we claim that the axiomatization of the Qualitative Bayes Rule that we provided can deal with iterated revision and satisfies the conceptual content of (K*7) and (K*8).

The set of postulates (K*1) through (K*6) is called the *basic set* of postulates for belief revision (Gärdenfors 1988 p. 55). The next proposition shows that our axioms imply that the basic set of postulates are satisfied.

PROPOSITION 21. Fix an arbitrary model and an arbitrary state α and let $K = \{\psi : \alpha \models B_0\psi\}$. Suppose that there is a formula ϕ such that $\alpha \models I\phi$ and define $K_\phi^* = \{\psi : \alpha \models B_1\psi\}$. If at α the following hypotheses are satisfied for all formulas ψ and χ

$\alpha \models I\psi \rightarrow B_1\psi$	Acceptance
$\alpha \models (I\psi \wedge B_1\chi) \rightarrow B_0(\psi \rightarrow \chi)$	Minimality
$\alpha \models (I\psi \wedge \neg B_0\neg\psi) \rightarrow (B_0\chi \rightarrow B_1\chi)$	Persistence
$\alpha \models B_1\chi \rightarrow \neg B_1\neg\chi$	Consistency of B_1 (axiom D)

then K_ϕ^* satisfies postulates (K^*1) to (K^*6).

Proof. (K*1): we need to show that K_ϕ^* is a belief set, that is, $K_\phi^* = [K_\phi^*]^{PL}$. Clearly, $K_\phi^* \subseteq [K_\phi^*]^{PL}$, since $\psi \rightarrow \psi$ is a tautology.

Thus we only need to show that $[K_\phi^*]^{\text{PL}} \subseteq K_\phi^*$. Let $\psi \in [K_\phi^*]^{\text{PL}}$, i.e. there exist $\phi_1, \dots, \phi_n \in K_\phi^*$ such that $(\phi_1 \wedge \dots \wedge \phi_n) \to \psi$ is a tautology. Then $\alpha \models B_1((\phi_1 \wedge \dots \wedge \phi_n) \to \psi)$. By definition of K_ϕ^*, since $\phi_1, \dots, \phi_n \in K_\phi^*, \alpha \models B_1(\phi_1 \wedge \dots \wedge \phi_n)$. Thus $\alpha \models B_1\psi$, that is, $\psi \in K_\phi^*$.

(K*2): we need to show that $\phi \in K_\phi^*$, that is, $\alpha \models B_1\phi$. This is an immediate consequence of our hypotheses that $\alpha \models I\phi$ and $\alpha \models I\phi \to B_1\phi$ (by the Acceptance axiom).

(K*3): we need to show that $K_\phi^* \subseteq [K \cup \{\phi\}]^{\text{PL}}$. Let $\psi \in K_\phi^*$, i.e. $\alpha \models B_1\psi$. By hypothesis, $\alpha \models (I\phi \wedge B_1\psi) \to B_0(\phi \to \psi)$ (by Minimality) and $\alpha \models I\phi$. Thus $\alpha \models B_0(\phi \to \psi)$, that is, $(\phi \to \psi) \in K$. Hence $\{\phi, (\phi \to \psi)\} \in K \cup \{\phi\}$ and, since $(\phi \wedge (\phi \to \psi)) \to \psi$ is a tautology, $\psi \in [K \cup \{\phi\}]^{\text{PL}}$.

(K*4): we need to show that if $\neg\phi \notin K$ then $[K \cup \{\phi\}]^{\text{PL}} \subseteq K_\phi^*$. Suppose $\neg\phi \notin K$, that is, $\alpha \models \neg B_0\neg\phi$. By hypothesis, $\alpha \models I\phi$ and $\alpha \models (I\phi \wedge \neg B_0\neg\phi) \to (B_0\psi \to B_1\psi)$ (by Persistence). Thus

$$(4) \qquad \alpha \models (B_0\psi \to B_1\psi), \text{ for every formula } \psi.$$

Let $\chi \in [K \cup \{\phi\}]^{\text{PL}}$, that is, there exist $\phi_1, \dots, \phi_n \in K \cup \{\phi\}$ such that $(\phi_1 \wedge \dots \wedge \phi_n) \to \chi$ is a tautology. We want to show that $\chi \in K_\phi^*$, i.e. $\alpha \models B_1\chi$. Since $(\phi_1 \wedge \dots \wedge \phi_n) \to \chi$ is a tautology, $\alpha \models B_0((\phi_1 \wedge \dots \wedge \phi_n) \to \chi)$. If $\phi_1, \dots, \phi_n \in K$, then $\alpha \models B_0(\phi_1 \wedge \dots \wedge \phi_n)$ and therefore $\alpha \models B_0\chi$. Thus, by (4), $\alpha \models B_1\chi$. If $\phi_1, \dots, \phi_n \notin K$, then w.l.o.g. $\phi_1 = \phi$ and $\phi_2, \dots, \phi_n \in K$. In this case we have $\alpha \models B_0(\phi_2 \wedge \dots \wedge \phi_n)$ and $\alpha \models B_0((\phi_2 \wedge \dots \wedge \phi_n) \to (\phi \to \chi))$ since $(\phi_1 \wedge \dots \wedge \phi_n) \to \chi$ is a tautology and it is equivalent to $(\phi_2 \wedge \dots \wedge \phi_n) \to (\phi \to \chi)$. Thus $\alpha \models B_0(\phi \to \chi)$. Hence, by (4) (with $\psi = (\phi \to \chi)$), $\alpha \models B_1(\phi \to \chi)$. From the hypotheses that $\alpha \models I\phi$ and $\alpha \models I\phi \to B_1\phi$ it follows that $\alpha \models B_1\phi$. Thus $\alpha \models B_1\chi$.

(K*5): we have to show that $K_\phi^* \neq \Phi$, unless ϕ is a contradiction. As noted above, the possibility of contradictory information is ruled out by the conjunction of Consistency of revised beliefs ($B_1\psi \to \neg B_1\neg\psi$) and Acceptance ($I\psi \to B_1\psi$). Thus we only need to show that $K_\phi^* \neq \Phi$. By hypothesis, $B_1\psi \to \neg B_1\neg\psi$; thus if $\psi \in K_\phi^*$ then $\neg\psi \notin K_\phi^*$ and therefore $K_\phi^* \neq \Phi$.

(K*6): we have to show that if $\phi \leftrightarrow \psi$ is a tautology then $K_\phi^* = K_\psi^*$. If $\phi \leftrightarrow \psi$ is a tautology, then $||\phi \leftrightarrow \psi|| = \Omega$, that is, $||\phi|| = ||\psi||$. Thus $\alpha \models I\phi$ if and only if $\alpha \models I\psi$. Hence, by definition, $K_\phi^* = K_\psi^*$.
□

6. ITERATED REVISION

As is well known[22], the AGM postulates are not sufficient to cover iterated belief revision, that is, the case where the individual receives a sequence of pieces of information over time. Only a limited amount of iterated revision is expressed by postulates (K*7) and (K*8), which require that the minimal change of K to include both information ϕ and information ψ (that is, $K^*_{\phi \wedge \psi}$) ought to be the same as the expansion of K^*_ϕ by ψ, so long as ψ does not contradict the beliefs in K^*_ϕ.

In our framework we model, at every state, only the information that is actually received by the individual and do not model how the individual would have modified his beliefs if he had received a different piece of information. Thus we cannot compare the revised beliefs the individual holds after receiving information ϕ with the beliefs he would have had if he had been informed of both ϕ and ψ. On the other hand, it is possible in our framework to model the effect of receiving first information ϕ and then information ψ. Indeed, any sequence of pieces of information can be easily modeled. In order to do this, we need to add a time index to the belief and information operators. Thus, for $t \in \mathbb{N}$ (where $\mathbb{N}$ denotes the set of natural numbers), we have a belief operator B_t representing the individual's beliefs at time t. In order to avoid confusion, we attach a double index $(t, t + 1)$ to the an information operator, so that $I_{t,t+1}$ represents the information received by the individual between time t and time $t + 1$. Thus the intended interpretation is as follows:

$B_t\phi$ at time t the individual believes that ϕ

$I_{t,t+1}\phi$ between time t and time $t + 1$ the individual is informed that ϕ

$B_{t+1}\phi$ at time $t + 1$ (in light of the information received between t and $t + 1$) the individual believes that ϕ.

Let $\mathcal{B}_t$ and $\mathcal{I}_{t,t+1}$ be the associated binary relations. The iterated version of the qualitative Bayes rule then is the following simple extension of QBR: $\forall \omega \in \Omega, \forall t \in \mathbb{N}$,

$$\text{if } \mathcal{B}_t(\omega) \cap \mathcal{I}_{t,t+1}(\omega) \neq \varnothing \text{ then } \mathcal{B}_{t+1}(\omega) = \mathcal{B}_t(\omega) \cap \mathcal{I}_{t,t+1}(\omega).$$
$$\text{(IQBR)}$$

The iterated Bayes rule plays an important role in game theory, since it is the main building block of two widely used solution concepts for dynamic (or extensive) games, namely Perfect Bayesian Equilibrium[23] and Sequential Equilibrium (Kreps and Wilson 1982). The idea behind these solution concepts is that, during the play of the game, a player should revise his beliefs by using Bayes' rule "as long as possible". Thus if an information set has been reached that had positive prior probability, then beliefs at that information set are obtained by using Bayes' rule (with the information being represented by the set of nodes in the information set under consideration). If an information set is reached that had zero prior probability, then new beliefs are formed more or less arbitrarily, but from that point onwards these new beliefs must be used in conjunction with Bayes' rule, unless further information is received that is inconsistent with those revised beliefs. This is precisely what IQBR requires.

Within this more general framework, a simple adaptation of Propositions 3 and 11 yields the following result:

PROPOSITION 22. (1) The Iterated Qualitative Bayes Rule (IQBR) is characterized by the conjunction of the following three axioms:

Iterated Qualified Acceptance: $(\neg B_t \neg \phi \wedge I_{t,t+1}\phi) \rightarrow B_{t+1}\phi$

Iterated Persistence: $(\neg B_t \neg \phi \wedge I_{t,t+1}\phi) \rightarrow$
$(B_t \psi \rightarrow B_{t+1}\psi)$

Iterated Minimality $(I_{t,t+1}\phi \wedge B_{t+1}\psi) \rightarrow B_t(\phi \rightarrow \psi)$.

(2) The logic obtained by adding the above three axioms to the straightforward adaptation of logic $\mathcal{L}$ to a multi-period framework is sound and complete with respect to the class of frames that satisfy the Iterated Qualitative Bayes Rule.

7. CONCLUSION

The simple modal language proposed in this paper has two advantages: (1) information is modeled directly by means of a modal operator I, so that (2) three operators are sufficient to axiomatize the qualitative version of Bayes' rule. Previous modal axiomatizations of belief revision required an infinite number of modal operators and captured information only indirectly through this infinite collection. We also showed that a multi-period extension of our framework allows one to deal with information flows and iterated belief revision.

While the belief operators B_0 and B_1 are normal modal operators, the information operator I is not normal in that the inference rule "from $\phi \to \psi$ to infer $I\phi \to I\psi$" does not hold.[24] This is a consequence of using a non-standard rule for the truth of $I\phi$ ($\omega \models I\phi$ if and only if $\mathcal{I}(\omega) = \|\phi\|$, whereas the standard rule would simply require $\mathcal{I}(\omega) \subseteq \|\phi\|$). However, the addition of the global or universal modality allowed us to obtain a logic of belief revision which is sound and complete with respect to the class of frames that satisfy the Qualitative Bayes Rule.

As pointed out in Remark 1, one might want to impose restrictions on the type of formulas that can constitute information (that is, on what formulas ϕ can be under the scope of the operator I). This is best done by imposing suitable axioms, rather than by restricting the syntax itself. For example, contradictory information is ruled out by imposing axiom $\neg I(\phi \wedge \neg\phi)$, which is characterized by seriality of $\mathcal{I}$ ($\forall \omega, \mathcal{I}(\omega) \neq \varnothing$).[25] Other axioms one might want to impose are:

$B_0\phi \to \neg I \neg B_0\phi$ (if you initially believed that ϕ then you cannot be informed that you did not believe that ϕ)[26], $\neg I(\phi \wedge \neg B_1\phi)$ (you cannot be informed that ϕ and that you will not believe that ϕ), etc. In this paper we have focused on characterization and completeness results and we leave the study of desirable refinements of the proposed logic for future work.

APPENDIX A

In this appendix we prove Propositions 10 and 11. First some preliminaries.

Let $\mathbb{M}$ be the set of maximally consistent sets (MCS) of formulas of $\mathcal{L}$. Define the following binary relation $\mathcal{A} \subseteq \mathbb{M} \times \mathbb{M}: \alpha \mathcal{A}\beta$ if and only if $\{\phi : A\phi \in \alpha\} \subseteq \beta$. Such a relation is well defined (see Chellas 1984, Theorem 4.30(1), p. 158) and is an equivalence relation because of axioms T_A and 5_A (Chellas 1984, Theorem 5.13(2) and (5), p. 175).

LEMMA 23. Let $\alpha, \beta \in \mathbb{M}$ be such that $\alpha \mathcal{A}\beta$ and let ϕ be a formula such that $I\phi \in \alpha$ and $\phi \in \beta$. Then, for every formula ψ, if $I\psi \in \alpha$ then $\psi \in \beta$, that is, $\{\psi : I\psi \in \alpha\} \subseteq \beta$.

Proof. Suppose that $\alpha A \beta$, $I\phi \in \alpha$ and $\phi \in \beta$. Fix an arbitrary ψ such that $I\psi \in \alpha$. Then $I\phi \wedge I\psi \in \alpha$. Since $(I\phi \wedge I\psi) \rightarrow A(\phi \leftrightarrow \psi)$ is a theorem, it belongs to every MCS, in particular to α. Hence $A(\phi \leftrightarrow \psi) \in \alpha$. Then, since $\alpha A \beta$, $\phi \leftrightarrow \psi \in \beta$. Since $\phi \in \beta$, it follows that $\psi \in \beta$. $\square$

Similarly to the definition of A, let the binary relations B_0 and B_1 on $\mathbb{M}$ be defined as follows: $\alpha B_0 \beta$ if and only if $\{\phi : B_0\phi \in \alpha\} \subseteq \beta$ and $\alpha B_1 \beta$ if and only if $\{\phi : B_1\phi \in \alpha\} \subseteq \beta$. It is straightforward to show that, because of axioms Incl_0 and Incl_1, both B_0 and B_1 are subrelations of A, that is, $\alpha B_0 \beta$ implies $\alpha A \beta$ and $\alpha B_1 \beta$ implies $\alpha A \beta$.

Let ω_0 be an arbitrary object such that $\omega_0 \notin \mathbb{M}$, that is, ω_0 can be anything but a MCS, Define the following relation $\mathcal{I}$ on $\mathbb{M} \cup \{\omega_0\}$: $\alpha \mathcal{I} \beta$ if and only if either for some ϕ, $I\phi \in \alpha$ and $\phi \in \beta$ and $\alpha A \beta$ (thus $\alpha, \beta \in \mathbb{M}$) or for all ϕ, $I\phi \notin \alpha$, $\alpha \in \mathbb{M}$ and $\beta = \omega_0$.

DEFINITION 24. An augmented frame is a quintuple $\langle \Omega, B_0, B_1, \mathcal{I}, A \rangle$ obtained by adding an equivalence relation A to a regular frame $\langle \Omega, B_0, B_1, \mathcal{I} \rangle$ with the additional requirements that $B_0 \subseteq A$ and $B_1 \subseteq A$.

The structure $\langle \mathbb{M} \cup \{\omega_0\}, B_0, B_1, \mathcal{I}, A \rangle$ defined above is an augmented frame. For every $\alpha \in \mathbb{M}$, let $A(\alpha) = \{\omega \in \mathbb{M} : \alpha A \omega\}$. Consider the canonical model based on this frame defined by $||p|| = \{\omega \in \mathbb{M} : p \in \omega\}$, for every atomic proposition p. For every formula ϕ define $||\phi||$ according to the semantic rules given in Section 2, with the following modified truth conditions for the operators I and A: $\alpha \models I\phi$ if and only if $\mathcal{I}(\alpha) = ||\phi|| \cap A(\alpha)$ and $\alpha \models A\phi$ if and only if $A(\alpha) \subseteq ||\phi||$. The proof of the following lemma is along the lines of Goranko and Passy (1992) (p. 25).

LEMMA 25. For every formula ϕ, $||\phi|| = \{\omega \in \mathbb{M} : \phi \in \omega\}$.

Proof. The proof is by induction on the complexity of ϕ. For the non-modal formulas and for the cases where ϕ is either $B_0\psi$ or $B_1\psi$ or $A\psi$, for some ψ, the proof is standard (see Chellas 1984, Theorem 5.7, p. 172). That proof makes use of rule of inference RK for the modal operators. Since this rule of inference does not hold for I (see Remark 8), we need a different proof for the case where $\phi = I\psi$ for some ψ. By the induction hypothesis, $||\psi|| = \{\omega \in \mathbb{M} : \psi \in \omega\}$. We need to show that $||I\psi|| = \{\omega \in \mathbb{M} : I\psi \in \omega\}$, that is, that

(1) if $\alpha \models I\psi$ (i.e. $\mathcal{I}(\alpha) = ||\psi|| \cap \mathcal{A}(\alpha)$) then $I\psi \in \alpha$, and
(2) if $I\psi \in \alpha$ then $\mathcal{I}(\alpha) = ||\psi|| \cap \mathcal{A}(\alpha)$ (i.e. $\alpha \models I\psi$).

For (1) we prove the contrapositive, namely that if $\alpha \in \mathbb{M}$ and $I\psi \notin \alpha$ then $\mathcal{I}(\alpha) \neq ||\psi|| \cap \mathcal{A}(\alpha)$. Suppose that $\alpha \in \mathbb{M}$ and $I\psi \notin \alpha$. Two cases are possible: (1.a) $I\chi \notin \alpha$ for every formula χ, or (1.b) $I\chi \in \alpha$ for some χ. In case (1.a), by definition of $\mathcal{I}, \mathcal{I}(\alpha) = \{\omega_0\}$. Since $\omega_0 \notin \mathbb{M}$ (and $\mathcal{A}(\alpha) \subseteq \mathbb{M}$) it follows that $\mathcal{I}(\alpha) \neq ||\psi|| \cap \mathcal{A}(\alpha)$. In case (1.b) it must be that $(I\chi \rightarrow I\psi) \notin \alpha$ (since $I\psi \notin \alpha$). By axiom I_2, $A(\chi \leftrightarrow \psi) \rightarrow (I\chi \rightarrow I\psi) \in \alpha$. Thus $A(\chi \leftrightarrow \psi) \notin \alpha$. Since α is a MCS, $\neg A(\chi \leftrightarrow \psi) \in \alpha$. Now, $\neg A(\chi \leftrightarrow \psi)$ is propositionally equivalent to $\neg A \neg \neg (\chi \leftrightarrow \psi)$, which in turn is equivalent to $\neg A \neg ((\chi \wedge \neg \psi) \vee (\psi \wedge \neg \chi))$. Thus this formula belongs to α. Hence there is a β such that $\alpha A \beta$ and either (1.b.1.) $(\chi \wedge \neg \psi) \in \beta$ or (1.b.2) $(\psi \wedge \neg \chi) \in \beta$. In case (1.b.1), $\chi \in \beta$ and $\psi \notin \beta$. By definition of $\mathcal{I}$, since $\alpha A \beta$ and $I\chi \in \alpha$ and $\chi \in \beta$, we have that $\beta \in \mathcal{I}(\alpha)$ while $\beta \notin ||\psi||$, since $\psi \notin \beta$ and, by the induction hypothesis, $||\psi|| = \{\omega \in \mathbb{M} : \psi \in \omega\}$. Thus $\mathcal{I}(\alpha) \neq ||\psi|| \cap \mathcal{A}(\alpha)$. In case (1.b.2), $\chi \notin \beta$ and $\psi \in \beta$, so that, by the induction hypothesis, $\beta \in ||\psi||$; furthermore, $\beta \in \mathcal{A}(\alpha)$. We want to show that $\beta \notin \mathcal{I}(\alpha)$, so that $\mathcal{I}(\alpha) \neq ||\psi|| \cap \mathcal{A}(\alpha)$. To see this, suppose by contradiction that $\beta \in \mathcal{I}(\alpha)$. Then by definition of $\mathcal{I}$, there is some ζ such that $I\zeta \in \alpha$ and $\zeta \in \beta$. By Lemma 23 $\{\theta : I\theta \in \alpha\} \subseteq \beta$, implying that $\chi \in \beta$, since, by hypothesis, $I\chi \in \alpha$. But this contradicts $\chi \notin \beta$. This completes the proof of (1).

Next we prove (2). Suppose that $I\psi \in \alpha$. First we show that $||\psi|| \cap \mathcal{A}(\alpha) \subseteq \mathcal{I}(\alpha)$. Fix an arbitrary $\beta \in ||\psi|| \cap \mathcal{A}(\alpha)$. Since $\beta \in ||\psi||$, by the induction hypothesis, $\psi \in \beta$ and, therefore, by definition of $\mathcal{I}, \beta \in \mathcal{I}(\alpha)$. Next we show that $\mathcal{I}(\alpha) \subseteq ||\psi|| \cap \mathcal{A}(\alpha)$. Fix an arbitrary $\beta \in \mathcal{I}(\alpha)$. By definition of $\mathcal{I}, \beta \in \mathcal{A}(\alpha)$ and there exists a χ such that $I\chi \in \alpha$ and $\chi \in \beta$. By Lemma 23, $\{\theta : I\theta \in \alpha\} \subseteq \beta$ and therefore, since $I\psi \in \alpha, \psi \in \beta$. By the induction hypothesis, $||\psi|| = \{\omega \in \mathbb{M} : \psi \in \omega\}$. Thus $\beta \in ||\psi|| \cap \mathcal{A}(\alpha)$. $\qquad\square$

PROPOSITION 26. Logic $\mathcal{L}$ is sound and complete with respect to the class of augmented frames $\langle \Omega, \mathcal{B}_0, \mathcal{B}_1, \mathcal{I}, \mathcal{A} \rangle$ under the semantic rules given in Section 2, with the following modified truth conditions for the operators I and A : $\alpha \models I\phi$ if and only if $\mathcal{I}(\alpha) = \|\phi\| \cap A(\alpha)$ and $\alpha \models A\phi$ if and only if $\mathcal{A}(\alpha) \subseteq \|\phi\|$, where $\mathcal{A}(\alpha) = \{\omega \in \Omega : \alpha \mathcal{A} \omega\}$.

Proof. (A) SOUNDNESS. It is straightforward to show that the inference rules MP and NEC$_A$ are validity preserving and axioms K$_0$, K$_1$, K$_A$, T$_A$, 5$_A$, Incl$_0$ and Incl$_1$, are valid in all augmented frames. Thus we only show that axioms I$_1$ and I$_2$ are valid in all augmented frames.

1. Validity of axiom I$_1$: $I\phi \wedge I\psi \rightarrow A(\phi \leftrightarrow \psi)$. Fix an arbitrary model, and suppose that $\alpha \models I\phi \wedge I\psi$. Then $\mathcal{I}(\alpha) = \|\phi\| \cap \mathcal{A}(\alpha)$ and $\mathcal{I}(\alpha) = \|\psi\| \cap \mathcal{A}(\alpha)$. Thus, $\|\phi\| \cap \mathcal{A}(\alpha) = \|\psi\| \cap \mathcal{A}(\alpha)$ and hence $\mathcal{A}(\alpha) \subseteq \|\phi \leftrightarrow \psi\|$, yielding $\alpha \models A(\phi \leftrightarrow \psi)$.

2. Validity of axiom, I$_2$: $A(\phi \leftrightarrow \psi) \rightarrow (I\phi \leftrightarrow I\psi)$. Fix an arbitrary model and suppose that $\alpha \models A(\phi \leftrightarrow \psi)$. Then $\mathcal{A}(\alpha) \subseteq \|\phi \leftrightarrow \psi\|$ and therefore, $\|\phi\| \cap \mathcal{A}(\alpha) = \|\psi\| \cap \mathcal{A}(\alpha)$. Thus, $\alpha \models I\phi$ if and only if $\mathcal{I}(\alpha) = \|\phi\| \cap \mathcal{A}(\alpha)$ if and only if $\mathcal{I}(\alpha) = \|\psi\| \cap \mathcal{A}(\alpha)$, if and only if $\alpha \models I\psi$. Hence $\alpha \models I\phi \leftrightarrow I\psi$.

(B) COMPLETENESS. Let ϕ be a formula that is valid in all augmented frames. Then ϕ is valid in the canonical structure $\langle \mathbb{M} \cup \{\omega_0\}, \mathcal{B}_0, \mathcal{B}_1, \mathcal{I}, \mathcal{A} \rangle$ defined above, which is an augmented frame. Thus ϕ is valid in the canonical model based on this frame. By Lemma 25, for every formula ψ, $\|\psi\| = \{\omega \in \mathbb{M} : \psi \in \omega\}$. Thus ϕ belongs to every MCS and therefore is a theorem of $\mathfrak{L}$ (Chellas, 1984, Theorem 2.20, p, 57).

To prove Proposition 10, namely that logic, $\mathfrak{L}$ is sound and complete with respect to the class of frames $\langle \Omega, \mathcal{B}_0, \mathcal{B}_1, \mathcal{I} \rangle$, we only need to invoke the result (Chellas 1984, Theorem 3.12, p, 97) that soundness and completeness with respect to the class of augmented frames (where $\mathcal{A}$ is an equivalence relation) implies soundness and completeness with respect to the generated sub-frames (where $\mathcal{A}$ is the universal relation). The latter are precisely what we called frames. In a frame where the relation $\mathcal{A}$ is the universal relation the semantic rule $\alpha \models I\phi$ if and only if $\mathcal{I}(\alpha) = \|\phi\| \cap \mathcal{A}(\alpha)$ becomes $\alpha \models I\phi$ if and only if $\mathcal{I}(\alpha) = \|\phi\|$ and the semantic rule $\alpha \models A\phi$ if and only if $\mathcal{A}(\alpha) \subseteq \|\phi\|$ becomes $\alpha \models A\phi$ if and only if $\|\phi\| = \Omega$, since $\mathcal{A}(\alpha) = \Omega$.

Next we turn to the **proof of Proposition 11**, namely that logic $\mathfrak{R}$ is sound and complete with respect to the class of frames $\langle \Omega, \mathcal{B}_0, \mathcal{B}_1 \mathcal{I} \rangle$ that satisfy the Qualitative Bayes Rule (QBR).

Proof. (A) SOUNDNESS. This follows from Propositions 3 and 10.

(B) COMPLETENESS. By Proposition 10 we only need to show that the frame associated with the canonical model is a QBR frame. First we show that

(5) $\forall \omega \in \mathbb{M}$, if $\mathcal{B}_0(\omega) \cap \mathcal{I}(\omega) \neq \varnothing$ then $\mathcal{B}_1(\omega) \subseteq \mathcal{I}(\omega)$.

Let $\beta \in \mathcal{B}_0(\alpha) \cap \mathcal{I}(\alpha)$. Since $\mathcal{B}_0(\alpha) \subseteq \mathbb{M}$, $\beta \in \mathbb{M}$ and therefore, by definition of $\mathcal{I}$, there exists a formula ϕ such that $I\phi \in \alpha$ and $\phi \in \beta$. Since $\beta \in \mathcal{B}_0(\alpha)$, $\neg B_0 \neg \phi \in \alpha$ (Chellas 1984, Theorem 5.6, p. 172). Thus $(I\phi \wedge \neg B_0 \neg \phi) \in \alpha$. Since Qualified Acceptance is a theorem, $(I\phi \wedge \neg B_0 \neg \phi) \rightarrow B_1\phi \in \alpha$. Thus $B_1\phi \in \alpha$. We want to show that $\mathcal{B}_1(\alpha) \subseteq \mathcal{I}(\alpha)$. Fix an arbitrary $\gamma \in \mathcal{B}_1(\alpha)$. Then, by definition of $\mathcal{B}_1$, $\{\psi : B_1\psi \in \alpha\} \subseteq \gamma$. In particular, since $B_1\phi \in \alpha$, $\phi \in \gamma$. By definition of $\mathcal{I}$ since $I\phi \in \alpha$ and $\phi \in \gamma$, $\gamma \in \mathcal{I}(\alpha)$,

Next we show that

(6) $\forall \omega \in \mathbb{M}$, if $\mathcal{B}_0(\omega) \cap \mathcal{I}(\omega) \neq \varnothing$ then $\mathcal{B}_1(\omega) \subseteq \mathcal{B}_0(\omega)$.

Let $\beta \in \mathcal{B}_0(\alpha) \cap \mathcal{I}(\alpha)$. As shown above, there exists a ϕ such that $I\phi \in \alpha$, $\phi \in \beta$ and $\neg B_0 \neg \phi \in \alpha$. By Persistence, for every formula, ψ, $(I\phi \wedge \neg B_0 \neg \phi) \rightarrow (B_0\psi \rightarrow B_1\psi) \in \alpha$. Thus

(7) $(B_0\psi \rightarrow B_1\psi) \in \alpha$.

Fix an arbitrary $\gamma \in \mathcal{B}_1(\alpha)$. Then, by definition of $\mathcal{B}_1$, $\{\psi : B_1\psi \in \alpha\} \subseteq \gamma$. We want to show that $\gamma \in \mathcal{B}_0(\alpha)$, that is, that $\{\psi : B_0\psi \in \alpha\} \subseteq \gamma$. Let ψ be such that $B_0\psi \in \alpha$. By (7) $B_1\psi \in \alpha$ and therefore $\psi \in \gamma$.

Finally we show that

(8) $\forall \omega \in \mathbb{M}$, $\mathcal{B}_0(\omega) \cap \mathcal{I}(\omega) \subseteq \mathcal{B}_1(\omega)$.

Fix an arbitrary $\alpha \in \mathbb{M}$. If $\mathcal{B}_0(\alpha) \cap \mathcal{I}(\alpha) = \varnothing$, there is nothing to prove. Suppose therefore that $\beta \in \mathcal{B}_0(\alpha) \cap \mathcal{I}(\alpha)$ for some β. Then there exists a ϕ such that $I\phi \in \alpha$ and $\phi \in \beta$. Fix an arbitrary $\gamma \in \mathcal{B}_0(\alpha) \cap \mathcal{I}(\alpha)$. We want to show that $\gamma \in \mathcal{B}_1(\alpha)$, that is, that $\{\psi : B_1\psi \in \alpha\} \subseteq \gamma$. Let ψ be an arbitrary formula such that $B_1\psi \in \alpha$. Then $(I\phi \wedge B_1\psi) \in \alpha$. By Minimality, $(I\phi \wedge B_1\psi) \rightarrow B_0(\phi \rightarrow \psi) \in \alpha$. Thus $B_0(\phi \rightarrow \psi) \in \alpha$. Since $\gamma \in \mathcal{B}_0(\alpha)$, $(\phi \rightarrow \psi) \in \gamma$. Since $I\phi \in \alpha$, $\mathcal{I}(\alpha) = \|\phi\|$. Thus, since $\gamma \in \mathcal{I}(\alpha)$, $\gamma \models \phi$ and, by Lemma 25, $\phi \in \gamma$. It follows from this and $(\phi \rightarrow \psi) \in \gamma$ that $\psi \in \gamma$.

[170]

NOTES

* I am grateful to two anonymous reviewers for helpful and constructive comments. A first draft of this paper was presented at the Sixth Conference on Logic and the Foundations of Game and Decision Theory (LOFT6), Leipzig, July 2004.

[1] For extensive surveys of the role of beliefs and rationality in game theory see Dekel and Gul (1997), Battigalli and Bonanno (1999) and vand der Hoek and Pauly (forthcoming).

[2] For an extensive overview see Gärdenfors (1988).

[3] There is an ongoing debate in the philosophical literature as to whether or not Bayes' rule is a requirement of rationality: see, for example, Brown (1976), Jeffrey (1983), Howson and Urbach (1989), Maher (1993) and Teller (1973).

[4] In a probabilistic setting, if P_0 is the prior probability measure representing the initial beliefs at state ω and P_1 is the posterior probability measure representing the revised beliefs at ω then $\mathcal{B}_0(\omega) = \mathrm{supp}(P_0)$ and $\mathcal{B}_1(\omega) = \mathrm{supp}(P_1)$.

[5] See, for example, Blackburn et al [5]. The connectives $\wedge$ (for "and"), $\rightarrow$ (for "if ... then ...") and $\leftrightarrow$ (for "if and only if") are defined as usual: $\phi \wedge \psi = \neg(\neg\phi \vee \neg\psi), \phi \rightarrow \psi = \neg\phi \vee \psi$ and $\phi \leftrightarrow \psi = (\phi \rightarrow \psi) \wedge (\psi \rightarrow \phi)$.

[6] In an interpersonal setting, however, information that pertains to beliefs (rather than merely to facts) ought to be allowed, at least to the extent that the information received by an individual be about the beliefs of *another* individual.

[7] More examples of problematic situations are: $I(\phi \wedge \neg B_1\phi)$ (the individual is informed that ϕ and that he will not believe ϕ), $B_0\phi \wedge I\neg B_1 B_0\phi$ (the individual initially believes ϕ and is informed that he will forget that he did), etc.

[8] In a probabilistic setting, where $\mathcal{B}_0(\omega)$ is the support of the probability measure representing the initial beliefs at ω, we would have that $\omega \models B_0\phi$ if and only if the individual assigns probability 1 to the event $\|\phi\|$. Similarly for $\omega \models B_1\phi$.

[9] The reason for this will become clear later. Intuitively, this allows us to distinguish between the content of information and its implications.

[10] As is well known (see Chellas 1984) consistency of initial beliefs is characterized by seriality of $\mathcal{B}_0$ ($\forall\omega \in \Omega, \mathcal{B}_0(\omega) \neq \varnothing$). If there is a state α such that $\mathcal{B}_0(\alpha) = \varnothing$ then $\alpha \models B_0\psi$ for every formula ψ.
To see that without consistency of initial beliefs Proposition 7 is not true, consider the following example. $\Omega = \{\alpha\}$, $\mathcal{B}_0(\alpha) = \varnothing$, $\mathcal{B}_1(\alpha) = \mathcal{I}(\alpha) = \{\alpha\}$. Then, for every formula ϕ, $\alpha \nvDash \neg B_0\neg\phi$ so that Persistence is trivially valid. It is also trivially true, for every ϕ and ψ, that $\alpha \models B_0(\phi \rightarrow \psi)$ so that Minimality is also valid. Let p be an atomic proposition such that $\alpha \models p$. Then $\alpha \models B_0 p \wedge I p \wedge B_0\neg p \wedge \neg B_1\neg p$ so that the No Change axiom is falsified at α with $\phi = p$ and $\psi = \neg p$.

[11] Consider the following frame: $\Omega = \{\alpha, \beta, \gamma\}$, and for every $\omega \in \Omega, \mathcal{B}_0(\omega) = \{\beta, \gamma\}$, $\mathcal{I}(\omega) = \{\omega\}$ and $\mathcal{B}_1(\omega) = \{\alpha, \beta\}$. By Lemma 5, Persistence is not valid in this frame (since $\mathcal{B}_0(\beta) \cap \mathcal{I}(\beta) \neq \varnothing$ and yet $\mathcal{B}_1(\beta) \nsubseteq \mathcal{B}_0(\beta)$. By Lemma 6, also Minimality is not valid (since $\mathcal{B}_0(\gamma) \cap \mathcal{I}(\gamma) \nsubseteq \mathcal{B}_1(\gamma)$). However, No Change is trivially valid in this frame. In fact, fix an arbitrary model and an arbitrary formula ϕ. It is easy to see that, for every $\omega \in \Omega, \omega \nvDash B_0\phi \wedge I\phi$. For example, if $\beta \models I\phi$ then $\|\phi\| = \mathcal{I}(\beta) = \{\beta\}$, so that $\gamma \notin \|\phi\|$, implying that $\beta \nvDash B_0\phi$.

[12] *Proof.*
  1. $B_1\phi \to B_1 B_1 \phi$ positive introspection axiom
  2. $I\phi \wedge B_1\phi \to I\phi \wedge B_1 B_1 \phi$ 1,PL
  3. $I\phi \wedge B_1 B_1 \phi \to B_0(\phi \to B_1\phi)$ instance of Minimality with $\psi = B_1\phi$
  4. $I\phi \wedge B_1\phi \to B_0(\phi \to B_1\phi)$ 2,3, PL.   □

[13] If ϕ is a valid formula, then $\|\phi\| = \Omega$. Let $\alpha \in \Omega$ be a state where $\mathcal{I}(\alpha) \neq \Omega$. Then $\alpha \nvDash I\phi$ and therefore $I\phi$ is not valid.

[14] Consider the following model: $\Omega = \{\alpha, \beta\}$, $\mathcal{I}(\alpha) = \{\alpha\}$, $\mathcal{I}(\beta) = \{\beta\}$, $\|p\| = \{\alpha\}$ and $\|q\| = \Omega$. Then $\|p \to q\| = \Omega$, $\|Ip\| = \{\alpha\}$, $\|Iq\| = \varnothing$ and thus $\|Ip \to Iq\| = \{\beta\} \neq \Omega$.

[15] *Proof.* Fix a frame, an arbitrary model and a state α. For it to be the case that $\alpha \vDash I(\phi \to \psi) \wedge I\phi$ we need $\mathcal{I}(\alpha) = \|\phi\|$ and $\mathcal{I}(\alpha) = \|\phi \to \psi\|$. Now, $\|\phi \to \psi\| = \|\neg\phi \vee \psi\| = \|\neg\phi\| \cup \|\psi\|$ and therefore we need the equality $\|\phi\| = \|\neg\phi\| \cup \|\psi\|$ to be satisfied. This requires $\|\phi\| = \|\psi\| = \Omega$. Thus if $\mathcal{I}(\alpha) = \|\phi\| = \|\psi\| = \Omega$. then $\alpha \vDash I(\phi \to \psi) \wedge I\phi \wedge I\psi$. In every other case, $\alpha \nvDash I(\phi \to \psi) \wedge I\phi$ and therefore the formula $I(\phi \to \psi) \wedge I\phi \to I\psi$ is trivially true at α.   □

[16] Furthermore, Self Trust is implied by a stronger property of beliefs, namely Negative Introspection ($\neg B_0\phi \to B_0 \neg B_0\phi$) which is characterized by euclideanness of $\mathcal{B}_0$ (if $\beta \in \mathcal{B}_0(\alpha)$ then $\mathcal{B}_0(\alpha) \subseteq \mathcal{B}_0(\beta)$).

[17] The frame also satisfies Positive Introspection of initial beliefs ($B_0\phi \to B_0 B_0\phi$) since $\mathcal{B}_0$ is transitive.

[18] Although, by Lemma 6, it does validate Minimality.

[19] "This language is called 'restricted' since (1) it does not allow for iterations of the B operator, and (2) it is not closed under truth-functional operations other than negation" (Hansson (1994), p. 22).

[20] For simplicity we consider the simplest case where the underlying logic is classical propositional logic.

[21] The expansion of K_ϕ^* by ψ is $[K_\phi^* \cup \{\psi\}]^{\mathrm{PL}}$

[22] See, for example, Rott (1991, p. 170).

[23] See, for example, Battigalli (1996), Bonanno (1993), Fudenberg and Tirole (1991).

[24] Furthermore, no formula of the type $I\phi$ or its negation is universally valid. Recall, however, that I trivially satisfies axiom $K : I(\phi \to \psi) \wedge I\phi \to I\psi$.

[25] *Proof.* Suppose $\mathcal{I}$ is serial and $\neg I(\phi \wedge \neg\phi)$ is not valid, that is, there is a state α and a formula ϕ such that $\alpha \vDash I(\phi \wedge \neg\phi)$. Then $\mathcal{I}(\alpha) = \|\phi \wedge \neg\phi\|$. But $\|\phi \wedge \neg\phi\| = \varnothing$, while by seriality $\mathcal{I}(\alpha) \neq \varnothing$. Conversely, suppose that $\mathcal{I}$ is not serial. Then there exists a state α such that $\mathcal{I}(\alpha) = \varnothing$. Since, for every formula ϕ, $\|\phi \wedge \neg\phi\| = \varnothing$, it follows that $\alpha \vDash I(\phi \wedge \neg\phi)$ so that $\neg I(\phi \wedge \neg\phi)$ is not valid.

Note that, given the non-standard validation rule for $I\phi$, the equivalence of axiom $D(I\phi \to \neg I\neg\phi)$ and seriality breaks down. It is still true that if $\mathcal{I}$ is serial then the axiom $I\phi \to \neg I\neg\phi$ is valid, but the converse is not true. Proof of the first part: assume seriality and suppose that the axiom is not valid, i.e. there is a formula ϕ such that $\alpha \vDash I\phi \wedge I\neg\phi$. Then $\mathcal{I}(\alpha) = \|\phi\|$ and $\mathcal{I}(\alpha) = \|\neg\phi\|$. By seriality, there exists a $\beta \in \mathcal{I}(\alpha)$. Then $\beta \vDash \phi \wedge \neg\phi$, which is impossible. Now, to see that the converse is not true, first note that the truth condition for $I\phi$ is equivalent to

$$\forall\beta, \text{ if } \beta \in \mathcal{I}(\alpha) \text{ then } \beta \vDash \phi, \text{ and } \forall\gamma, \text{ if } \gamma \vDash \phi \text{ then } \gamma \in \mathcal{I}(\alpha).$$

Thus $\alpha \models \neg I \neg \phi$ iff $\alpha \not\models I \neg \phi$ iff $not\,(\forall \beta, \beta \in \mathcal{I}(\alpha) \Longrightarrow \beta \models \neg \phi$ and $\forall \gamma, \gamma \models \neg \phi \Longrightarrow \gamma \in \mathcal{I}(\alpha))$ which is equivalent to

either $\exists \beta \in \mathcal{I}(\alpha)$ such that $\beta \models \phi$ or $\exists \gamma$ such that $\gamma \models \neg \phi$ and $\gamma \notin \mathcal{I}(\alpha)$.

Now, suppose that $\mathcal{I}(\alpha) = \varnothing$. Then, for every formula ϕ either $\|\phi\| \neq \varnothing$, in which case $\alpha \not\models I\phi$ and therefore $\alpha \models I\phi \to \psi$ for every formula ψ (in particular for $\psi = \neg I \neg \phi$) or $\|\phi\| = \varnothing$, in which case $\alpha \models I\phi$ and, since $\alpha \models \neg \phi$ and $\alpha \notin \mathcal{I}(\alpha), \alpha \models \neg I \neg \phi$. Thus validity of $I\phi \to \neg I \neg \phi$ does not guarantee seriality of $\mathcal{I}$ (let $\mathcal{I}$ be empty everywhere, then the axiom is valid!).

[26] Indeed, one might want to go further and impose memory axioms: $B_0 \phi \to B_1 B_0 \phi$ (if in the past you believed ϕ then later on you remember this) and $\neg B_0 \phi \to B_1 \neg B_0 \phi$ (at a later time you remember what you did *not* believe in the past).

REFERENCES

Alchourron, C., P. Gärdenfors and D. Makinson: 1985, 'On the Logic of Theory Change; Partial Meet Contraction and Revision Functions', *The Journal of Symbolic Logic* **50**, 510–530.

Battigalli, P.: 1996, 'Strategic Independence and Perfect Bayesian Equilibria', *Journal of Economic Theory* **70**, 201–234.

Battigalli, P. and G. Bonanno: 1992, 'The Logic of Belief Persistence', *Economics and Philosophy* **13**, 39–59.

Battigalli, P. and G. Bonanno: 1999, 'Recent Results on Belief, Knowledge and the Epistemic Foundations of Game Theory', *Research in Economics* **53**, 149–225.

Blackburn, P., M. de Rijke, and Y. Venema: 2001, *Modal Logic,* Cambridge University Press.

Board, O.: 2004, 'Dynamic Interactive Epistemology', *Games and Economic Behavior* **49**, 49–80.

Bonanno, G.: 1993, 'Rational Belief Equilibria', *Economic Notes* **22**, 430–463.

Brown, P. M.: 1976, 'Conditionalization and Expected Utility', *Philosophy of Science* **43**, 415–419.

Chellas, B.: 1984, *Modal Logic; An Introduction*, Cambridge University Press.

Dekel, E. and F. Gul: 1997, 'Rationality and Knowledge in Game Theory', in: D. M. Kreps, and K. F. Wallis (eds.), *Advances in Economic Theory, Seventh World Congress*, Cambridge University Press, 87–172.

van Ditmarsch, H. and W. van der Hoek: 2004, Dynamic epistemic logic, working paper.

Friedman, N. and J. Halpern: 1999, 'Belief Revision: A Critique', *Journal of Logic, Language, and Information* **8**, 401–420.

Fudenberg, D. and J. Tirole: 1991, 'Perfect Bayesian Equilibrium and Sequential Equilibrium', *Journal of Economic Theory* **53**, 236–260.

Fuhrmann, A.: 1989, 'Reflective Modalities and Theory Change', *Synthese* **81**, 115–134.

Fuhrmann, A.: 1991, 'On the Modal Logic of Theory Change', in A. Fuhrmann (ed.), *The logic of theory change,* Lecture notes in Artificial Intelligence No. 465, Springer-Verlag, Berlin, 259–281.

Gärdenfors, P.: 1988, *Knowledge in Flux: Modeling the Dynamics of Epistemic States*, MIT Press.

Goranko, V. and S. Passy: 1992, 'Using the Universal Modality: Gains and Questions', *Journal of Logic and Computation* **2**, 5–30.

Hansson, S. O.: 1994, 'Taking Belief Bases Seriously', in D. Prawitz and D. Westerståhl (eds.), *Logic and Philosophy of Science in Uppsala*, Kluwer, 13–28.

Harman, G.: 1986, *Change in View: Principles of Reasoning*, MIT Press.

Hintikka, J.: 1962, *Knowledge and Belief*, Cornell University Press.

van der Hoek, W.: 1993, 'Systems for Knowledge and Belief', *Journal of Logic and Computation*, **3**, 173–195.

van der Hoek, W. and M. Pauly: forthcoming 'Game Theory', in: J. van Benthem, P. Blackburn, and F. Wolter (eds.), *Handbook of Modal Logic*.

Howson, C, and P. Urbach: 1989, *Scientific Reasoning*, Open Court.

Jeffrey, R.: 1983, *The Logic of Decision*, 2nd ed. University of Chicago Press.

Kreps, D. and R, Wilson: 1982, 'Sequential Equilibria', *Econometrica*, **50**, 863–894.

Kripke, S.: 1963, A Semantical Analysis of Modal Logic I: Normal Propositional Calculi', *Zeitschrift für Mathematische Logik and Grundlagen der Mathematik* **9**, 67–96.

Levi, I.: 1988, 'Iteration of conditionals and the Ramsey test', *Synthese* **76**, 49–81.

Liau, C.-J.: 2003, 'Belief, Information Acquisition, and Trust in Multi-Agent Systems – A Modal Logic Formulation, *Artificial Intelligence* **149**, 31–60.

Maher, P.: 1993, *Betting on Theories*, Cambridge University Press.

Rott, H.: 1991, 'Two Methods of Constructing Contractions and Revisions of Knowledge Systems', *Journal of Philosophical Logic* **20**, 149–173.

Segerberg, K.: 1999, 'Two traditions in the Logic of Belief: Bringing them Together', in: H. J, Ohlbach and U. Reyle (eds.), *Logic, Language and Reasoning*, Kluwer Academic Publishers. pp. 135–147.

Teller, P.: 1973, 'Conditionalization and Observation', *Synthese*, **26**, 218–258.

Department of Economics
University of California
Davis, CA 95616-8578
U.S.A.
E-mail: gfbonanno@ucdavis.edu

HANS P. VAN DITMARSCH

PROLEGOMENA TO DYNAMIC LOGIC FOR BELIEF REVISION

ABSTRACT. In 'belief revision' a theory $\mathcal{K}$ is revised with a formula φ resulting in a revised theory $\mathcal{K} * \varphi$. Typically, $\neg \varphi$ is in $\mathcal{K}$, one has to give up belief in $\neg \varphi$ by a process of retraction, and φ is in $\mathcal{K} * \varphi$. We propose to model belief revision in a dynamic epistemic logic. In this setting, we typically have an information state (pointed Kripke model) for the theory $\mathcal{K}$ wherein the agent believes the negation of the revision formula, i.e., wherein $B\neg\varphi$ is true. The revision with φ is a program $*\varphi$ that transforms this information state into a new information state. The transformation is described by a dynamic modal operator $[*\varphi]$, that is interpreted as a binary relation $[\![*\varphi]\!]$ between information states. The next information state is computed from the current information state and the belief revision formula. If the revision is successful, the agent believes φ in the resulting state, i.e., $B\varphi$ is then true. To make this work, as information states we propose 'doxastic epistemic models' that represent both knowledge and degrees of belief. These are multi-modal and multi-agent Kripke models. They are constructed from preference relations for agents, and they satisfy various characterizable multi-agent frame properties. Iterated, revocable, and higher-order belief revision are all quite natural in this setting. We present, for an example, five different ways of such dynamic belief revision. One can also see that as a non-deterministic epistemic action with two alternatives, where one is preferred over the other, and there is a natural generalization to general epistemic actions with preferences.

1. INTRODUCTION

Both belief revision and knowledge change have independently been on the research agenda for quite a while (Alchourrón et al. 1985; Kraus et al. 1990; Fagin et al. 1995; van Benthem 1996).

Belief revision has been studied from the perspective of structural properties of reasoning about changing beliefs (Gärdenfors 1988), from the perspective of changing, growing and shrinking knowledge bases, and from the perspective of models and other structures of belief change wherein such knowledge bases may be interpreted, or that satisfy assumed properties of reasoning about beliefs. Such models are the starting point of our investigations. A typical approach involves preferential orders to express increasing or

Synthese (2005) 147: 229–275
Knowledge, Rationality & Action 175–221
DOI 10.1007/s11229-005-1349-7

decreasing degrees of belief (Kraus et al. 1990; Meyer et al. 2000; Meyer 2001; Ferguson and Labuschagne 2002). Within this tradition multi-agent belief revision has also been investigated, e.g., belief merging (Konieczny and Pérez 2002).

Knowledge change has been extensively studied in what is by now called dynamic epistemic logic, with seminal publications of (Plaza 1989; Gerbrandy and Groeneveld 1997; Baltag et al. 1998; Gerbrandy 1999; van Ditmarsch 2000, 2002; Baltag 2002; ten Cate 2002; Kooi 2003). These investigations are often on the level of arbitrary modal operators, so that such 'knowledge change' also includes belief change. The typical perspective is multi-agent, with one belief or knowledge operator per agent, and 'change' always is *growth* of knowledge or *strengthening* of belief, and therefore not revisions unless they are simple expansions. This research has been mainly driven by the attempt to model higher-order belief change phenomena, and was initially motivated by the attempt to model the so-called 'unsuccessful updates', as in the well-known muddy children problem (Moses et al. 1986): from a public update with 'nobody knows whether he/she is muddy', the muddy children may learn that they are muddy. Such research still has to be carried to the stage of integrated frameworks for factual, knowledge and belief change. In particular, in the current approaches it is not possible to revise your beliefs, only to expand them. Our work is a proposal for integrated knowledge and belief change, including belief revision.

Interactions between these two different research directions have been somewhat limited, notwithstanding outstanding expositions that relate them (Segerberg 1999b), and apart from a tradition of modeling belief revision but in *non-dynamic* modal logic (Board 2004; Asheim and Søvik 2005). This may partly be due to a different choice of semantic primitives. In belief revision, preference relations such as templated (preferential), smooth, faithful orders, play a major part (Alchourrón et al. 1985; Meyer et al. 2000). In dynamic knowledge update, the primitives are binary accessibility relations between factual states, expressing that these states cannot be distinguished from one another. This difference in focus can be bridged elegantly. Another reason for the lack of interaction appears to be the supposed difficulty and assumed complexity of higher-order belief change: apart from revising or updating one's beliefs about facts, one may also want to revise one's beliefs about other agent's beliefs, including different degrees of belief. In Segerberg (1999b) dynamic belief revision is restricted to factual revision, and the related (Lindström and Rabinowicz 1999) describes in detail why higher-order revision is problematic. Higher-order belief change has been incorporated in dynamic epistemics, but at the cost of (at least) giving up the analogue of the AGM postulate of 'success'. We also address that challenge.

We propose relational structures where each agent has not one but a set of associated accessibility relations, based on Lewis (1973), and similar to those in Board (2004). The set of accessibility relations corresponds to the preferences of the agent. A language for degrees of belief and knowledge is interpreted on such structures. Higher-order belief revision then becomes very natural. Higher-order belief revision is with formulas that themselves express beliefs, of that agent or of other agents. In the remainder of this introduction we illustrate this by an example.

EXAMPLE 1. Consider one (anonymous) agent and two facts p and q that the agent has some uncertainty about; and, as a matter of fact, both p and q are false (see Figure 1). There are four states of the world, $\{00, 01, 10, 11\}$. Atom p is only true in $\{10, 11\}$, and atom q is only true in $\{01, 11\}$. The actual state is 00. The agent has preferences among these states. He considers it most likely that 11 is the actual state, i.e., that both p and q are true, slightly less likely that 01 or 10 are the actual state, and least likely that 00 is the actual state. (And we assume that these preferences are, in this particular case, the same whatever the actual state.) We write

$$11 < 01 = 10 < 00$$

The agent *believes* propositions when they hold in the most likely world. For example, he believes that p and q are true. This is

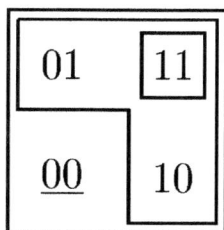

Figure 1. A model representing different degrees of belief in p and q. The actual state 00 is underlined.

described as

$$B(p \land q)$$

or as $B^0(p \land q)$, where 0 stands for 'degree of belief 0', i.e., 'most preferred'. His belief in the slightly weaker proposition p or q, is slightly stronger than his belief in p and q. Note that p or q are true in all three of 11, 01, and 10, i.e., including state 11 in level 0. For that, we write

$$B^1(p \lor q)$$

In B^1, the 1 means 'slightly less preferred than 0'. His strongest beliefs, or knowledge, in this case only involve tautologies such as $p \lor \neg p$. This is described as

$$K(p \lor \neg p)$$

or as $B^2(p \lor \neg p)$.[1] His strong beliefs are also about his preferences. For example, he knows that he believes p and q

$$K B(p \land q)$$

This is, because whatever the actual state of the world (even though it is 00), $B(p \land q)$ is true.

EXAMPLE 2. In the example above, imagine that the agent wants to revise his current beliefs. He believed that p and q are both true, but he has been given sufficient reason to be willing to revise his beliefs with $\neg p$ instead. We can accomplish that when we allow an information state transformation. On the right in Figure 2 the agent believes that p is false and that q is true. So in particular, in modal terms, $B\neg p$ is true. Therefore, the revision was successful. This can already be expressed in the information state on the left, by using

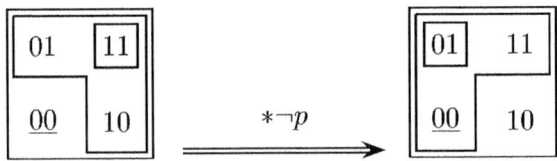

Figure 2. The agent changes his belief in p and q by revising with $\neg p$. After the revision, the agent believes $\neg p$ instead. He still believes q.

a dynamic modal operator $[*\neg p]$ for the relation induced by the program "belief revision with $\neg p$", followed by what should hold after that program is executed. On the left, it is true that the agent believes p and that after belief revision with $\neg p$ the agent believes that $\neg p$. In the language that we propose this is described as $Bp \wedge [*\neg p]B\neg p$.

In 'belief revision', the initial beliefs in the model on the left in the figure are described by a theory (with belief base) $\{p, q\}$, excluding preferences. In this case, the preferences can be measured as preferences on sets of formulas, namely $\{p, q\} < \{\neg p, q\} = \{p, \neg q\} < \{\neg p, \neg q\}$. The theory is revised with $\neg p$, resulting in a theory $\{\neg p, q\}$ that comes with different preferences.

Example 2 illustrates how 'standard' and 'dynamic' belief revision compare. In 'standard' belief revision a theory $\mathcal{K}$ is revised with a formula φ resulting in a revised theory $\mathcal{K} * \varphi$. Typically, $\neg \varphi$ is in $\mathcal{K}$, one has to give up belief in $\neg \varphi$ by a process of retraction, and φ is in $\mathcal{K} * \varphi$. In 'dynamic' belief revision we have an information state (pointed Kripke model) for the theory $\mathcal{K}$ wherein the agent believes the negation of the revision formula, i.e., wherein $B\neg \varphi$ is true. The revision with φ is a 'program' $*\varphi$ that transforms this information state into a new information state. The transformation is described by a dynamic modal operator $[*\varphi]$, that is interpreted as a binary relation $[\![*\varphi]\!]$ between information states. After successful belief revision, the revision formula φ is believed in the new information state, i.e., $B\varphi$ is now true.

The suitability of such a dynamic modal revision operator can be assessed by matching the AGM requirements for belief revision to corresponding properties of dynamic logical semantics. A few general observations can already be made at this stage.

AGM requirements concerning inconsistency are implicit in this approach. They are now related to the executability of (belief revision) programs. If a program $*\varphi$ is executable, an information state results from its execution, that is a model of a non-trivial theory. If it is not executable, this means that the revision cannot take place. Given the semantics of a modal 'necessity'- or $\square$-type operator such as $[*\varphi]$, we then have that $[*\varphi]\psi$ is true for all ψ; in other words, the trivial theory of all formulas results. It is implicit, because we do not reach an information state wherein all formulas are true (an impossibility).

'Success' can no longer be required for all formulas. This because the revision formula is not necessarily a propositional formula, and for some epistemic formulas success is impossible. A typical example is to attempt 'revision' with $p \wedge \neg Bp$: after that, you believe that p so it is no longer the case that you don't believe it. Instead, one may require success for belief revision on *propositional* formulas. In general, the question becomes for which formulas (i.e., for what fragment of the logical language) belief revision is successful.

Another way of using dynamic logic for belief revision, and more particularly belief expansion, is outside the (Segerberg 1999b) setting. Instead of a revision where (a model that satisfies) $B\neg\varphi$ is revised to $B\varphi$, one could conceive revision where $\neg B\varphi$ is revised to $B\varphi$. For that, for the more specific case of knowledge instead of belief, see van Ditmarsch et al. (2004). In this setting, revision of facts is not possible, but for a non-trivial fragment of epistemic formulas revision is successful.

There are obvious relations between the *qualitative* reasoning about plausibilities and degrees of belief, and *quantitative* reasoning about probabilities. Realistic accounts of belief change also need to incorporate the latter. We consider such interactions, and the combination of probability, knowledge and belief, and change, outside the scope of our contribution. In part this is an independent area of investigation, and for another part it may be seen as an extension of a qualitative approach, namely wherein we retain preferences between states (worlds), but add a probability distribution that assigns weights to individual states. See Halpern (2003) for an overview.

In Section 2 we introduce and motivate our perspective on preference relations. In Section 3 we introduce the doxastic epistemic states, and the language and semantics of doxastic epistemic logic. This is the logic for reasoning about different degrees of belief and knowledge, that is interpreted on these structures. The section contains summary observations on the relation to probability. In Section 4 we introduce dynamic belief revision, including five example belief revision operators. In Section 5 we present a generalization of these ideas to epistemic actions. Section 6 investigates and summarizes to what extent **AGM** postulates are fulfilled. In the concluding Section 7 we give suggestions for further research.[2] Appendix A on belief revision by conditional modal belief operators contains a detailed comparison of our approach to that of (Board 2004).

2. PREFERENCES

Given an 'actual' state, with which we associate a factual description of the world, an agent may be uncertain which of a set of different states is actually the case. This is the set of *plausible* states, given the actual state. The actual state itself may or may not be included in the set of plausible states.[3] Any of those plausible states may have its own associated set of plausible states, relative to *that* state, etc. It is customary to assume a 'set of all states', or *domain,* such that the actual state is in the domain and the set of its plausible states is a subset of the domain. We also assume that all states are 'about' the same facts, and that the agent is merely ignorant about which facts are true and which false, in a given state. In 'belief revision' it is often assumed that for all states, the set of plausible states is the *entire* domain. In 'dynamic epistemics' the set of plausible states is typically not the entire domain but a *subset of* the domain, and 'plausible' means 'accessible'.

Given two plausible states, the agent may think that one is more likely to be the actual state than the other, in other words, the agent has a preference among states. This view originates with (Lewis 1973) and is also propagated in Grove (1988) and Spohn (1988). Instead of preferences one may think of 'systems of spheres' or 'order relations between worlds'. Preferences are assumed to be partially ordered. Specific partial orders have been investigated in the belief revision community (total, smooth, faithful, modular, ...) (Alchourrón et al. 1985). In particular, one does not *have* to assume that states that are equally preferable are comparable (Alchourrón et al. 1985; Meyer et al. 2000). We assume that all plausible states are totally ordered, but some of our results are generalizable to such partial orders.

The implausible states are less preferable than any plausible state. As the plausible states are totally ordered, one *could* therefore extend the order to the entire domain. This is partly a matter of taste, and we prefer not to extend it like that. This can be justified by, firstly, the *qualitative* difference between plausible and implausible states: though some plausible states are preferred over other plausible states, the implausible states are all equally implausible. Also, a reasonable assumption is that states that have once been excluded should always be excluded; it should not be possible to change your mind on what is plausible and what not. Finally, in a setting that is both dynamic and multi-agent any *finite* number of different preferences seems counterintuitive. But if there are infinitely many degrees of belief, no matter how much you increase your belief in a proposition, it would be surprising if you could make it knowledge by doing *just* that: the assumption that 'knowledge' is merely a 'high degree of belief' has counterintuitive consequences, even when the actual state is *known* to be plausible.[4]

We write $<^s$ for an agent's preference relation on the set of plausible states given s. For example, $t <^s t'$ means that t is preferable to t', from the perspective of state s (and *not* the other way round). For more agents, write $<^s_n$ for the preference relation given state s and agent n.

In belief revision a totally ordered set of plausible states typically has least or most preferred elements. These represent the 'normally believed' states. The relation between beliefs and states is as follows: if the denotation of a(n) (objective) proposition contains all states in level 0, we say that this proposition is 'normally believed', or simply, 'believed'. These are the states that determine the belief set or belief base of the agent. The higher levels play a part in the belief *revision* process. In order to ensure that the revision with some objective (propositional) formula φ is successful (which means that that formula is believed after the revision), the contraction part of the revision process consists of 'expanding the bottom level to the lowest level above it for which the negation of φ is no longer 'normally believed'. This means that the contraction part of the revision depends on the order among plausible states, and it also means that the contraction may change the order among plausible states, thus possibly affecting the results of iterated belief revision. The contraction is followed by a different phase of the belief revision, called expansion. The 'dynamic logic' approach to belief revision achieves this result by globally adjusting all preferences, so that contraction and expansion are, so to speak, simultaneously executed.

We therefore also assume that each total order $<^s$ has a least element. Without loss of generality we further assume that for all states s in the domain (and for all agents), a $<^s$-totally ordered set of plausible states is isomorphic to a subset of the same total order $\langle \chi, < \rangle$ – in case of doubt, take the one induced by $\bigcup_s <^s$. This total order hereby becomes a parameter of the class of structures, and of the logical language. As the language includes postconditions of belief revision, this requirement determines that the 'background total order' remains unchanged throughout iteration of belief revision. This is relevant: *without* this requirement, it would be unclear in what logical language one describes *arbitrary* postconditions.

If there is only one agent, a finite number of preferences is always sufficient to distinguish factual state descriptions. If there are more agents, an agent can also have preferences about another agent's preferences. These preferences should therefore be seen as not so much between 'isolated' states, that can be identified with factual descriptions, but between 'information states', i.e., pointed Kripke-models. Even when reasoning about a finite model, a infinite number of preferences is then needed: Given the initial model, from dynamic developments (belief revision) more complex models may result. Such resulting models may have arbitrarily many preference distinctions.

3. DOXASTIC EPISTEMIC LOGIC

For sake of a simple exposition, assume that there is one agent only; the obvious multi-agent versions of definitions are given at the end of this section, in subsection 3.6. Throughout this section, we assume a set of atoms P, and a total order $\langle \chi, < \rangle$ with least element 0. The doxastic epistemic structures proposed below are (modulo nonessential differences) the same as the *belief revision structures* in the original publication (Board 2004), but the proposed logical language and the treatment of knowledge is rather different. See Appendix A for a comparison.

DEFINITION 3 (Doxastic epistemic model). A *doxastic epistemic model* is a triple $\langle S, <, V \rangle$. The set S is a *domain* of factual states, and *valuation* V is a function $V: P \to \mathcal{P}(S)$ such that each V_p is a subset of S. The *preference function* $<: S \to \mathcal{P}(S \times S)$ defines a *preference relation* $<^s$ for each $s \in S$. The subset $\mathsf{domain}(<^s) \cup \mathsf{range}(<^s)$ of the domain S is the set $Plaus^s$ of *plausible states* given s. There are two requirements on $Plaus^s$: it should contain a least element, and there must be a *degree function* $<^s: \langle Plaus^s, <^s \rangle \hookrightarrow \langle \chi, < \rangle$ that is an injection; note the overloading of the notation $<^s$. A *doxastic epistemic state* is a pointed doxastic epistemic model $(\langle S, <, V \rangle, s)$, with $s \in S$. A *doxastic epistemic frame* is a pair $\langle S, < \rangle$. For 'doxastic epistemic' we also write 'doxep'.

The set of plausible states $Plaus^s$ is therefore also totally ordered and (as such) is isomorphic to a subset of χ. The degree function $<^s$ determines the degree $<^s(t) \in \chi$ of a plausible state $t \in Plaus^s$. The reflexive closure of a preference relation $<^u$ is $\leq^u$. Define $s =^u t$ if and only if $s \leq^u t$ and $t \leq^u s$. Two states t and t' are *comparable* if there is an s such that $t, t' \in Plaus^s$, i.e., if they are both plausible given the same state. If t and t' are comparable (given s), either $t <^s t'$, or $t' <^s t$, or $t =^s t'$.

Write $s < s'$ for "there is a t such that $s <^t s'$." This $<$ is *not* a total or even partial order, as it need not be transitive. In Example 14, later, we have that $s <^s t$ and $t <^t s$, but obviously not $s <^s s$. Subject to additional structural constraints, $<$ will become a partial order, which justifies the notation. Note that we overload the notation $<$ and also use it for the preference function.

The totally ordered χ can be seen as a 'background' or 'reference' set of preferences for a given doxastic epistemic state. (And it remains that, after a state transition induced by belief revision.) Typical total orders $\langle \chi, < \rangle$ that we have in mind are the booleans $\langle \{0, 1\}, < \rangle$, the natural numbers $\langle \mathbb{N}, < \rangle$, and the closed interval from 0 to 1 of rational or real numbers $\langle [0, 1], < \rangle$. The first allows a distinction between belief and knowledge. The last may facilitate future comparisons with reasoning about probability and knowledge.

We required that $\langle \chi, < \rangle$ is total and has a least element 0. We did not require that $\langle \chi, < \rangle$ is well-founded, because we want to include $\langle [0, 1], < \rangle$ as a special case. A generalization to partial orders, in view of applying this semantic framework to orders that are common in standard belief revision, such as smooth and faithful orders, seems appropriate and does not appear to pose any difficulties.

Given a preference relation $<$ we can define a set of accessibility relations to interpret degrees of belief and knowledge.

DEFINITION 4 (Accessibility relations for belief and knowledge). Accessibility $\rightarrow^x$ is defined as $s \rightarrow^x t$ iff $<^s(t) \leq x$. Accessibility $\rightarrow^\chi$ is defined as $\bigcup_{x \in \chi} \rightarrow^x$.

Therefore, the accessibility relations $\rightarrow^0, \ldots, \rightarrow^x, \ldots, \rightarrow^\chi$ on a domain S are induced by the (combined) preferences $<^s$ for all $s \in S$. We emphasize that the preferences are relative to a state s, but that, somewhat surprisingly, the accessibility relations $\rightarrow^x$ are *not* relative to that s: it is now the first of a pair in that accessibility relation.

FACT 5. If χ is finite, then $\rightarrow^{\text{Max}\{x \in \chi\}} = \rightarrow^{\chi}$. Let $x, y \in \chi$, then $x \leq y$ iff $\rightarrow^x \subseteq \rightarrow^y$. Let $x \in \chi$, then $\rightarrow^x \subseteq \rightarrow^{\chi}$.

DEFINITION 6 (Language of doxastic epistemic logic).

$$\varphi ::= p \,|\, \neg\varphi \,|\, \varphi \wedge \psi \,|\, \Box^x \varphi \,|\, \Box^\chi \varphi$$

DEFINITION 7 (Semantics of doxastic epistemic logic). Given are a doxastic epistemic model $M = \langle S, <, V \rangle$ and an $s \in S$.

$$
\begin{aligned}
M, s &\models p && \text{iff } s \in V_p \\
M, s &\models \neg\varphi && \text{iff } M, s \not\models \varphi \\
M, s &\models \varphi \wedge \psi && \text{iff } M, s \models \varphi \text{ and } M, s \models \psi \\
M, s &\models \Box^x \varphi && \text{iff for all } s' : s \rightarrow^x s' \text{ implies } M, s' \models \varphi \\
M, s &\models \Box^\chi \varphi && \text{iff for all } s' : s \rightarrow^\chi s' \text{ implies } M, s' \models \varphi
\end{aligned}
$$

Formula φ is *valid* on M, iff $M, s \models \varphi$ for all $s \in S$. Formula φ is *valid*, if it is valid on M for arbitrary M.

We have some multi-modal frame correspondence 'for free' (see Fact 5).

FACT 8. Schema $\Box^y \varphi \rightarrow \Box^x \varphi$ is valid iff $\rightarrow^x \subseteq \rightarrow^y$ (iff $x \leq y$). Schema $\Box^\chi \varphi \rightarrow \Box^x \varphi$ is valid.

We introduced $\Box^\chi \varphi$ as a primitive in the language, defined by its semantics. Alternatively, one can define $\Box^\chi \varphi$ as the possibly infinite (or even uncountable) conjunction of all $\Box^x \varphi$:

$$\Box^\chi \varphi := \wedge_{x \in \chi} \Box^x \varphi$$

So $\Box^\chi \varphi$ can also be defined as an infinitary modal operator in this language. We find this an interesting observation, because, subject to additional relational constraints, $\Box^\chi$ becomes 'knowledge' K. In other words, individual knowledge K is an infinitary modal operator. Epistemic logicians are used to have K as a primitive in the language, and common knowledge C as an infinitary operator. But $\Box^\chi$ is not knowledge yet. In the following subsections we gradually make the transition to 'knowledge'.

The logic is obviously not compact, for example, for $\langle \chi, < \rangle = \langle \mathbb{N}, < \rangle$, and, for readability, $\Box = B$ and $\Box^\chi = K$, we have

$$\{B^0 p, B^1 p, B^2 p, \ldots\} \models Kp$$

Although there are possibly uncountably many modal operators, namely one $\Box^x$ for each $x \in \chi$, a formula in the language always only contains a finite subset of such modal operators. Therefore, completeness of a proof system for the logic (outside this contribution) does not pose any difficulty.

EXAMPLE 9. The model in Figure 1 can be described as a doxastic epistemic (doxep) state $(M, 00)$. Given is the set of atoms $\{p, q\}$ and the total order $\langle\{0, 1, 2\}, <\rangle$. The doxep state $(M, 00)$ has underlying doxep model $M = \langle\{00, 01, 10, 11\}, <, V\rangle$ with preferences

$$11 <^{00} \ 01 =^{00} \ 10 <^{00} \ 00$$
$$11 <^{01} \ 01 =^{01} \ 10 <^{01} \ 00$$
$$11 <^{10} \ 01 =^{10} \ 10 <^{10} \ 00$$
$$11 <^{11} \ 01 =^{11} \ 10 <^{11} \ 00$$

and valuation

$$Vp = \{10, 11\}$$
$$Vq = \{01, 11\}$$

The accessibility relations computed from these preferences are:

$$\to^0 = \{(00, 11), (01, 11), (10, 11), (11, 11)\}$$
$$\to^1 = \to^0 \cup \{(00, 01), (01, 01), (10, 01), (11, 01)\} \cup$$
$$\{(00, 10), (01, 10), (10, 10), (11, 10)\}$$
$$\to^2 = \to^1 \cup \{(00, 00), (01, 00), (10, 00), (11, 00)\}$$
$$= \{00, 01, 10, 11\} \times \{00, 01, 10, 11\}$$
$$\to^{\{0,1,2\}} = \to^2$$

The following hold throughout the model

$$M \vDash \Box^0 (p \wedge q)$$
$$M \vDash \Box^1 (p \vee q)$$
$$M \vDash \Box^2 (p \vee \neg p)$$
$$M \vDash \Box^{\{0,1,2\}} (p \vee \neg p)$$

Instead of $M \vDash \Box^0 (p \wedge q)$ we may say that the agent believes $p \wedge q$, i.e., $M \vDash B(p \wedge q)$, and instead of $M \vDash \Box^{\{0,1,2\}} (p \vee \neg p)$, or $M \vDash \Box^2 (p \vee \neg p)$, we may say that the agent knows $p \vee \neg p$, i.e., $M \vDash K(p \vee \neg p)$.

As an example we prove that $M, 00 \vDash \square^1 (p \vee q)$. Note that $00 \rightarrow^1$ $10, 00 \rightarrow^1 01$, and $00 \rightarrow^1 11$, and that $M, 10 \vDash p \vee q$, and $M, 01 \vDash p \vee q$, and $M, 11 \vDash p \vee q$. It holds that $M, 10 \vDash p \vee q$, because $M, 10 \vDash p$, because $10 \in V_p = \{10, 11\}$. Etc. As $p \vee q$ holds in all three states that are $\rightarrow^1$-accessible from state 00, we have by definition that $M, 00 \vDash \square^1 (p \vee q)$.

There is no obvious relation between plausibilities and probabilities. Note that $\square^x \varphi$ cannot be identified with $Pr(\varphi) \leq z$ for some value $z \in [0, 1)$. Clearly, $\square^x (\varphi \wedge \psi)$ is equivalent to ($\square^x \varphi$ and $\square^x \psi$), but if both $Pr(\varphi) \leq z$ and $Pr(\psi) \leq z$, we would expect $Pr(\varphi \wedge \psi)$ to be between z^2 and z. One solution is to provide a probability distribution that ensures that normally believed propositions are always believed with probability 1, and uses conditional (posterior) probabilities so that this remains the case for other degrees of belief (or for conditional belief) (Stalnaker 1996; Board 2004). Alternatively, lexicographic probability has been proposed, that provides a different probability distribution for each degree of belief, see Halpern (2001) for an overview and for prior references. Such settings of conditional reasoning for belief revision are unrelated to the dynamic belief revision that we propose (see Appendix A).

3.1. *Preferences from Accessibility Relations*

Instead of defining accessibility relations $\rightarrow^x$ given preferences $<^s$, we could also have defined preference relations given accessibility relations for belief and knowledge. The difference is a mere matter of taste.

DEFINITION 10 (Preferences from access). Given are accessibility relations $\rightarrow^0, \ldots, \rightarrow^x, \ldots, \rightarrow^X$ such that for all $x < y : \rightarrow^x \subseteq \rightarrow^y$, and $\rightarrow^X = \bigcup_{x \in \chi} \rightarrow^x$, and such that, for arbitrary s and t, all subsets of χ of the form $\{x \in \chi \mid s \rightarrow^x t\}$ have a least element. Define $<^s (t) :=$ $\mathsf{Min}\{x \in \chi \mid s \rightarrow^x t\}$, and $t <^s t'$ iff $<^s (t) <<^s (t')$.

The requirement that all subsets $\{x \in \chi \mid s \rightarrow^x t\}$ have a least element, corresponds to the requirement on preference functions that all orders $<^s$ have a least element. Given a multi-modal structure $\langle S, \{\rightarrow^0, \ldots, \rightarrow^x, \ldots, \rightarrow^X\}, V \rangle$ with access satisfying the requirements in Definition 10, and a belief function $<$ computed from that access such that for each state $s \in S$, $<^s$ is as in that definition. So:

FACT 11. Structure $\langle S, <, V \rangle$ is a doxep model.

How much information is contained in a doxep state? To this question is a clear answer. Consider a doxep model $M = \langle S, <, V \rangle$ and the set of accessibility relations $\rightarrow^0, \ldots, \rightarrow^x, \ldots, \rightarrow^\chi$ induced by $<$. We can think of M as an $M^\rightarrow = \langle S, \{\rightarrow^0, \ldots, \rightarrow^x, \ldots, \rightarrow^\chi\}, V \rangle$. Two doxep states (M, s) and (M', s') *contain the same information* if and only if $(M^\rightarrow, s)$ and $(M'^\rightarrow, s')$ are *bisimilar*.[5]

3.2. *Belief*

We assumed that $\langle \chi, < \rangle$ has a least element 0 and that each totally ordered $<^s$ has a least element. A least element t of $<^s$ does not necessarily have degree 0. If each order $<^s$ has a least element of degree 0, the relation $\rightarrow^0$ is *serial*. If we additionally require that $\rightarrow^0$ is transitive and euclidean, then $\square^0$ corresponds to 'normal belief'. If we additionally require that, for all other $x \in \chi$, the accessibility relations $\rightarrow^x$ are transitive and euclidean, then all other modal operators $\square^x$ correspond to 'standard' belief operators. Their seriality follows from the assumption that $\rightarrow^0$ is serial, and from $\rightarrow^0 \subseteq \rightarrow^x$ for arbitrary $x \in \chi$. We call a relation a *belief relation* if and only if it is serial, transitive, and euclidean.

PROPOSITION 12. If $\rightarrow^0$ is serial and if all accessibility relations $\rightarrow^x$ are transitive and euclidean, then $\rightarrow^\chi$ is a belief relation.

Proof. We prove that $\rightarrow^\chi$ is serial, transitive and euclidean.

Serial: This is because $\rightarrow^0 \subseteq \rightarrow^\chi$.

Transitive: Suppose $s \rightarrow^\chi t$ and $t \rightarrow^\chi u$. Then there are $x, y \in \chi$ such that $s \rightarrow^x t$ and $t \rightarrow^y u$. Let $z = \mathsf{Max}(x, y)$, then $s \rightarrow^z t$ and $t \rightarrow^z u$. As $\rightarrow^z$ is transitive, $s \rightarrow^z u$. Therefore $s \rightarrow^\chi u$.

Euclidean: Suppose $s \rightarrow^\chi t$ and $s \rightarrow^\chi u$. Then there are $x, y \in \chi$ such that $s \rightarrow^x t$ and $s \rightarrow^y u$. Let $z = \mathsf{Max}(x, y)$, then $s \rightarrow^z t$ and $s \rightarrow^z u$. As $\rightarrow^z$ is euclidean, $t \rightarrow^z u$. Therefore $t \rightarrow^\chi u$.

We conclude that not just all $\square^x$ but also $\square^\chi$ satisfy the standard properties of belief. Therefore, from now on, write B^x for $\square^x$ and write B^χ for $\square^\chi$. We are getting closer to knowledge (and conviction), but we are not yet there. We close this section with one more relevant result:

[188]

PROPOSITION 13. (χ-reflexivity for plausible states). Assume all accessibility relations $\to^x$ are euclidean. If a state s is plausible, then $s \to^x s$.

Proof. Let $s \in S$, and suppose there is a $t \in S$ and a $x \in \chi$ such that $t \to^x s$. From $t \to^x s$ and $t \to^x s$ and euclidicity follows $s \to^x s$. Therefore $s \to^x s$.

Of course, some states may be implausible given *any* state whatsoever. So that in particular, they are not χ-accessible from themselves.

3.3. *Global Preferences*

A natural restriction for preferences seems to be, that a preference of state s over state t should be independent of which of these is the actual state. More precisely – because neither of the two may be actually the case – the same preference should apply if one of those states is actually the case. This is not yet so:

EXAMPLE 14. Consider the doxep model $\langle \{s, t\}, <, V \rangle$ for set of atoms $\{p\}$ and total order $\langle \{0, 1\}, < \rangle$. The domain consists of two states s and t. Atom p is only true in state t. The preferences are that $s <^s t$ and $t <^t s$, or, in other words: if s is actually the case, then the agent considers s more likely than t, whereas if t is actually the case, then the agent considers t more likely than s. The associated accessibility relations are the identity for $\to^0$ and the universal relation for $\to^1$, and $\to^{\{0,1\}} = \to^1$. So all are equivalence relations, and therefore belief relations as well.

A different wording of this restriction is, that we may have various degrees of uncertainty over which is the actual state, but that we are highly confident of our preferences. This is expressed by the multimodal schemata $B^x \varphi \to B^\chi B^x \varphi$ and $\neg B^x \varphi \to B^\chi \neg B^x \varphi$, that say, so to speak, that 'you are convinced of what you believe and what not'. The obvious frame correspondence is as follows. (See also Figure 3.)

PROPOSITION 15. Schema $B^x \varphi \to B^y B^x \varphi$ of arbitrary positive introspection corresponds to frame property (for all $s, s', s'' \in S$: $s \to^y s'$ and $s' \to^x s''$ implies $s' \to^x s''$) of arbitrary transitivity.

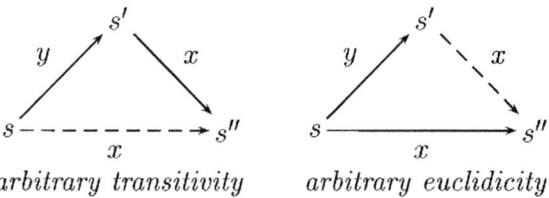

Figure 3. Frame correspondence expressing conviction of preference.

Schema $\neg B^x \varphi \to B^y \neg B^x \varphi$ of *arbitrary negative introspection* corresponds to frame property (for all $s, s', s'' \in S : s \to^y s'$ and $s \to^x s''$ implies $s' \to^x s''$) of *arbitrary euclidicity.*

For $x = 0$, and anticipating on the later identification of B^χ with K, $B^x \varphi \to B^\chi B^x \varphi$ and $\neg B^x \varphi \to B^\chi \neg B^x \varphi$ become the familiar (Kraus and Lehmann 1988) :

$$B\varphi \to K B\varphi$$
$$\neg B\varphi \to K \neg B\varphi$$

DEFINITION 16 (Belief function). The preference function of a belief state where all induced access are belief relations and satisfy arbitrary transitivity and euclidicity, is called a *belief function.*

In a doxep state with a belief function, the preferences are independent from our perspective of the 'actual state'.

LEMMA 17 (χ-symmetry for plausible states). Let $\langle S, <, V \rangle$ be a doxep state with belief function $<$, and let $s, t \in S$ be plausible. If $s \to^\chi t$, then $t \to^\chi s$.

Proof. Assume that $s \to^\chi t$. From $s \to^\chi t$ follows that $s \to^x t$ for some $x \in \chi$. As s is plausible, by Proposition 13 it is plausible from itself. So there is a $y \in \chi$ such that $s \to^y s$. From $s \to^x t$ and $s \to^y s$ and euclidicity follows $t \to^y s$. So $t \to^\chi s$.

PROPOSITION 18. In a doxep state with belief function, the degree of a plausible state is unique.

Proof. Let $<^s (t) = x$ and $<^{s'} (t) = y$, and suppose towards a contradiction that $x \neq y$. Without loss of generality, assume that $x > y$.

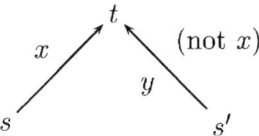

Figure 4.   This situation cannot occur, because in a doxep model with belief function, the degree of a plausible state is unique.

Then we have that $s \rightarrow^x t$ and $s' \rightarrow^y t$, and *not* $s' \rightarrow^x t$. See Figure 4. From twice $s \rightarrow^x t$ and arbitrary euclidicity follows (as in Lemma 13) $t \rightarrow^x t$. From $s' \rightarrow^y t, t \rightarrow^x t$ and arbitrary transitivity follows $s' \rightarrow^x t$. Contradiction.

PROPOSITION 19. Let $\langle S, <, V \rangle$ be a doxep state with belief function $<$. The relation $<$ (with the same name) is a partial order on S.

Proof. We remind the reader that $<$, seen as a relation, was defined as: $s < s'$ iff there is a t such that $s <^t s'$. Let $s < s'$ and $s' < s''$. Then there are t, t' such that $s <^t s'$ and $s' <^{t'} s''$. From $s <^t s'$ and the previous follows $s <^{s'} s'$, from $s' <^{t'} s''$ and the previous follows $s' <^{s'} s''$. From $s <^{s'} s'$ and $s' <^{s'} s''$ follows $s <^{s'} s''$. From $s <^{s'} s''$ follows by definition $s < s''$.

Proposition 19 therefore justifies writing $<$ for that relation. It follows from Proposition 18 that: $s < s'$ iff for all t, either $s, s' \notin Plaus^t$ or $s <^t s'$. In other words, we can always write $<$ instead of $<^t$. Our preferences are no longer local, depending on an actual state, but global, irrespective of the actual state for which comparable states are plausible.

There may still be states that are implausible from any state. Given two such implausible states, it follows from Proposition 18 that if but a single state is plausible from (the perspective of) both, then they have the same preferences on the entire domain. (Note that in the proof of Proposition 18, s and s' may be such implausible states. And we proved that the degree x of a shared plausible t must be unique.)

The meaning of the operator B^x now appears to be the notion of *conviction* in Lenzen (2003) and others. Work in progress by Guillaume Aucher explores the notion in depth. Also Asheim and Søvik (2005) investigate similar relations between preferences and belief operators. We will not explore the notion of conviction any further, but instead make one more requirement such that 'knowledge' results.

3.4. *Knowledge*

We still have situations where $\to^\chi$ is not an equivalence relation, and therefore conviction B^χ not the same as knowledge K.

EXAMPLE 20. Consider the doxep state $((\{s, t\}, <, V), s)$ for set of atoms $\{p\}$ and total order $\langle\{0\}, <\rangle$. The domain consists of two states s and t. There is one atom p that is only true in state t. The preferences are that $t =^s t$ and $t =^t t$. Therefore, the induced access is that $\to^0 = \{(s, t), (t, t)\}$. (As the order is finite, $\to^{\{0\}} = \to^0$.) We now have that

$$\langle\{s, t\}, <, V\rangle, s \vDash B^{\{0\}} p \wedge \neg p$$

In other words: the agent is convinced that p, but p is false after all.

Example 20 is a simple case where $\to^\chi$ is not reflexive. If we require that $\to^\chi$ is reflexive, B^χ becomes knowledge. If $\to^\chi$ is reflexive, all states are plausible. This is similar to a requirement that, for all states $s \in S :< (s) \in \chi$. This is because $< (s) \in \chi$ means by definition that there is a state t such that $<^t (s) \in \chi$. From that, it follows that $t \to^{<^t(s)} s$. From twice that follows, with euclidicity, that $s \to^{<^t(s)} s$, i.e., $s \to^\chi s$.

PROPOSITION 21. Let $\langle S, <, V \rangle$ be a doxep state with belief function $<$. If all states are plausible, $\to^\chi$ is an equivalence relation.

Proof. Let $s \in S$. Because s is plausible, $s \to^\chi s$ (Proposition 13). It was already shown that $\to^\chi$ is a belief function (Proposition 12).

In such doxep states, the relation $<$ is called a *knowledge function*, instead of $s \to^\chi t$ we write $s \sim t$, and instead of B^χ we write K.

'All states are plausible' does not mean that they are all plausible for the same state, it only means that they are plausible for *some* state. Specifically, they are plausible for any state in their equivalence class.

PROPOSITION 22. If all states are plausible, arbitrary transitivity follows from arbitrary euclidicity.

Proof. Assume that arbitrary transitivity does *not* hold, that arbitrary euclidicity holds, and that all states are plausible. Suppose $s \rightarrow^y s'$ and $s' \rightarrow^x s''$. To prove: $s \rightarrow^x s''$. As s is plausible, there is a $z \in \chi$ such that $s \rightarrow^z s$ (Proposition 13). From $s \rightarrow^y s'$, $s \rightarrow^z s$, and arbitrary euclidicity follows $s' \rightarrow^z s$. From $s' \rightarrow^z s$ and $s' \rightarrow^x s''$ follows with arbitrary euclidicity $s \rightarrow^x s''$.

In other words, within the setting of knowledge, frame axiom "for all $x, y : B^x \varphi \rightarrow B^y B^x \varphi$" (arbitrary positive introspection) follows from "for all $x, y : \neg B^x \varphi \rightarrow B^y \neg B^x \varphi$" (arbitrary negative introspection) and "there is a x such that $B^x \varphi \rightarrow \varphi$".

3.5. *Visualizing Information*

For doxep models with a belief or with a knowledge function, such that preferences are global, we can resort to an elegant visualization. The domain of a doxep model with a knowledge function is split up in a number of equivalence classes. Each equivalence class consists of a (possibly uncountable) number of degrees, or levels. It can be seen as a 'copy' of the order $\langle \chi, < \rangle$. (With some allowance for layers to be empty, as formally the equivalence class is isomorphic to a subset of χ.) In other words, the domain of the model is entirely covered by a set of disjoint total orders $\langle \chi, < \rangle$. We can visualize such a model as a 'bag of onions', where each 'onion' stands for such an epistemic class and each onion peel corresponds to a level in the total order. The 'innermost peel', that is non-empty, contains the normally believed states (Figure 5).

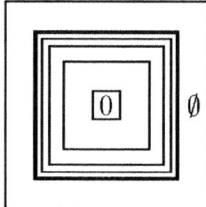 

Figure 5. On the left, we see *one* totally ordered subset of the domain, visualizing the preferences given some state. There is an emptly layer of 'unreachable' or implausible states. The preferences given a different state, may be totally unrelated, and intersect with this 'system of squares'. But if the preference function is a knowledge function, and the preferences are therefore global, the 'unreachable' part ∅ of the domain is entirely covered by similar copies of that total order. The copies do *not* intersect.

This visualization also applies to the notion of 'conviction' in the previous section. Apart from mutually disjoint total orders of plausible states (onions), we then have 'implausible states' outside onions. Given an implausible state and an agent, one (and only one) onion may be accessible (but for another agent that same state may be linked to another onion). In other words, we now have onions with roots.

The 'running' Example 1, wherein one agent has different beliefs in atoms p and q, is a doxep state with a knowledge function, where degrees of belief indeed approach knowledge. Note that in the formal description of this doxep state in Example 9 the preferences are indeed independent from the choice of actual state. In this example, there is only one epistemic class: whatever the actual state, all states in the domain are plausible given that state. We close this section with another example, illustrating more than one epistemic class. The next section contains a multi-agent example. This may help the reader to appreciate fully the information modeling opportunities of these structures.

EXAMPLE 23. Consider a model M' where the agent may be uncertain about the truth of two facts p and q. In state 01, p is false and q is true, etc. Let p stand for 'the fan is on' and let q stand for 'the light is on.' In the model of Figure 6, if the agent *knows* that the fan is off, he *believes* that the light is off too; and if he knows that the fan is on, he believes that the light is on too. This is described in the formula $(Kp \to Bq) \land (K\neg p \to B\neg q)$. For example, all the following hold:

$$M', 00 \vDash B(\neg p \land \neg q)$$
$$M', 00 \vDash K\neg p$$
$$M', 10 \vDash B(p \land q)$$
$$M', 10 \vDash Kp$$
$$M' \vDash (Kp \to Bq) \land (K\neg p \to B\neg q)$$

3.6. *More Agents*

We present multi-agent versions of the relevant definitions. As there are no multi-agent interaction axioms, belief relations and belief and knowledge functions and all results are as before. The outcome is a multi-agent language for reasoning about knowledge and belief.

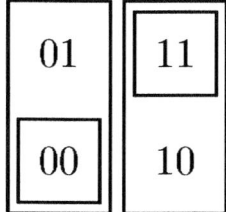

Figure 6. A doxep model consisting of two equivalence classes.

The (often implicit) assumption in a multi-agent epistemic setting is that agents are completely aware of each other's epistemic distinctions (i.e., the accessibility relations). In the current setting it means that they are completely aware of each other's preferences. This is part of the background knowledge, so to speak. Or, yet differently said, the agents know the entire structure of the model, and the only thing that they do (may) not know, is what the actual state of the model is.

DEFINITION 24 (Multi-agent doxastic epistemic model). A *doxastic epistemic model* (doxep model) is a triple $\langle S, <, V \rangle$. The set S is a *domain* of factual states, and *valuation* V is a function $V : P \to \mathcal{P}(S)$ such that each V_p is a subset of S. The *preference* function $< : N \to S \to \mathcal{P}(S \times S)$ defines a preference relation $<_n^s$; for each agent $n \in N$ and for each $s \in S$. The subset $\mathsf{domain}\,(<_n^s) \cup \mathsf{range}(<_n^s)$ of the domain S is the set $Plaus_n(s)$ of *plausible states* for agent n given s. There are two requirements on $Plaus_n(s)$: it should contain a least element, and there must be a *degree* function $<_n^s : \langle Plaus_n(s), <_n^s \rangle \hookrightarrow \langle \chi, < \rangle$ that is an injection.

DEFINITION 25 (Language of multi-agent doxastic epistemic logic).

$$\varphi ::= p \,|\, \neg\varphi \,|\, \varphi \wedge \psi \,|\, \Box_n^x \varphi \,|\, \Box_n^\chi \varphi$$

DEFINITION 26 (Semantics of multi-agent doxastic epistemic logic). Let $\langle S, <, V \rangle$ be a multi-agent doxep state and $s \in S$. Then:

$$M, s \vDash \Box_n^x \varphi \quad \text{iff} \quad \text{for all } s' : s \to_n^x s' \text{ implies } M, s' \vDash \varphi$$
$$M, s \vDash \Box_n^\chi \varphi \quad \text{iff} \quad \text{for all } s' : s \to_n^\chi s' \text{ implies } M, s' \vDash \varphi$$

EXAMPLE 27. Consider the setting of Example 23, only now there are two agents, operators Anne (a) and Bill (b), say, that have different access to the state of the fan and the light. In Figure 7, access for a is solid and access for b is dashed. This doxep model M'' represents a situation where Anne knows whether the fan is on, whereas Bill knows whether the light is on. Also, as in Example 23, if Anne knows that the fan is on, she believes that the light is on, and if she knows that the fan is off, she believes that the light is off.

We can now evaluate various statements. In the state where the fan and light are both off, Anne believes that, and believes that Bill knows that the light is off. In the state where the fan is off and the light is on, Anne knows that the fan is off, and she (incorrectly) believes that the light is off; she even believes that Bill knows that the light is off, even though Bill actually knows that the light is on. Formally:

$$M'', 00 \models B_a(\neg p \wedge \neg q) \wedge B_a K_b \neg q$$
$$M'', 01 \models \neg p \wedge K_a \neg p \wedge B_a \neg q \wedge B_a K_b \neg q \wedge K_b q$$

One can of course expand this multi-agent language with notions for common knowledge and collective belief (Fagin et al. 1995). There are interesting corresponding (dynamic) notions of merged belief revision for subgroups of the public. We have not explored that in depth yet (but see Liu (2004) for an investigation).

4. DYNAMIC BELIEF REVISION

We expand the language with a dynamic operator $[*\varphi]$ expressing belief revision with a formula φ. Again, for simplicity, we give the single-agent version. Given are a set of atoms P and a total order

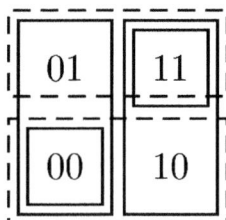

Figure 7. Two operators Anne and Bill have different access to the state of a fan and a light.

$\langle \mathcal{X}, < \rangle$, as before. Explanations and justifications follow the definitions, and the example after that.

DEFINITION 28 (Language of dynamic belief revision).

$$\varphi ::= p \,|\, \neg\varphi \,|\, \varphi \wedge \psi \,|\, \Box^x \varphi \,|\, \Box^{\mathcal{X}} \varphi \,|\, [*\varphi]\psi$$

DEFINITION 29 (Semantics of belief revision). Given are a doxep model $M = \langle S, <, V \rangle$ and an $s \in S$; $M^* = \langle S^*, <^*, V^* \rangle$ is an arbitrary doxep model for (the same) atoms P, agents N, and order $\langle \mathcal{X}, < \rangle$.

$$M, s \vDash [*\varphi]\psi \text{ iff for all } (M^*, s^*) : (M, s)\llbracket *\varphi \rrbracket (M^*, s^*)$$
$$\text{implies } M^*, s^* \vDash \psi$$

The formula $[*\varphi]\psi$ reads as 'after revision with φ, ψ holds'. The semantics that we propose is typical for a dynamic modal operator: a state transformer $[*\varphi]$ induces a binary relation $\llbracket *\varphi \rrbracket$ between doxep states. As revision with a formula φ is a deterministic process, the relation $\llbracket *\varphi \rrbracket$ is functional. (Also) Because the value of facts in states does not change, we may choose $s^* = s$. For most sorts of 'dynamic belief revision $*$' that we consider the function $\llbracket *\varphi \rrbracket$ is also total, and $S^* = S$, and $V^* = V$. We call this *type a* belief revision. In that case the definition becomes

$$\langle S, <, V \rangle, s \vDash [*\varphi]\psi \quad \text{iff} \quad \langle S, <^*, V \rangle, s \vDash \psi$$

where $<^*$ is a revised preference function for order $\langle \mathcal{X}, < \rangle$ that is computed from $<$ and (the value of) φ (in M). Apart from that we will also consider a also tentatively named *type b* belief revision where some states may become implausible, specifically: where $S^* \subset S$.

DEFINITION 30 (Successful belief revision). Belief revision $*$ is *successful on* φ if $\neg\Box^{\mathcal{X}}\neg\varphi \to [*\varphi]\Box^0\varphi$ is valid. Belief revision $*$ is *propositionally successful* if it is successful for all propositional φ.

EXAMPLE 31. We define a (type a) belief revision operator $*^1$. The effect of two consecutive $*^1\neg p$ revisions is pictured in Figure 8. It is not successful after the first revision, but only after the second. In the leftmost doxep state $(M, 00)$, our running example, the

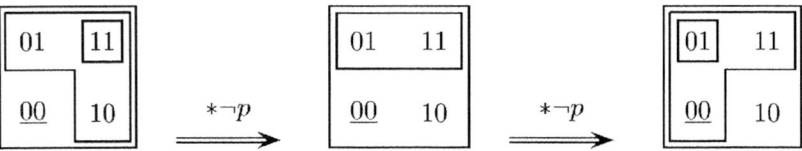

Figure 8. The agent changes his belief in p and q by revising with $\neg p$. After the revision, the agent no longer believes p, but the revision is not successful, because he does not believe $\neg p$ either. After applying the same revision once more, it is successful.

(global) order on the states is $11 < 01 = 10 < 00$. (It is inessential in this example which is the actual state.) In other words:

$$< (11) = 0$$
$$< (01) = 1$$
$$< (10) = 1$$
$$< (00) = 2$$

Revision $*^1$ with some formula φ first makes the states where the formula does *not* hold one degree *less* likely. I.e., $<^{*1} (s) = < (s)$ if $M, s \vDash \varphi$ but $<^{*1} (s) = < (s) + 1$ if $M, s \nvDash \varphi$. Then, we "start counting levels from 0" again and we remove gaps between levels. This is called normalization. If we revise $(M, 00)$ with $\neg p$ according to this recipe, we first get

$$< (11) = 0 + 1 = 1$$
$$< (01) = 1$$
$$< (10) = 1 + 1 = 2$$
$$< (00) = 2$$

Then we normalize and get

$$< (11) = 0$$
$$< (01) = 0$$
$$< (10) = 1$$
$$< (00) = 1$$

This revision is the leftmost transition pictured in the figure. The revision is not successful, because even though the agent no longer believes p, he has not come to believe $\neg p$. If we apply the revision once more, the rightmost doxep state results, wherein the agent believes $\neg p$. (See also subsection 4.1.)

In the semantics of $[*\varphi]\psi$, how $<^*$ is computed from $<$ and φ in type a revision, needs precision. We give four different ways to perform that computation, tentatively named belief revision $*^1$, $*^3$, $*^4$, and $*^5$. In type a semantics, belief revision $*\varphi$ is an always executable and deterministic program, corresponding to the *tentative* public announcement of φ. It consists of changing preferences *only*, therefore both the domain S and the valuation V remain the same. To change a preference function $<^*$ means that each $<^s(t)$ may be assigned a new degree in $\langle \mathcal{X}, < \rangle$. This new degree is the value of $(<^*)^s(t)$ in the 'new' preference function $<^*$. Such dynamic revision $*\varphi$ can be seen as the *tentative public announcement* of φ, or 'soft update' with φ. The announcer assumes that both φ and $\neg\varphi$ may be true, but considers it more likely that φ is true.

When do we expect the revision $*\varphi$ not to be executable, unlike type a? For each state $s \in S$, the preference relation $<^s$ has two features: what the subset *Plaus*^{s} of the domain is for which the preferences are given, and how the degrees $x \in \mathcal{X}$ are distributed over that subset. Therefore, there are two ways in which one might want to adjust the preference function $<$: change the degrees of the plausible states *Plaus*^{s} given s, or constrain the set *Plaus*^{s}. If *Plaus*^{s} becomes the empty set, $<^s$ no longer corresponds to a non-empty total order: as there are no accessible states, the agent believes everything, in other words, he has "gone mad". The solution to such a problem is to remove state s from the domain S, but if s happens to be the actual state, we cannot do that. Therefore, in that case we do not want $*\varphi$ to be executable. This is the belief revision that we consider to be of *type b* of which we propose an example $*^2$ revision – in that case $*\varphi$ is executable whenever φ is true. See subsection 4.2 for further details.

Other realistic forms of such belief revision, where we adjust *both* the set of plausible states relative to a state *and* the degrees among the plausible states, may well be conceived.

Revision with φ is successful, if the revision formula is believed by the agent after the revision has taken place, i.e., if $\square^0\varphi(B\varphi)$ is true in the resulting doxep state. A precondition for that, is that the revision formula was plausible for that agent, i.e., that before the revision $\neg\square^{\mathcal{X}}\neg\varphi(\neg K\neg\varphi)$ holds. In this dynamic setting we cannot require that belief revision is *always* successful. This is because, first, we do not want implausible states to become plausible again. If we *know* $\neg\varphi$ already, a fortiori we *believe* $\neg\varphi$, but belief revision

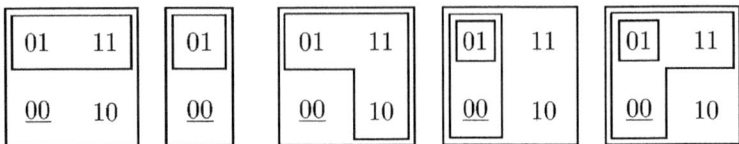

Figure 9. The five doxastic epistemic states in this Figure depict the result of belief revision with $\neg p$ in the doxastic epistemic state $(M, 00)$ of Figure 1, for, from left to right, five different forms of belief revision $*^1, *^2, *^3, *^4, *^5$. Note that only $*^2, *^4, *^5$ are successful. Typical examples of formulas valid in the original model M are $[*^1\neg p]Bq, [*^2\neg p]K\neg p, [*^3\neg p]B(p \vee q), [*^4\neg p]\neg B^1 p, [*^5\neg p]B(\neg p \wedge q)$. Details are found in the text.

$*\varphi$ should in that case not result in an information state where we believe φ. This explains the condition $\neg\Box^{\mathcal{X}}\neg\varphi$ for success. Second, for some formulas, such as 'Moore'-sentences $p \wedge \neg Bp$, belief revision is always unsuccessful. See Section 6 for more details.

For the remainder of this section, assume that all states are all doxastic epistemic models come with knowledge functions, in other words, that all states are plausible, that we have 'standard' degrees of belief B^x and knowledge K, and that preferences are global. We also assume that $\langle \mathcal{X}, < \rangle = \langle \mathbb{N}, < \rangle$, and that all epistemic classes map to prefixes of $\mathbb{N}$, or in other words, that there are no gaps between levels (apparently for some realistic revision scenarios this assumption cannot be made). To keep this feature after belief revision, we 'normalize' the computed preferences, i.e., we remove gaps that may have appeared between degrees and make the least degree 0, as in Example 31. Normalization can be made formal by a simple algorithm that we omit from this presentation. For expository purposes, we have left it implicit in the following definitions. An overview of the effect of the five example belief revision operators $*^1, *^2, *^3, *^4, *^5$ on the running example doxep state is found in Figure 9.[6]

4.1. *Minimal Belief Revision*

$$<^{*^1}(s) = <(s) \quad \text{if } M, s \vDash \varphi, \text{ and else } <^{*^1}(s) = <(s) + 1$$

See Figure 10. Both Bp and Bq hold (in the model) before revision $*^1\neg p$. After the revision, Bp no longer holds, but only Bq. The revision was not successful, as $B\neg p$ has not been achieved.

In minimal belief revision, the states where the revision formula does not hold are made one degree less likely. The degrees of the

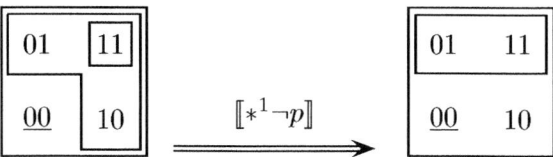

Figure 10. Minimal belief revision.

remaining states, i.e., the states where the revision formula holds, do not change. A normalization may have to take place, so that the least level is again 0. We call this belief revision minimal, because an even smaller increase in likelihood is not possible in this discrete setting. We do not have a reduction axiom that relates belief before and after the revision.

We think that $*^1$, even though it is not propositionally successful, is a realistic form of belief revision: A typical but, we think, often implicit assumption in 'belief revision' is that one *only* revises one's beliefs after being convinced that the new information is more reliable than the old information. In $*^1$ belief revision this assumption is not made. The new information is considered just as reliable as the old information. Therefore a $[*^1\varphi]$ operator is not idempotent – as in Figure 31. New information φ may reach us from different sources or agents. Each time it strengthens our $[*^1\varphi]$-induced support for φ. The assumption behind this sort of belief revision appears to be more common from a viewpoint of *belief merging*. A possibly less appealing consequence of $*^1$-revision is that also information from the same source increases in strength by mere repetition – a phenomenon with which parents with young children are not entirely unfamiliar. This may be considered of interest for modeling bounded rationality.

Though not necessarily successful, belief revision $*^1$ is indeed *eventually* propositionally successful. If φ is propositional and if $\neg K \neg \varphi$ holds (plausibility is a precondition for success, see Definition 30), then after some finite number of revisions with φ, $B\varphi$ will hold. In other words, if we were to define a more complex action language with a Kleene-$*$ operation on belief revision programs, such that $(*^1\varphi)^*$ stands for 'iterative belief revision with φ for some finite number' – we cannot avoid notational confusion here – then $\neg K \neg \varphi \rightarrow [(*^1\varphi)^*]B\varphi$ is a validity for propositional φ.

Belief revision $*^1$ is revocable. Valid is

$$\psi \rightarrow [*^1\varphi][*^1\neg\varphi]\psi$$

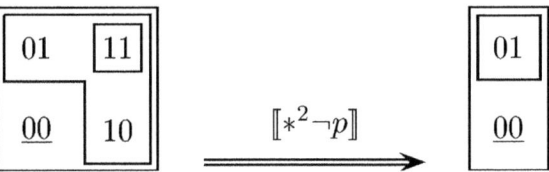

Figure 11.   Maximal belief revision.

In other words, by revising with the negation of the revision formula, the original doxastic epistemic state results. Belief revision $*^1$ is somewhat reminiscent of belief revision • in Darwiche and Pearl (1997), but there are important differences, e.g., the φ-states are treated differently in •.

4.2. *Maximal Belief Revision*

$$<^{*2}(s) = <(s) \text{ if } M, s \vDash \varphi \text{ and else } s \text{ is implausible.}$$

See Figure 11. Before the revision, both Bp and Bq hold, afterwards, $B\neg p$ and Bq. Note that in this case, it is essential that the actual state is one of 00 and 01 (namely 00), and not one of the remaining states, because they have become implausible. (For the other four belief revisions, it does not matter what the actual state is in this model, as from every state in the model the entire domain is plausible.) After $*^2\neg p$ revision, the degrees have to be normalized: the degree of state 00 was 2 and becomes 1; the degree of state 01 was 1 and becomes 0.

We call $*^2$ belief revision 'maximal belief revision', because the states where the revision formula is false are removed from the domain. As an operation on epistemic states – i.e., epistemic classes only without preferences – this form of 'belief revision' is known as 'truthful public announcement' (Plaza 1989; Baltag et al. 1998; Baltag and Moss, 2004). A reformulation of the standard definition for doxep models is as follows. In the definition, $[\![\varphi]\!]_M$ stands for $\{s \in S | (M, s) \vDash \varphi\}$.

DEFINITION 32 (Semantics of public announcement).

$$M, s \vDash [*^2\varphi]\psi \quad \text{iff} \quad M, s \vDash \varphi \text{ implies } M^{*2}, s \vDash \psi$$

where $M^{*2} = \langle S^{*2}, <^{*2}, V^{*2} \rangle$ is defined as (modulo normalization of $<^{*2}$)

$$S^{*2} \quad = [\![\varphi]\!]_M$$
$$<^{*2}(s) = <(s) \cap ([\![\varphi]\!]_M \times [\![\varphi]\!]_M)$$
$$V_p^{*2} \quad = V_p \cap [\![\varphi]\!]_M$$

Public announcement is propositionally successful. The interaction between belief revision operator $[*^2\varphi]$ and knowledge operator K is

$$[*^2\varphi]K\psi \leftrightarrow (\varphi \rightarrow K[*^2\varphi]\psi)$$

The interaction between belief revision operator $[*^2\varphi]$ and normal belief operator B is unclear (but should not be hard to find – though one has to take normalization into account). Public announcement is irrevocable.

4.3. *Majoring Belief Revision*

$$<^{*3}(s) = <(s) \text{ if } M, s \vDash \varphi, \text{ and else } <^{*3}(s) = \text{Max}(<(s), 1)$$

See Figure 12. Revision $*^3$ is not propositionally successful. Revision $*^3$ can be seen as a special case – namely for a two-point epistemic action – of a rather straightforward generalization – namely for more than one modal operator per agent – of an adaptation of the framework for epistemic action models of (Baltag and Moss 2004) – namely for epistemic actions with preferences between them, similar to those between states. As a revision with φ is a tentative announcement of φ, this can in principle be seen as a nondeterministic action consisting of two parts, and where the agent considers it more likely that we announce φ than that we announce $\neg\varphi$. He therefore gives the first action degree 0 and the second degree 1. In the definition of $<^{*3}$, we have taken the *maximum* of the degrees of these actions and the degrees of the states wherein they are executable. The reason for that will only become clear in Section 5, wherein we discuss this framework in more detail.

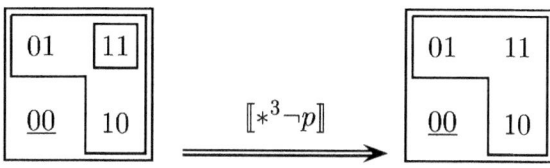

Figure 12.   Majoring belief revision.

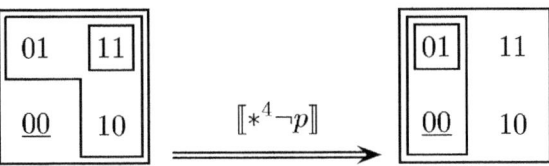

Figure 13.   Biased belief revision.

4.4. *Biased Belief Revision*

$$<^{*4}(s) = <(s) \text{ if } M, s \vDash \varphi, \text{ and else}$$
$$<^{*4}(s) = \mathsf{Max}\{<(t)|M, t \vDash \varphi\} + 1$$

See Figure 13. Belief revision $*^4$ is called biased, because the belief distinctions between φ-states are retained, whereas those between $\neg\varphi$-states are 'forgotten'. The set of all the $\neg\varphi$-states is made one degree less likely than the least likely of the φ-states. Belief revision $*^4$ was suggested in van Benthem (2003), in a different context of conditional implication, and it is not entirely faithful to that proposal. Because of the discrepancy, the reduction axiom suggested in van Benthem (2003) does not apply. Revision $*^4$ is successful.

4.5. *Successful Minimal Belief Revision*

$$<^{*5}(s) = <(s) - \mathsf{Min}\{<(t)|M, t \vDash \varphi\} \quad \text{if } M, s \vDash \varphi \text{ and else}$$
$$<^{*5}(s) = <(s) + 1 - \mathsf{Min}\{<(t)|M, t \vDash \neg\varphi\}$$

See Figure 14. Revision $*^5$ is a specific action that fits the more general framework by Aucher (2003), which is a dynamic epistemic setting according to Baltag and Moss (2004) of a proposal in Spohn (1988), and which is for a finite number of degrees of belief. We will give more details in Section 5. Belief revision $*^5$ is propositionally successful. The interaction axiom for belief and revision in Aucher (2003) is rather complex. It explicitly refers to all degrees of preference, and therefore does not appear to apply in the current setting with $\mathbb{N}$ degrees. Note that the transition in Figure 2 is the same.

5. DOXASTIC EPISTEMIC ACTIONS

In this section we outline some further generalizations. Just as for doxastic epistemic models, there are obvious multi-agent versions

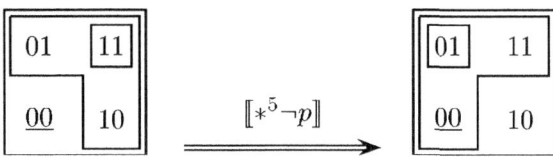

Figure 14. Successful minimal belief revision.

for dynamic belief revision. They are, in a concise presentation, as follows. A set of atoms P, an order $\langle \mathcal{X}, < \rangle$, and a set of agents N, are given.

DEFINITION 33 (Multi-agent dynamic belief revision). The language of multi-agent dynamic belief revision:

$$\varphi ::= p \mid \neg\varphi \mid \varphi \wedge \psi \mid \Box_n^x \varphi \mid \Box_n^{\mathcal{X}} \varphi \mid [\![*\varphi]\!]\psi$$

The semantics of multi-agent belief revision, for a given doxep model $M = \langle S, <, V \rangle$ and an $s \in S$:

$$M, s \vDash [\![*\varphi]\!]\psi \text{ iff for all } (M^*, s^*) : (M, s) [\![*\varphi]\!] (M^*, s^*)$$
$$\text{implies } M^*, s^* \vDash \psi$$

where $M^* = \langle S^*, <^*, V^* \rangle$ and $s^* \in S^*$ as before, and $<^*$ is a preference function for order $\langle \mathcal{X}, < \rangle$ that is computed from $<$ and φ by way of revising, *for each agent,* all $<_n^x (s)$. Belief revision $*$ is *successful on φ for agent n* if $\neg\Box_n^{\mathcal{X}}\neg\varphi \to [\![*\varphi]\!]\Box_n^0\varphi$ is valid. Belief revision $*$ is *propositionally successful for agent n* if it is successful for that agent for all propositional φ.

Similarly, one can define that belief revision $*$ is *successful on* φ for a group $G \subseteq N$ of agents, iff it is successful on φ for all agents in that group. Obviously, a belief revision mechanism $*$ may be successful on a formula for *some* agents but not for others.

One can go beyond that. The reader may recall that we primarily envisaged dynamic belief revision $*\varphi$ as a tentative public announcement of φ, where, so to speak, it was unclear whether φ or $\neg\varphi$ was actually the case. A different perspective on this action is that it consists of *two* actions, that cannot be told apart by the agent. The more likely one is an announcement of φ, and a less likely one 'casts doubt' on the first, and is an announcement of $\neg\varphi$. More complex scenarios where actions consist of more than two parts can easily be constructed, with or without a multi-agent setting. For example, a card player may show one of cards clubs, spades, and hearts, to another player, with the remaining players observing that a card is being shown, but not which card. Anne may then think it more likely that Bill plays clubs than spades, and spades more likely than hearts, whereas Cath may think it more likely that Bill plays spades than clubs. This calls for a generalized setting of doxastic epistemic actions, to be described in a 'dynamic doxastic epistemic logic'. Various approaches to dynamic epistemic logic, as already mentioned in the introduction, can be adjusted to that purpose. For an example, we outline an 'action model' approach as in Baltag and Moss (2004) and a 'relational action' approach as in van Ditmarsch et al. (2003). The action model with preferences approach originates with (Aucher 2003; Liu 2004), our setup is only a slight generalization of that.

An action model is like a Kripke model, but it is a dynamic version of that. Instead of a *valuation* of the points in the (static) domain, that determines which facts are true in the state for which this point stands, we now have *a precondition* for each point in the (dynamic) domain, that determines where the action for which that point stands can be executed – i.e., in which states of the (static) domain. (Instead of a precondition, one can also see this as the denotation of a precondition, if one wishes to separate syntax and semantics clearly.) A doxastic epistemic action is a pointed action model. The execution of the action consists of computing a restricted modal product of a doxastic epistemic state with the action, and this is how we can envisage 'dynamic belief revision'.

EXAMPLE 34. Given the 'running example' doxep state $(M, 00)$, we model belief revision with $*^1 \neg p$ as a doxep action. The action model consists of two 'basic actions' np and p with preconditions $\neg p$ and p, respectively, that are both considered plausible, and where np is considered more likely than p. (And where np really takes place, because the actual state is 00 wherein $\neg p$ is true.) Therefore, the degree of the action np is 0, and the less likely action p receives degree 1. Executing the action means computing a *restricted* modal product. The domain of the resulting doxep state consists of pairs (state, basic action). The product is restricted, because the domain is restricted to pairs where the precondition of the action is true in the state. See Figure 15 (and compare this with Figure 10). For example, $(01, np)$ is a pair in the domain of

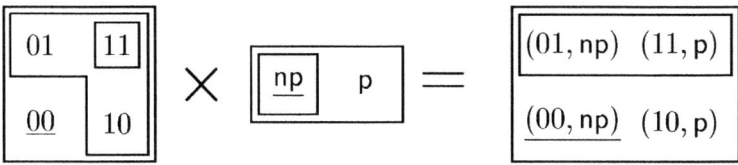

Figure 15. Belief revision as an action model. Performing the belief revision means computing a restricted modal product of a doxep state and a doxep action model. A 'state' in the resulting model is a pair (state, action) such that the precondition of 'action' is satisfied in 'state'. To compute its degree in the resulting model, add the degrees of 'state' and 'action', and normalize (start counting from 0 again).

the resulting epistemic state, because $(M, 01) \vDash \neg p$. The valuation of atoms in the new state $(01, \text{np})$ remains that of 01.

The degree, or weight, of a (state, action) pair in the resulting doxep model is a function of the degree of the state and the degree of the action. In this $*^1$-revision example, we simply *add* the degrees, and normalize. For example: $<^{*1} (11, \text{p}) = < (11) + < (\text{p}) = 0 + 1 = 1$ and $<^{*1} (01, \text{np}) = < (01) + < (\text{np}) = 1 + 0 = 1$. Similarly, the two remaining states get degree 2. Normalization results in $<^{*1}$ $(11, \text{p}) <^{*1} (01, \text{np}) = 0$, just as before, etc. This completes the computation of the new doxep state.

DEFINITION 35 (Doxastic epistemic action model). A *doxastic epistemic action model* **A** is a triple $\langle A, <, \text{pre} \rangle$. The set A is a domain of basic actions, and for each $a \in A$, $\text{pre}(a)$ is a formula that is the *precondition* for executing basic action a. The *preference function* $<: A \to \mathcal{P}(A \times A)$ defines *a preference relation* $<^a$ for each $a \in A$. The set *Plaus^a* of plausible actions given a should contain a least element, and there must be a *degree function* $<^a: \langle Plaus^a, <^a \rangle \hookrightarrow \langle \mathcal{X}, < \rangle$ that is an injection. A *doxastic epistemic action* is a pointed doxastic epistemic action model $(\langle A, <, V \rangle, a)$.

DEFINITION 36 (Doxastic epistemic action execution). Given are a *doxastic epistemic state* $(M, s) = (\langle S, <, V \rangle, s)$ and a *doxastic epistemic action* $\alpha = (\mathbf{A}, a) = (\langle A, <, \text{pre} \rangle, a)$. Assume that $M, s \vDash \text{pre}(a)$. The result of executing α in (M, s) is the doxastic epistemic state with underlying model $M \otimes A = \langle S^*, <^*, V^* \rangle$ that is defined as follows. Domain: all pairs (t, b) such that $M, t \vDash \text{pre}(b)$. Valuation: a fact p is true in (t, b) iff it is true in t, formally: $(t, b) \in V_p^*$ iff $t \in V_p$. Preference function $<^*$ is defined as a function $*$ of the preference function $<$ of M and the preference function $<$ of $\mathbf{A}$, i.e., $(<^*)^{(t',b')}(t,b) = *(<^{t'}(t), <^{b'}(b))$. The point of the resulting doxep state is (s,a).

In the remainder, assume that $\langle \chi, < \rangle = \langle \mathbb{N}, < \rangle$, and that the preference functions for doxep models and doxep actions are knowledge functions. Three different ways $*1$, $*3$, and $*5$ to define belief revision with epistemic actions are defined as:

$$<^{*1}(s,a) = <(s) + <(a)$$
$$<^{*3}(s,a) = \mathsf{Max}(<(s), <(a))$$
$$<^{*5}(s,a) = <(s) + <(a) - \mathsf{Min}\{<(t) | M, t \vDash \mathsf{pre}(a)\}$$

As usual, assume normalization of the $<^{*1}$, $<^{*3}$, $<^{*5}$ thus defined. These are natural generalizations of the dynamic belief revisions $<^{*1}$, $<^{*3}$, $<^{*5}$ in the previous section.

Belief revision $<^{*3}$ corresponds to the computation of modal access in the restricted product. To put it in the terminology of our setting, the way to compute access is by way of

$$(s,a) \to^x (t,b) \quad \text{iff } s \to^x t \text{ and } a \to^x b$$

For levels $x \leq y \in \chi$ we use that $\to^x \subseteq \to^y$ and we get

$$(s,a) \to^{\mathsf{Max}(x,y)} (t,b) \quad \text{if } s \to^x t \text{ and } a \to^y b$$

If we do that for the minimum degree x given s,t, respectively, y given a,b, we get

$$(s,a) \to^{\mathsf{Max}(x,y)} (t,b) \quad \text{iff } s \to^x t \text{ and } a \to^y b$$

This directly corresponds to

$$<^{*3}(s,a) = \mathsf{Max}(<(s), <(a))$$

Belief revision $<^{*5}$ is the one presented in Aucher (2003) (which is similar to Aucher 2005a), and is based on Spohn (1988). Aucher also provides a completeness result for the logic. See also Liu (2004). Just as for dynamic belief revision $*\varphi$ with a single formula, also for this generalization of dynamic belief revision there are many other options than $*^1$, $*^3$, and $*^5$: two more proposals for belief revision can be based on two other suggestions also found in Spohn (1988), that can be reformulated in the terms presented here. Another interesting approach, possibly allowing reformulation in our terms as well, is based on similarity of valuations (Herzig et al. 2005).

An alternative to the 'action model' approach is the 'relational action' approach in van Ditmarsch (2002) and van Ditmarsch et al. (2003). We outline the semantics of 'belief revision' in that setting, too. Belief revision is now seen as a *basic action constructor* $*\varphi$. This is defined as a (doxep) state transformer, that results in changed preferences, as before. But unlike in the Definitions 29 and 33 we *postpone* computing agent access from those preferences. In those definitions the assumption was that 'the agent' or 'all agents' are aware of the revision taking place. Now, instead, the awareness of the agents of the revision may vary. To express that, there are other constructs in the language, called 'learning' operators L_G (for a group of agents G). The preferences computed in $*\varphi$ are 'activated' by such learning operators. Public learning – for all agents – is L_N. Belief revision $*\varphi$ as in Definition 33 is, from that perspective, described as $L_N * \varphi$ – as this is observed by all agents. Surprisingly, one appears to be able to plug in all the previously distinguished notions $*^1, \dots, *^5$ again, in this relational setting. The constructor $*\varphi$ is a generalization of the 'test' operator $?\varphi$ in the relational action language, and test $?\varphi$ has now become the special case $*^2\varphi$. In this more general language, public announcement $*^2\varphi$ is described as $L_N *^2 \varphi$ – for 'everybody learns (L_N) that φ is true $(*^2\varphi)$', in terms of this language: as $L_N?\varphi$. And, for another example, the specific revision with $*^1\neg p$ in our running example is now described as $L *^1 \neg p$ – with just L for a single anonymous agent. The transition is, now for the last time, visualized in Figure 16. In a multi-agent setting more appealing examples can be found, such as the 'card show action with preferences' above.

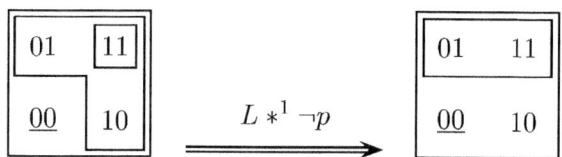

Figure 16. Example of a relational doxastic epistemic action.

6. THE AGM REQUIREMENTS IN A DYNAMIC LOGICAL SETTING

The standard reference for the well-known eight AGM postulates for belief revision of a theory $\mathcal{K}$ with a formula φ is Alchourrón et al. (1985). A *theory* $\mathcal{K}$ is a deductively closed set of formulas in the logical language. The *trivial theory* $\mathcal{K}_\perp$ is the set of all formulas. Details of AGM belief expansion (as in $\mathcal{K} + \varphi$, in postulate *3) are not given; for that, see also Alchourrón et al. (1985).

$*_{\text{agm}}1$	$\mathcal{K} * \varphi$ is a theory	type
$*_{\text{agm}}2$	$\varphi \in \mathcal{K} * \varphi$	success
$*_{\text{agm}}3$	$\mathcal{K} * \varphi \subseteq \mathcal{K} + \varphi$	upper bound
$*_{\text{agm}}4$	if $\neg\varphi \notin \mathcal{K}$, then $\mathcal{K} + \varphi \subseteq \mathcal{K} * \varphi$	lower bound
$*_{\text{agm}}5$	$\mathcal{K} * \varphi = \mathcal{K}_\perp$ iff φ is inconsistent	triviality
$*_{\text{agm}}6$	if φ is equivalent to ψ then $\mathcal{K} * \varphi = \mathcal{K} * \psi$	extensionality
$*_{\text{agm}}7$	$\mathcal{K} * (\varphi \wedge \psi) \subseteq (\mathcal{K} * \varphi) + \psi$	iteration upper bound
$*_{\text{agm}}8$	if $\neg\psi \notin \mathcal{K} * \varphi$, then $(\mathcal{K} * \varphi) + \psi \subseteq \mathcal{K} * (\varphi \wedge \psi)$	iteration lower bound

In the semantic setting of dynamic logic, the phrasing of the AGM postulates tends to be somewhat different, and a clear unequivocal equivalent is not necessarily available. But the main idea is that the postulates can be required for propositional formulas. For details of that we refer to their discussion in Segerberg 1999a, b; Lindström and Rabinowicz 1999), as their dynamic setup is not dissimilar to ours, for propositional formulas. We therefore focus on the non-propositional, purely epistemic formulas, for which the postulates *cannot* be universally required (this is also partly discussed in Lindström and Rabinowicz 1999). The multi-agent setting for the AGM postulates does not pose extra technical or conceptual complications. Therefore we restrict ourselves, as usual, to the single-agent situation.

In the remainder, let (M, s) be a doxep state (pointed Kripke-model) such that for all $\psi \in \mathcal{K}$, $(M, s) \vDash \psi$, and let, as before, φ be the revision formula. We have not given postulates for expansion (which features in $*_{\text{agm}}7$ and $*_{\text{agm}}8$): we merely need to assume that at least $B\varphi$ should hold in a doxep state resulting from expansion with φ.[7]

6.1. *Postulate* $*_{\text{agm}}1$

In case $(M, s) \nvDash \varphi$ by definition $(M, s) \vDash [*\varphi]\psi$ for all formulas ψ, so in particular for all ψ of the form $B\chi$ – where again χ can be an arbitrary formula. So the trivial theory of all formulas is then the result of belief revision, and this is a belief set. This can of course only be the case for belief revision operations that are partial functions, such as 'public announcement', i.e., $*^2$ belief revision. In the remaining $*^i$ example belief revisions, the revision is a total function, so that the theory of which the resulting doxep state (M^*, s) is a model, is obviously non-trivial.

The concept of a set of formulas that is not deductively closed, makes no sense in this semantic setting – so in that respect we need have no fears whether $K * \varphi$ is a 'theory'.

6.2. *Postulate* $*_{\mathsf{agm}}2$

This postulate cannot be generally required in a dynamic logical setting. As already mentioned in the introductory section, some revisions are never successful, the poignant example is $p \wedge \neg Bp$. This is because for modal belief operators B, the formula $B(p \wedge \neg Bp)$ is inconsistent.[8]

Success of revisions with propositional formulas may be required (see also the discussion ad $*_{\mathsf{agm}}3$ and $*_{\mathsf{agm}}4$). In $*^1$, as we have seen, this is not necessarily the case, by design; but in $*^2$, $*^4$, and $*^5$, it is.

6.3. *Postulates* $*_{\mathsf{agm}}3$ *and* $*_{\mathsf{agm}}4$

Together, these postulates describe 'minimal change', and in particular, that contraction is only required when contrary beliefs are held. In a higher-order belief setting, this is – once more – a problematic requirement, because knowledge and belief cannot properly said to increase. This is, because ignorance ('disbelief') of the revision formula φ always has to be retracted in order to avoid inconsistency, unless φ was already believed, so that revision is trivial. For example, if we revise with atom p and $M, w \vDash B\neg Bp$ holds before revision, then $B\neg Bp$ cannot *also* be true after (successful) revision after which Bp is true and therefore also BBp.

Still, some aspects of 'minimal change' correspond to requirements on modalities. In case the agent believes $\neg \varphi$, a contraction is required prior to expansion with φ. In our terms, this is therefore required when $M, s \vDash B\neg \varphi$. The condition $\neg \varphi \notin K$ in $*_{\mathsf{agm}}4$ translates into $M, s \nvDash B\neg \varphi$, i.e., $M, s \vDash \neg B\neg \varphi$. This says, that there is at least one state among the normally believed states (worlds) that satisfies φ. In that case, a revision with φ 'should not' require a contraction prior to expansion – so that $K + \varphi \subseteq K * \varphi$. Even though this requirement cannot – again – be made in general, it is reasonable to require this for *propositional* revision formulas. We may require that no contraction of facts should be needed if they are already 'considered normally possible' (in the above sense). This is indeed satisfied by all proposed revision operators: this is, because the truth of a propositional φ depends on single states *only*, because such states are (trivially) contained in the extension of that φ in (M, s), and because the φ-states never become *less* plausible in all proposed $*^i$.

There is yet another respect in which dynamic belief revision models 'minimal change', even though this is not expressed in the $*_{\mathsf{agm}}3$ and $*_{\mathsf{agm}}4$ postulates. We model belief *and* knowledge, and knowledge is irrevocable. If φ is *known* to be false, revision with φ should 'have no effect', even if it can be executed in some way. In modal terms this, is the requirement (already expressed in Definition 30 namely) that $(M, s) \vDash \neg K \neg \varphi \leftrightarrow [*\varphi]B\varphi$: a necessary and sufficient requirement for successful (propositional) belief revision is that φ is 'considered plausible', even when not (normally) believed.[9]

6.4. *Postulate* $*_{\mathsf{agm}}5$

If φ is inconsistent, $(M, s) \vDash \varphi$ is false, and therefore $(M, s) \vDash [\varphi]B\chi$ is true for all formulas χ (see the discussion ad postulate $*_{\mathsf{agm}}1$), so indeed $\mathcal{K} * \varphi = \mathcal{K}_\perp$.

6.5. *Postulate* $*_{\mathsf{agm}}6$

If φ is equivalent to ψ, then $(M, s) \vDash \varphi$ iff $(M, s) \vDash \psi$. So, whatever the proposed revision operator, for all $\chi : (M, s) \vDash [*\varphi]B\chi$ iff $(M, s) \vDash [*\psi]B\chi$. In other words: extensionality is trivially satisfied, as all belief revision operators operate on the value of the revision formula in the current doxep state.

6.6. *Postulates* $*_{\mathsf{agm}}7$ *and* $*_{\mathsf{agm}}8$

In this case we encounter similar problems as for $*_{\mathsf{agm}}3$ and $*_{\mathsf{agm}}4$: reasonable corresponding requirements are satisfied for propositional formulas, but not in general. And we run into new problems as well: suppose the revision formula $\varphi \wedge \psi$ in postulates $*_{\mathsf{agm}}7$ and $*_{\mathsf{agm}}8$ is $\neg Bp \wedge p$, and suppose $M, s \vDash \neg Bp \wedge p$, so that the revision is executable and a doxep state (and therefore a non-trivial theory) results. Further, assume that $M, s \nvDash B \neg p$. For the theory $\mathcal{K}$ of which doxep state (M, s) is a model we now have: $\mathcal{K} * \neg Bp$ is non-trivial, but $(\mathcal{K} * \neg Bp) + p$ is trivial (it contains both $\neg Bp$ and Bp. So $*_{\text{agm}}7$ is satisfied. The premise of $*_{\text{agm}}8$ is also satisfied. We can see this as follows. Given that $M, s \nvDash B \neg p$, this will persist after revision $* \neg Bp$ on grounds of *minimal* change. Therefore also $M, s \nvDash [* \neg Bp]B \neg p$, in other words: $\neg p \notin \mathcal{K} * \neg Bp$. But on the other hand $(\mathcal{K} * \neg Bp) + p \nsubseteq (\mathcal{K} * (\neg Bp \wedge p))$. So the conclusion of $*_{\text{agm}}8$ is *not* satisfied, so $*_{\text{agm}}8$ is not satisfied. Ouch.

Similar concerns exist for generalizations of the $[*\varphi]$ belief revision to general epistemic actions, for a discussion of the AGM postulates in this context, see Aucher (2003). If revision with propositional formulas satisfies the AGM postulates, but revision with arbitrary formulas not, the obvious question is for which language fragment the postulates are still satisfied. There are several interesting fragments to consider, such as the *positive formulas* (Meyer and van der Hoek 1995), the *preserved formulas* (van Benthem 2002), and the *successful formulas* (van Ditmarsch and Kooi 2005). We think these questions are not easy to answer, e.g., for the successful formulas a syntactic characterization is not even known.

7. CONCLUSIONS AND FURTHER RESEARCH

We have provided a dynamic approach to belief revision. Belief revision with φ is modeled as a dynamic operator $[*\varphi]$ that is interpreted as a binary relation between information states for knowledge and degrees of belief. As such information states we proposed doxastic epistemic states, that can be built from preferences between plausible states. Thus, we can reason about degrees of belief and knowledge, and change of belief and knowledge. This includes higher-order belief change, iterated belief revision, and revocable belief revision. We have given multi-agent versions of such belief revision, and we have outlined a generalization of belief revision to epistemic actions with preferences.

There seems to be a wealth of opportunities for further research. The logics of the various proposals in this contribution are yet unclear, the generalization of the semantics to include epistemic actions is only outlined, a generalization to preferences that are not totally but only partially ordered is on the agenda and seems quite feasible, modal operators for groups such as common knowledge and collective belief can easily be added to the language, it is unclear which language fragments precisely satisfy the AGM postulates for a given revision operator, and – last but by no means least – there are various concerns on the relation between plausibility and probability.

ACKNOWLEDGMENTS

I thank for their comments, during and after various presentations on this topic, and on different versions of this manuscript: Geir Asheim, Guillaume Aucher, Johan van Benthem, Oliver Board, Giacomo Bonanno, Jan Broersen, Balder ten Cate, Jan van Eijck, Valentin Goranko, Johannes Heidema, Andreas Herzig, Wiebe van der Hoek, Barteld Kooi, Erik Krabbe, Willem Labuschagne, Jérôme Lang, Maurice Pagnucco, Steven Shapiro, Allard Tamminga, Sieuwert van Otterloo, Robert van Rooij, Hans Rott, Rineke Verbrugge, Frank Wolter, and – last but not least – an anonymous referee supplied by the Journal Knowledge, Rationality, an Action.

Apart from that, specific (not otherwise referenced) credits go the following persons. Willem Labuschagne started my involvement in this subject matter. The setting of Example 23 is one of his all-time favorites – need I say more. Johan van Benthem's interest inspired me, and he involved me in Guillaume Aucher's MSc. supervision. Guillaume Aucher suggested Example 14, corrected a crucial error in Definition 10 and also states that relation in work in progress, established Proposition 18 independently by a different proof. Barteld Kooi gave brilliant opposition during the Changing Minds workshop. Giacomo Bonanno wisely filled an important gap in the references for this contribution; this resulted in many changes, and most importantly in the addition of the appendix containing the detailed comparison with (Board 2004). Constructive suggestions by the anonymous referee resulted in the addition of a section listing the (partial) fulfilment of AGM requirements. Wiebe van der Hoek suggested that $*^1$ belief revision is eventually successful. Hans Rott observed the relation between $*^1$ belief revision and • belief revision in Darwiche and Pearl (1997).

APPENDIX A: BELIEF REVISION AS CONDITIONAL MODAL OPERATORS

A different way to model belief revision in modal logic, but that falls outside the dynamic modal framework that we present, is to interpret belief revision as conditional or counterfactual reasoning. The proposition "ψ holds after belief revision with φ" is then interpreted as "in case φ were true (even though we currently may not have evidence for that), we would believe ψ (even though we currently may believe something else)". In an epistemic logic this can be described as $B^\varphi \psi$. What is the difference between $[*\varphi]\psi$ and $B^\varphi \psi$?

Work that falls under the scope of conditional belief revision includes Stalnaker (1996), Board (2004), Asheim and Søvik (2005) and Bonanno (2005). In fact, Stalnaker (1996) does not introduce a logical language, and in Bonanno (2005) the dependence of new beliefs on received information is by the intermediation of yet another modal operator $\mathcal{I}$, but all these approaches have in common that a belief in ψ *after* revision can be evaluated by somehow localizing (by modal access, or otherwise) a subset of the domain in the information state (Kripke model) *before* revision where that formula ψ should hold.

This philosophy is rather different from that of dynamic logic, where "ψ holds after belief revision with φ" is interpreted as "processing the information φ changes the information state, and in the resulting information state ψ holds". Apart from the references in the introductory section we mention as well (Aucher 2003, 2005a, b; van Ditmarsch and Labuschagne 2003; Herzig et al. 2005; Cantwell 2005).

The difference between these approaches can be seen in terms similar to those with which physicists view their experiments: conditional belief revision takes the non-interfering 'observer's' point of view, whereas dynamic belief revision can be said to take the invasive 'participant's' point of view. Typically (but not necessarily, Bonanno (2005) is an exception) iterated belief revision cannot be modeled in conditional belief revision, because it is unclear what the *new* conditions are upon which such an iteration should be based, as there is no update of preferences. But in dynamic belief revision this is a matter of course: we have arrived in a new information state, that contains all features to perform another belief revision: all preferences in all states have been updated.

A poignant example that explains the difference between $[*\varphi]\psi$ and $B^\varphi\psi$, is when we revise with $p \wedge \neg Bp$ in a model where this formula is already valid, e.g., in the two-point model where the agent cannot distinguish between p and $\neg p$, and when we take public announcement as dynamic belief revision. We then have that

$$M, s \not\models [p \wedge \neg Bp](p \wedge \neg Bp)$$

The postcondition of the announcement is evaluated in a *different* model, wherein p is now believed. And as Bp is then true, $(p \wedge \neg Bp)$ must be false. On the other hand, in conditional belief revision we have

$$M, s \models B^{p \wedge \neg Bp}(p \wedge \neg Bp)$$

As the condition was already satisfied throughout the model, it still is in the restriction of the model to the condition. It should, as such conditional belief revision is successful anyway.

We now list some specific differences between the conditional approaches mentioned and ours. The structures proposed in Stalnaker (1996) are similar to doxep structures with belief functions (i.e., *with* global preferences that induce KD45 belief operators). Stalnaker models knowledge as *justified true belief* – which results in an $S4.3$ knowledge operator. We, instead, model knowledge as *true belief,* which results in an $S5$ knowledge operator. This is the stance typical in computer science, and for modeling multi-agent systems, and it is a generalization (namely for different degrees of belief) of logics that combine belief and knowledge (Kraus and Lehmann 1988). Board's belief revision structures (Board 2004) are the same as our doxastic epistemic structures (i.e., *without* global preferences), and his careful setup introducing conditions R1, R2, R3, and R4 and (part of) their modal correspondence, is also strikingly similar to our approach, introducing first global preferences and then knowledge. So in this respect, what we have to say is already sufficiently and beautifully expressed in Board (2004). There are also some striking differences, which, we think, validates our alternative exposition. Unlike (Board 2004), we do not require that preferences are well-ordered (a restriction enforced by his conditional view of belief revision), and in this we stay closer to the 'limit assumption' in Lewis (1973); we only require that each totally ordered $<^s$ has a minimal element. We tend to find our definition of knowledge as interpreted by $\rightarrow^x$ original. Board follows Stalnaker, concerning knowledge. Unlike Board, because of the *dynamic* nature of our belief revision, we need a total order χ to be a *parameter* for doxep structures, ensuring that doxep structures resulting from belief revision adhere to the same parameter.

Obviously, operators B^φ do not correspond in any obvious way to dynamic revision operators $[*\varphi]$, as already explained above. But, interestingly, they correspond to *degrees of belief* $B^x\varphi$.[10] For a simple case, take $x = 0$, for most normal belief, and $\varphi = p$. We then have that:

$$B_n^0 p \quad \text{iff} \quad B_n^{\text{true}} p$$
$$B_n^1 p \quad \text{iff} \quad B_n^{\text{true}} p \wedge B_n^{\neg B_n^{\text{true}} p} p$$

which may help to understand the following global embedding t of the *static* part of our language (for the special case of an enumerable total order) into Board's. All clauses are trivial, except:

$$t(B_n^0 \varphi) = B_n^{\text{true}} t(\varphi)$$
$$t(B_n^{x+1} \varphi) = t(B_n^x \varphi) \wedge B_n^{\neg t(B_n^x \varphi)} t(\varphi)$$

This correspondence makes immediately clear that our axioms of *arbitrary positive introspection* and *arbitrary negative introspection* are a special case of Board's axioms TPI and TNI. We do not know whether Board's language can be similarly embedded into the *static* part of our language. We conjecture that our full language, i.e., including the *dynamic* part, i.e., the dynamic belief revision operators, is more expressive than Board's. This would require demonstrating that our language can distinguish two doxep structures that cannot be distinguished (are bisimilar) in Board's language – an interesting but fairly formal exercise that goes beyond this exposition. The conjecture appears corroborated by results that show that dynamic epistemic logics are more expressive than their non-dynamic counterparts (Baltag et al. 1998).

Board's conditional *individual* belief operators B_n^φ seem to be generalizable to conditional *common* belief operators C_G^φ, for arbitrary sets of agents. He currently only introduces 'ordinary' common belief C which is C^{true}, for the group of all agents. Such a generalization might enhance the expressivity of his language. Recent work in van Benthem et al. (2005) shows that conditional *common* knowledge is more expressive than certain dynamic epistemic logics.

NOTES

[1] The correspondence between the K in $K\varphi$, for 'Knowing that φ', and the K of a theory $\mathcal{K}$, a set of formulas that is deductively closed, is a coincidence. We have simply followed tradition.

[2] This contribution is titled 'Prolegomena to Dynamic Logic for Belief Revision' and not just 'Dynamic Logic for Belief Revision'. This is because we venture to see it as an outline for research in dynamic logic for belief revision (see the detailed suggestions in Section 7), by providing a sound semantic foundation. Therefore, they are proper 'prolegomena' in the sense of what should be said before starting the real work. Although there are many Prolegomena, the one that we have in mind and that inspired us is (Ibn Khaldun 1938), praised throughout the centuries as a great work by the scholar Ibn Khaldun. He taught at the Al-Azhar University in Cairo, around 1400 AD. The Al-Azhar may well be the oldest university of the world in the sense of having a continuous tradition of learning – the mosque of the same name to which this center of learning is associated was inaugurated in 972 AD. It may be interesting to observe, and justify the inclusion of this digression, that Ibn Khaldun discusses the nature of human knowledge, and also (Aristotelian) logic (see (Ibn Khaldun 1938), 2:433ff. and 3:149ff.). Both positive introspection, and knowledge as justified belief, appear to be mentioned. Unfortunately we did not find a reference smelling of negative introspection. Positive introspection serves as an alternative proof of the existence of the human soul. To put it in contemporary terminology: when we observe facts p describing the world, as in Kp, we are not different from the animals, that also make such observations, But when we observe knowledge, as in KKp, what is the object of our observations? We then observe our spiritual nature, and this proves the existence of the human soul.

[3] This assumption is not universal in the literature. In a somewhat different perspective – namely of similarity to the actual world – the actual state is always plausible, and always in the innermost circle or bottom level of 'most normal' states at that. See Lewis (1973).

[4] We owe our view on implausible states to Ferguson and Labuschagne (2002). (Lewis 1973) gives similar considerations for and against extending ordered preferences to the entire domain. He appears to remain in doubt on what is best: "An alternative method would be to let $\leq_i$ [the preference relation] be an ordering not of all worlds [states], but only of accessible worlds [plausible states], so that S_i [the set of plausible states] could be defined as the field of the relation $\leq_i$; but this method is even clumsier" (Lewis 1973, p. 50).

[5] One may wonder what happens after belief revision. It seems quite straightforward to show that the dynamic belief revision operators defined in later sections preserve bisimilarity of (the multi-modal correspondent of) doxastic epistemic states, but we do not address that matter in detail.

[6] We only incidentally give reduction axioms relating $[*\varphi]$ and B^x and K. We expect proof systems for a logic with belief revision $[*\varphi]$ not to be much more complex than the logic without. We conjecture that only such reduction axioms need to be added, after which one may well be able to reduce derivability in this logic to that in the logic without belief revision, just as in the logic of public announcements without common knowledge (Plaza 1989).

[7] The AGM postulates for expansion also prove the so-called 'Gärdenfors Impossibility Theorem', which states that expansions cannot always be added in a non-trivial way, not even if they are separately non-trivial. See Gärdenfors (1986), and, generalizing these results, (Segerberg 1989). These results appear to hold similarly for the dynamic setting of belief revision, as also in this setting such postulates at least apply to propositional formulas.

[8] It entails both Bp and $B\neg Bp$; from Bp follows BBp, and from $B\neg Bp$ and BBp follows $B(Bp \wedge \neg Bp)$. In other words, $B_\perp$, but given seriality $B\varphi \to \neg B\neg\varphi$ this entails $\neg B\neg\perp$; that says that a doxep state is considered possible wherein $\perp$ is true, a contradiction.

[9] This constraint is also given in conditional belief revision approaches where knowledge and belief are both modeled (Stalnaker 1996; Board 2004).

[10] Board does not distinguish degrees of belief, and in particular not 'conviction'. For that, see the setup for 'certain belief' in Asheim and Søvik (2005). Asheim et al. use, again, fairly though less strikingly similar structures.

REFERENCES

Alchourrón, C., P. Gärdenfors, and D. Makinson: 1985, 'On the Logic of Theory Change: Partial Meet Contraction and Revision Functions', *Journal of Symbolic Logic* **50**, 510–530.

Asheim, G. and Y. Søvik: 2005, 'Preference-Based Belief Operators', *Mathematical Social Sciences* **50**(1), 61–82.

Aucher, G.: 2003, 'A Combined System for Update Logic and Belief Revision', Master's thesis, ILLC, University of Amsterdam, Amsterdam, the Netherlands.

Aucher, G.: 2005a, 'A Combined System for Update Logic and Belief Revision'. In M. Barley and N. Kasabov (eds.), *Intelligent Agents and Multi-Agent Systems – 7th Pacific Rim International Workshop on Multi-Agents (PRIMA 2004)*. pp. 1–17, Springer. LNAI 3371.

Aucher, G.: 2005b, 'How Our Beliefs Contribute To Interpret Actions', To appear in the Proceedings of CEEMAS, see www.ceemas.org/ceemas05/.

Baltag, A.: 2002, 'A Logic for Suspicious Players: Epistemic Actions and Belief Updates in Games', *Bulletin of Economic Research* **54**(1), 1–45.

Baltag, A. and L. Moss: 2004, 'Logics for Epistemic Programs', *Synthese* **139**, 165–224. *Knowledge, Rationality and Action* 1–60.

Baltag, A., L. Moss, and S. Solecki: 1998, 'The Logic of Common Knowledge, Public Announcements, and Private Suspicions'. In I. Gilbao (ed.), *Proceedings of the 7th conference on theoretical aspects of rationality and knowledge (TARK 98)*, pp. 43–56.

Board, O.: 2004, 'Dynamic Interactive Epistemology', *Games and Economic Behaviour* **49**, 49–80.

Bonanno, G.: 2005, 'A Simple Modal Logic for Belief Revision', *Knowledge, Rationality and Action*, this volume.

Cantwell, J.: 2005, 'A Formal Model of Multi-Agent Belief-Interaction', *Journal of Logic, Language, and Information*. To appear.

Darwiche, A. and J. Pearl: 1997, 'On the Logic of Iterated Belief Revision', *Artificial Intelligence* **89** (1–2), 1–29.

Fagin, R., J. Halpern, Y. Moses, and M. Vardi: 1995, *Reasoning about Knowledge*, MIT Press, Cambridge MA.

Ferguson, D. and W. Labuschagne: 2002, 'Information-Theoretic Semantics for Epistemic Logic', In *Proceedings of LOFT 5*. Turin, Italy, ICER.

Gärdenfors, P.: 1986, 'Belief Revisions and the Ramsey test for Conditionals', *The Philosophical Review* **XCV**(1), 81–93.

Gärdenfors, P.: 1988, *Knowledge in Flux: Modeling the Dynamics of Epistemic States*, Bradford Books, MIT Press, Cambridge, MA.

Gerbrandy, J.: 1999, 'Bisimulations on Planet Kripke', Ph.D. thesis, University of Amsterdam. ILLC Dissertation Series DS-1999-01.

Gerbrandy, J. and W. Groeneveld: 1997, 'Reasoning about Information Change', *Journal of Logic, Language, and Information* **6**, 147–169.

Grove, A.: 1988, 'Two Modellings for Theory Change', *Journal of Philosophical Logic* **17**, 157–170.

Halpern, J.: 2001, 'Lexicographic Probability, Conditional Probability, and Non-standard probability', In *Proceedings of the Eighth Conference on Theoretical Aspects of Rationality and Knowledge (TARK 8)*, pp. 17–30.

Halpern, J.: 2003, *Reasoning about Uncertainty*. MIT Press, Cambridge MA.

Herzig, A., J. Lang, and P. Marquis: 2005, 'Revision and Update in Multiagent Belief Structures', Manuscript, also presented at the LOFT 6 conference, see http://www.econ.ucdavis.edu/faculty/bonanno/LOFT6.html.

Ibn Khaldun: 1938, *Les prolégomènes d'Ibn Khaldoun*, traduits en français et commentés par M. de Slane. Paris: Librairie Orientaliste Paul Geuthner. Three volumes, published in 1934, 1936, and 1938.

Konieczny, S. and R. P. Pérez: 2002, 'Merging Information under Constraints: A Logical Framework', *Journal of Logic and Computation* **12**(5), 773–808.

Kooi, B.: 2003, 'Knowledge, Chance, and Change', Ph.D. thesis, University of Groningen. ILLC Dissertation Series DS-2003-01.

Kraus, S. and D. Lehmann: 1988, 'Knowledge, Belief and Time', *Theoretical Computer Science* **58**.

Kraus, S., D. Lehmann, and M. Magidor: 1990, 'Nonmonotonic Reasoning, Preferential Models and Cumulative Logics', *Artificial Intelligence* **44**, 167–207.

Lenzen, W.: 2003, 'Knowledge, Belief, and Subjective Probability: Outlines of a Unified System of Epistemic/Doxastic Logic', In V. Hendricks, K. Jorgensen, and S. Pedersen (eds.), *Knowledge Contributors*, Dordrecht, pp. 17–31, Kluwer Academic Publishers, Synthese Library Volume 322.

Lewis, D.: 1973, *Counterfactuals*, Harvard University Press, Cambridge MA.

Lindström, S. and W. Rabinowicz: 1999, 'DDL Unlimited: Dynamic Doxastic Logic for Introspective Agents', *Erkenntnis* **50**, 353–385.

Liu, F.: 2004, 'Dynamic Variations: Update and Revision for Diverse Agents', Master's thesis, ILLC, University of Amsterdam, Amsterdam, the Netherlands.

Meyer, J.-J. and W. van der Hoek: 1995, *Epistemic Logic for AI and Computer Science*, Cambridge Tracts in Theoretical Computer Science Vol. 41. Cambridge University Press, Cambridge.

Meyer, T.: 2001, 'Basic Infobase Change', *Studia Logica* **67**, 215–242.

Meyer, T., W. Labuschagne, and J. Heidema: 2000, 'Refined Epistemic Entrenchment', *Journal of Logic, Language, and Information* **9**, 237–259.

Moses, Y. O., D. Dolev, and J. Y. Halpern: 1986, 'Cheating Husbands and Other Stories: A Case Study in Knowledge, Action, and Communication', *Distributed Computing* **1**(3), 167–176.

Plaza, J.: 1989, 'Logics of Public Communications', In M. Emrich, M. Pfeifer, M. Hadzikadic, and Z. Ras (eds.), *Proceedings of the 4th International Symposium on Methodologies for Intelligent Systems*. pp. 201–216.

Segerberg, K.: 1989, 'A Note on an Impossibility Theorem of Gärdenfors', *Noûs* **23**, 351–354.

Segerberg, K.: 1999a, 'Default Logic as Dynamic Doxastic Logic', *Erkenntnis* **50**, 333–352.

Segerberg, K.: 1999b, 'Two Traditions in the Logic of Belief: Bringing them Together', In H. Ohlbach and U. Reyle (eds.), *Logic, Language, and Reasoning*, Kluwer Academic Publishers, Dordrecht, pp. 135–147.

Spohn, W.: 1988, 'Ordinal Conditional Functions: A Dynamic Theory of Epistemic States', In W. Harper and B. Skyrms (eds.), *Causation in Decision, Belief Change, and Statistics*, Vol. II, pp. 105–134.

Stalnaker, R.: 1996, 'Knowledge, Belief and Counterfactual Reasoning in Games', *Economics and Philosophy* **12**, 133–163.

ten Cate, B.: 2002, 'Internalizing Epistemic Actions', In M. Martinez (ed.), *Proceedings of the NASSLLI-2002 Student Session*, Stanford University.

van Benthem, J.: 1996, *Exploring Logical Dynamics*. CSLI Publications.

van Benthem, J.: 2002, 'One is a Lonely Number: On the Logic of Communication', Technical report, ILLC, University of Amsterdam. Report PP-2002-27 (material presented at the Logic Colloquium 2002).

van Benthem, J.: 2003, 'Qualitative Belief Revision'. Manuscript.

van Benthem, J., J. van Eijck, and B. Kooi: 2005, 'Logics of Communication and Change'. Manuscript.

van Ditmarsch, H.: 2000, 'Knowledge Games'. Ph.D. thesis, University of Groningen. ILLC Dissertation Series DS-2000-06.

van Ditmarsch, H.: 2002, 'Descriptions of Game Actions', *Journal of Logic, Language and Information* **11** 349–365.

van Ditmarsch, H. and B. Kooi: 2005, 'The Secret of My Success', *Synthese*. To appear.

van Ditmarsch, H. and W. Labuschagne: 2003, 'A Multimodal Language for Revising Defeasible Beliefs', In E. Álvarez, R. Bosch, and L. Villamil (eds.), *Proceedings of the 12th International Congress of Logic, Methodology, and Philosophy of Science (LMPS)*, Oviedo University Press, pp. 140–141.

van Ditmarsch, H., W. van der Hoek, and B. Kooi: 2003, 'Concurrent Dynamic Epistemic Logic', In V. Hendricks, K. Jørgensen, and S. Pedersen (eds.), *Knowledge Contributors*, Kluwer Academic Publishers, Dordrecht, pp. 45–82, *Synthese Library Volume 322*.

van Ditmarsch, H., W. van der Hoek, and B. Kooi: 2004, 'Public Announcements and Belief Expansion', In R. Schmidt, I. Pratt-Hartmann, M. Reynolds, and H. Wansing (eds.), *Proceedings of AiML-2004 (Advances in Modal Logic)*, University of Manchester, pp. 62–73.

Hans P. van Ditmarsch
University of Otago
Dunedin, New Zealand
E-mail: hans@cs.otago.ac.nz

NOËL LAVERNY and JÉRÔME LANG

FROM KNOWLEDGE-BASED PROGRAMS TO GRADED BELIEF-BASED PROGRAMS, PART I: ON-LINE REASONING*

ABSTRACT. Knowledge-based programs (KBPs) are a powerful notion for expressing action policies in which branching conditions refer to implicit knowledge and call for a deliberation task at execution time. However, branching conditions in KBPs cannot refer to possibly erroneous beliefs or to graded belief, such as

"if my belief that φ holds is high
then do some action α
else perform some sensing action β".

The purpose of this paper is to build a framework where such programs can be expressed. In this paper we focus on the execution of such a program (a companion paper investigates issues relevant to the off-line evaluation and construction of such programs). We define a simple graded version of doxastic logic KD45 as the basis for the definition of belief-based programs. Then we study the way the agent's belief state is maintained when executing such programs, which calls for revising belief states by observations (possibly unreliable or imprecise) and progressing belief states by physical actions (which may have normal as well as exceptional effects).

1. INTRODUCTION

Knowledge-based programs, or KBPs (e.g. Fagin et al. 1995) are a powerful notion for expressing action policies in which branching conditions refer to implicit knowledge and call for a deliberation task at execution time: informally speaking, branching in KBPs has the following form:

$$\text{if } \mathbf{K}\varphi \text{ then } \pi \text{ else } \pi'$$

where $\mathbf{K}$ is an epistemic (generally S5) modality, and π, π' are subprograms. However, branching conditions in KBPs cannot refer to possibly erroneous beliefs or to graded belief, such as in

while I have no strong belief about the direction of the railway station
do ask someone

Synthese (2005) 147: 277–321
Knowledge, Rationality & Action 223–267
DOI 10.1007/s11229-005-1350-1

The purpose of this paper is to build a framework for such *belief-based programs* (BBPs). While knowledge states in KBPs are expressed in epistemic logic (usually S5), BBPs need a logic of *graded belief*, where different levels of uncertainty or entrenchment can be expressed. We therefore have to commit to a choice regarding the nature of uncertainty we wish to handle. Rather than reasoning with probabilistic belief states (and therefore introducing probabilistic modalities), which would take us far from usual logics of knowledge or belief such as S5 and KD45,[1] we choose to define belief states as *ordinal conditional functions* (OCF) (Spohn 1988) – also called *kappa-functions*. Introducing OCFs in logic is technically unproblematic (see [Goldszmidt and Pearl 1992; Boutilier et al. 1998, 1999] for logical frameworks of dynamicity and uncertainty based on OCFs); besides, OCFs are expressive enough in many situations where there exists only a small number of "belief degrees"; therefore they are a good trade-off between simplicity and expressivity, as well as between ordinality and cardinality, since they allow for an approximation of probabilities without the technical difficulties raised by the integration of logic and probability. Thus, unsurprisingly, OCFs have been used in several places for building logical frameworks of dynamicity and uncertainty (Goldszmidt and Pearl 1992; Boutilier 1998; Boutilier et al. 1998).

Then, many difficulties arise when considering the way a belief state should be progressed by an action. As in most logical frameworks for reasoning about action we distinguish between *pure sensing actions* who leave the state of the world unchanged and act only on the agent's mental state by giving her some feedback about the actual world, and *purely ontic* (or *physical*) actions, aiming at changing the state of the world without giving any feedback to the agent. This partition can be made without loss of generality (see e.g. [Scherl and Levesque 1993; Herzig et al. 2000), since complex actions (with both ontic effects and feedback) can be sequentially decomposed in two actions, the first being purely ontic and the second one being a pure sensing action.

Let us first consider sensing actions. In S5-based KBPs, observations provided by sensing actions are considered fully reliable; they are taken into account by a pure belief expansion operation. What we need is suitable handling of uncertain initial beliefs, uncertain and partially unreliable observations, and a belief revision operation for incorporating observations into the current belief state. As to ontic actions, BBPs, are intended to cope with the distinction between normal effects and more or less exceptional effects.

We start by defining a graded version of **KD45** (Section 2). In Section 3 we show how belief states are *revised* by possibly unreliable observations produced by sensing actions. In Section 4 we show how belief states are *progressed* (or updated) when the agent performs *(physical) actions* which may have alternative effects, some of which being more exceptional than others. Belief-based programs and their relationship to partially observable Markov decision processes are the subject of Section 5. Section 6 discusses further research directions. Since related work pertains to several different areas (depending on whether it relates to graded modalities, revision with uncertain inputs, or progression), we discuss it in the corresponding sections of the document, rather than having a specific Section on related work.

2. KD45$_G$

2.1. *Graded Beliefs and BBPs*

Our goal being to allow for branching conditions referring to implicit and graded beliefs, we start by generalizing the well-known doxastic logic **KD45** so as to allow for *graded belief modalities*.

Let *PS* be a finite set of propositional symbols, The (non-modal) language L_{PS} is defined in the usual way as the propositional language generated from *PS*, the usual connectives, and the Boolean constants $\top$ and $\bot$. Now, we define the language $\mathcal{L}_{PS}^O$ of graded doxastic logic **KD45**$_G$.

DEFINITION 1. The language $\mathcal{L}_{PS}^O$ generated from a set of propositional symbols *PS* is defined as follows:

- if φ is an objective formula of L_{PS} then $\mathbf{B}_1\varphi, \mathbf{B}_2\varphi, \ldots, \mathbf{B}_\infty\varphi$ are formulas of $\mathcal{L}_{PS}^O$;
- if φ is an objective formula of L_{PS} then $\mathbf{O}_1\varphi, \mathbf{O}_2\varphi, \ldots, \mathbf{O}_\infty\varphi$ are formulas of $\mathcal{L}_{PS}^O$;
- if Φ and Ψ are formulas of L_{PS} then $\neg\Phi, \Phi\vee\Psi, \Phi\wedge\Psi$ are formulas of $\mathcal{L}_{PS}^O$.

$\mathbf{B}_i\varphi$, for $i \in \overline{\mathbb{N}} = \mathbb{N}\cup\{\infty\}$, intuitively means that the agent believes φ with strength i. The larger i, the stronger the belief expressed by

$\mathbf{B}_i$, and $\mathbf{B}_\infty$ is a *knowledge* modality and may be denoted more simply by $\mathbf{K}$ (belief with infinite strength is true knowledge). Modalities $\mathbf{O}_1$, $\mathbf{O}_2$, $\mathbf{O}_n$ and $\mathbf{O}_\infty$ are *only belief* modalities, generalizing *only knowing* (Levesque and Lakemeyer 2000). Intuitively, $\mathbf{O}_{i\varphi}$ means that *all the agent believes to the degree at least i is φ*.

Note that the language $\mathcal{L}_{PS}^O$ considers only *subjective* and *flat* formulas. Neither formulas with nested modalities, nor formulas such as $\varphi \wedge \mathbf{B}_i \psi$, where φ, ψ are both objective, are formulas of $\mathcal{L}_{PS}^O$. This restriction is made for the sake of simplicity; it would be possible to consider a full modal language, and then prove, as it is the case in **KD45**, that each formula is equivalent to a flat formula, but we leave this technical issue aside since it has little relevance to the issues dealt with in this paper. Likewise, combinations of objective and subjective formulas do not play any role either as far as expressing and interpreting BBPs are concerned. Formulas of **KD45$_G$** are denoted by capital Greek letters Φ, Ψ etc. while objective formulas are denoted by small Greek letters φ, ψ etc.

A BBP is built up from the set of primitive actions *ACT* and usual program constructors. Given a set *ACT* of primitive actions, a BBP is defined inductively as follows:

- the empty plan λ is a BBP;
- for any $\alpha \in ACT$, α is a BBP;
- if π and π' are BBP then $\pi; \pi'$ is a BBP;
- if π and π' are BBP and Φ is a formula of $\mathcal{L}_{PS}^O$, then if Φ then π else π' and while Φ do π are BBPs.

Thus, a BBP is a program *whose branching conditions are doxastically interpretable* (since formulas of $\mathcal{L}_{PS}^O$ are subjective): the agent can decide whether she *believes* to a given degree that a formula is true (whereas she is generally unable to decide whether a given objective formula is true in the actual world). For instance, the agent performing the BPP

$$\pi = \text{while } \neg(\mathbf{B}_2 r \vee \mathbf{B}_2 \neg r) \text{ do } ask;$$
$$\text{if } \mathbf{B}_2 r \text{ then } goright \text{ else } goleft$$

performs the sensing action *ask* until she has a belief firm enough (namely of degree 2) about the way to follow (we'll see in Section 5 that if the *ask* action does not give fully reliable and informative outcomes then this program is not guaranteed to stop).

2.2. Semantics

We now give a semantics for interpreting formulas of $\mathcal{L}^O_{PS}$. Let $S = 2^{PS}$ be the (finite) set of *states* associated with PS. States are denoted by s, s' etc. Rather than writing a state with a subset of PS, for sake of clarity, we prefer to write them by listing all propositional symbols with a bar on the symbol when it is false in the state: for instance, if $PS = \{a, b, c, d\}$, then instead of $s = \{b, d\}$ we write $s = \bar{a}b\bar{c}d$; instead of $s = \emptyset$ we write $s = \bar{a}\bar{b}\bar{c}\bar{d}$ etc. If φ is objective then we note $\mathrm{Mod}(\varphi) = \{s \in S | s \models \varphi\}$. For $A \subseteq S$, $\mathrm{Form}(A)$ is the objective formula (unique up to logical equivalence) such that $\mathrm{Mod}(\mathrm{Form}(A)) = A$. If $A = \{s\}$ then we write $\mathrm{Form}(s)$ instead of $\mathrm{Form}(\{s\})$.

DEFINITION 2 (Belief states). An OCF (Spohn 1988), also called a belief state, is a function $\kappa: S \mapsto \overline{\mathbb{N}}$ such that $\min_{s \in S} \kappa(s) = 0$, κ is extended from states to objective formulas by $\kappa(\varphi) = \min\{\kappa(s) | s \models \varphi\}$.

Intuitively, $\kappa(s)$ is the *exceptionality degree* of s, $\kappa(s)$ is usually interpreted in terms of infinitesimal probabilities; $\kappa(s) = k < +\infty$ is then understood as $\mathrm{prob}(s) = o(\varepsilon^k)$, where ε is infinitely small. In particular:

- $\kappa(s) = 0$ means that s is a *normal state* (a normal state is not exceptional, to any degree).
- $\kappa(s) = 1$ means that s is "simply exceptional";
- $\kappa(s) = 2$ means that s is "doubly exceptional';
- $\kappa(s) = +\infty$ means that s is truly impossible. Any state s such that $\kappa(s) < \infty$ is called *a possible state*.

The *normalization constraint* $\min_{s \in S} \kappa(s) = 0$ imposes that there exists at least one normal state. The *void belief state* κ_{void} is defined by $\kappa_{\mathrm{void}}(s) = 0$ for all s.

We now define satisfaction of a $\mathcal{L}^O_{PS}$ formula by a belief state.

DEFINITION 3. A model for $\mathsf{KD45}_G$ is simply a an OCF κ. The satisfaction of a formula of $\mathcal{L}_{PS}$ in a model κ is defined by:

- for φ objective and $i \in \overline{\mathbb{N}}$, $\kappa \models \mathbf{B}_i \varphi$ iff $\kappa(\neg\varphi) \geq i$;
- for φ objective and $i \in \overline{\mathbb{N}}$, $\kappa \models \mathbf{O}_i \varphi$ iff $\forall s \in S$, $s \models \neg\varphi \Leftrightarrow \kappa(s) \geq i$
- $\kappa \models \Phi \vee \Psi$ iff $\kappa \models \Phi$ or $\kappa \models \Psi$
- $\kappa \models \neg\Phi$ iff $\kappa \not\models \Phi$.

The connectives $\wedge, \rightarrow, \leftrightarrow$ are defined from $\vee$ and $\neg$ in the usual way. Φ is *valid* (resp. *satisfiable*) *iff* it is satisfied in any model (resp, in at least one model). Ψ is a *consequence* of Φ (denoted by $\Phi \models \Psi$) *iff* for any $\kappa, \kappa \models \Phi$ implies $\kappa \models \Psi$. Φ and Ψ are equivalent (denoted by $\Phi \equiv \Psi$) *iff* $\Phi \models \Psi$ and $\Psi \models \Phi$.

Let us briefly comment the definitions.

- $\kappa \models \mathbf{B}_i \varphi$ holds as soon as any model of $\neg\varphi$ is exceptional at least to the degree i (i.e., is such that $\kappa(s) \geqslant i$), or, equivalently, all states such that $\kappa(s) < i$ (i.e., at most $i - 1$-exceptional) satisfy φ. In particular, $\mathbf{B}_1 \varphi$ is satisfied when all normal states satisfy φ, and $\mathbf{B}_\infty \varphi$ is satisfied when all possible states (to any degree) are models of φ.
- $\kappa \models \mathbf{O}_i \varphi$ holds in κ as soon as the states exceptional at least to the degree i are *exactly* the countermodels of φ, or equivalently, the states exceptional at most to degree $i - 1$ are *exactly* the models of φ. In particular, $\mathbf{O}_1 \varphi$ is satisfied when all normal states satisfy φ, and all models of φ are normal.

Importantly, $\mathbf{O}_1\top$, means that the agent does not believe anything to the degree 1, therefore nothing either to the degree 2, etc. The only κ satisfying $\mathbf{O}_1\top$ is κ_{void}.

It can be shown easily that each $\mathbf{B}_i$ is a **KD45** modality restricted to flat formulas:

PROPOSITION 1. For all φ, ψ in L_{PS} and all i, the following formulas are valid in **KD45**$_G$:

1. $\mathbf{O}_i \varphi \rightarrow \mathbf{B}_i \varphi$;
2. $\mathbf{B}_j \varphi \rightarrow \mathbf{B}_i \varphi$ whenever $j \geqslant i$;
3. $\mathbf{B}_i (\varphi \wedge \psi) \leftrightarrow \mathbf{B}_i \varphi \wedge \mathbf{B}_i \psi$;
4. $\neg \mathbf{B}_i \bot$.

Proof.

1. Let κ such that $\kappa \models \mathbf{O}_i \varphi$, which, by definition of the satisfaction relation, is equivalent to $\forall s \in S, s \models \neg\varphi$ *iff* $\kappa(s) \geqslant i$. This implies $\min\{\kappa(s) \mid s \models \neg\varphi\} \geqslant i$, that is, $\kappa(\neg\varphi) \geqslant i$, therefore $\kappa \models \mathbf{B}_i \varphi$.
2. Assume $j \geqslant i$. $\kappa \models \mathbf{B}_j \varphi$ is equivalent to $\kappa(\neg\varphi) \geqslant j$, which implies $\kappa(\neg\varphi) \geqslant i$, i.e., $\kappa \models \mathbf{B}_i \varphi$.

3. $\kappa \models \mathbf{B}_i(\varphi \wedge \psi)$ is equivalent to $\kappa(\neg(\varphi \wedge \psi)) \geqslant i$. Now, $\kappa(\neg(\varphi \wedge \psi)) = \kappa(\neg\varphi \vee \neg\psi) = \min(\kappa(\neg\varphi), \kappa(\neg\psi))$. Therefore, $\kappa \models \mathbf{B}_i(\varphi \wedge \psi)$ is equivalent to $\min(\kappa(\neg\varphi), \kappa(\neg\psi)) \geqslant i$ i.e., $\kappa(\neg\varphi) \geqslant i$ and $\kappa(\neg\psi) \geqslant i$, which is equivalent to $\kappa \models \mathbf{B}_i\varphi$ and $\kappa \models \mathbf{B}_i\psi$, i.e., $\kappa \models \mathbf{B}_i\varphi \wedge \mathbf{B}_i\psi$.

4. Let κ a belief state. Since there exists a s such that $\kappa(s) = 0$, we get $\kappa(\top) = 0$, hence for all $i \geqslant 1$, $\kappa \models \neg\mathbf{B}_i \bot$. $\qquad\square$

Remark that due to (3), $\mathbf{B}_i\varphi \to \mathbf{B}_i\psi$ is valid whenever $\varphi \models \psi$. Remark also that (2) and (3) fail to be valid if we replace $\mathbf{B}_i$ by $\mathbf{O}_i$.

EXAMPLE 1. Let κ defined by $\kappa(a\overline{b}) = 0$, $\kappa(ab) = 1$, $\kappa(\overline{a}b) = 1$ and $\kappa(\overline{a}\overline{b}) = \infty$. Then

- $\kappa \models \mathbf{B}_1 a \wedge \neg\mathbf{B}_2 a$: the agent believes a to the degree 1 (because the (single) normal state, i.e, $a\overline{b}$, satisfies a), but this belief is no firmer than that: a is not believed to the degree 2, because there is a $\neg a$-state s such that $\kappa(s) = 1$, namely $\overline{a}b$.
- $\kappa \models \mathbf{K}(a \vee b)$, because all possible states (namely, $a\overline{b}, ab$ and $\overline{a}b$) satisfy $a \vee b$;
- $\kappa \models \neg\mathbf{B}_1 b$, because the normal state $a\overline{b}$ does not satisfies b.
- $\kappa \models \mathbf{O}_1(a \wedge \neg b)$, because $a \wedge \neg b$ is all the agent believes in the normal states;
- $\kappa \models \mathbf{O}_\infty(a \vee b)$.

The meaning of $\kappa \models \mathbf{O}_i\varphi$ is better understood by the following simple result:

PROPOSITION 2. The two following statements are equivalent:

1. $\kappa \models \mathbf{O}_i\varphi$
2. for every objective formula ψ, $\kappa \models \mathbf{B}_i\psi$ iff $\varphi \models \psi$.

Proof.

(1) $\Rightarrow$ (2) Let $\kappa \models \mathbf{O}_i\varphi$.

(a) Let ψ such that $\varphi \models \psi$. By Proposition 1, $\kappa \models \mathbf{O}_i\varphi$ implies $\kappa \models \mathbf{B}_i\varphi$, therefore, by Proposition 1, $\kappa \models \mathbf{B}_i\psi$.

(b) Let ψ such that $\varphi \not\models \psi$, which entails that there exists a state s such that $s \models \varphi \wedge \neg\psi$. Now, $s \models \varphi$ and $\kappa \models \mathbf{O}_i\varphi$ together imply $\kappa(s) < i$, which in turn implies $\kappa(\neg\psi) < i$ and therefore $\kappa \not\models \mathbf{B}_i\psi$.

(2) $\Rightarrow$ (1) Assume (1) false, i.e., $\kappa \models \neg \mathbf{O}_i \varphi$; then either (c) there is an s such that $s \models \neg \varphi$ and $\kappa(s) < i$, or (d) there is an s such that $s \models \varphi$ and $\kappa(s) \geqslant i$. If (c) holds, then $\kappa \not\models \mathbf{B}_i \varphi$ and then taking $\psi = \varphi$ falsifies (2). If (d) holds, then take $\psi = \neg \text{Form}(s)$. We have $\kappa(\psi) = \kappa(s) \geqslant i$, and yet $\varphi \not\models \psi$, which falsifies(2). $\square$

Syntactically, since the number of states is finite, $\mathbf{O}_i \varphi$ can be defined from the $\mathbf{B}_i$ modalities by the following formula (which is finite only when PS is finite):

PROPOSITION 3.

$$\mathbf{O}_i \varphi \equiv \mathbf{B}_i \varphi \wedge \bigwedge_{s \models \varphi} \neg \mathbf{B}_i (\varphi \wedge \neg \text{Form}(s))$$

Proof.

- We start by showing $\mathbf{O}_i \varphi \models \mathbf{B}_i \varphi \wedge \bigwedge_{s \models \varphi} \neg \mathbf{B}_i (\varphi \wedge \neg \text{Form}(s))$. Let κ such that $\kappa \models \mathbf{O}_i \varphi$, which, by definition of the satisfaction relation, is equivalent to (a) $\forall s \models \varphi$, $\kappa(s) < i$ and (b) $\forall s \models \neg \varphi$, $\kappa(s) \geqslant i$. From point 1 of Proposition 1 we have $\kappa \models \mathbf{B}_i \varphi$. Now, let $s \models \varphi$, which by (a) implies $\kappa(s) < i$. $\kappa(s) < i$, together with $s \not\models \varphi \wedge \neg \text{Form}(s)$, imply $\kappa(\neg(\varphi \wedge \neg \text{Form}(s))) < i$, therefore $\kappa \models \neg \mathbf{B}_i (\varphi \wedge \neg \text{Form}(s))$. This being true for all $s \models \varphi$, and the set of states being finite, we get (d) $\kappa \models \bigwedge_{s \models \varphi} \neg \mathbf{B}_i (\varphi \wedge \neg \text{Form}(s))$. From (c) and (d) we get $\kappa \models \mathbf{B}_i \varphi \wedge \bigwedge_{s \models \varphi} \neg \mathbf{B}_i (\varphi \wedge \neg \text{Form}(s))$.
- Now, we show $\mathbf{B}_i \varphi \wedge \bigwedge_{s \models \varphi} \neg \mathbf{B}_i (\varphi \wedge \neg \text{Form}(s)) \models \mathbf{O}_i \varphi$. Let $\kappa \models \neg \mathbf{O}_i \varphi$. Then, either (e) there is a state s such that $s \models \neg \varphi$ and $\kappa(s) < i$ or (f) there is a state s such that $s \models \neg \varphi$ and $\kappa(s) \geqslant i$. If (e) holds, then $\kappa(\varphi) < i$ and therefore $\kappa \models \neg \mathbf{B}_i \varphi$ and *a fortiori* $\kappa \models \neg(\mathbf{B}_i \varphi \wedge \bigwedge_{s \models \varphi} \neg \mathbf{B}_i (\varphi \wedge \neg \text{Form}(s)))$. If (f) holds, then for this state s it holds $\kappa(\varphi \wedge \neg \text{Form}(s)) \geqslant i$, therefore $\kappa \models \mathbf{B}_i (\varphi \wedge \neg \text{Form}(s))$, which entails that $\kappa \models \neg(\bigwedge_{s \models \varphi} \neg \mathbf{B}_i (\varphi \wedge \neg \text{Form}(s)))$. In both cases (e) and (f) we have $\kappa \models \neg(\mathbf{B}_i \varphi \wedge \bigwedge_{s \models \varphi} \neg \mathbf{B}_i (\varphi \wedge \neg \text{Form}(s)))$. This being true for all $\kappa \models \neg \mathbf{O}_i \varphi$, we have $\neg \mathbf{O}_i \varphi \models \neg(\mathbf{B}_i \varphi \wedge \bigwedge_{s \models \varphi} \neg \mathbf{B}_i (\varphi \wedge \neg \text{Form}(s)))$, which is equivalent to $\mathbf{B}_i \varphi \wedge \bigwedge_{s \models \varphi} \neg \mathbf{B}_i (\varphi \wedge \neg \text{Form}(s)) \models \mathbf{O}_i \varphi$. $\square$

EXAMPLE 2. Let $PS = \{x, y\}$; we have $\mathbf{O}_2(x \vee y) \equiv \mathbf{B}_2(x \vee y) \wedge \neg \mathbf{B}_2 x \wedge \neg \mathbf{B}_2 y \wedge \neg \mathbf{B}_2(x \wedge \neg y \vee \neg x \wedge y)$. The formula $\mathbf{O}_1 x \wedge \mathbf{O}_2 x \wedge \mathbf{O}_3 \top$ means that the agent believes only x to the degree 2, that he does not believe more to the degree 1 and that he does not believe anything to a degree > 2.

2.3. *Normal Forms*

We now introduce some useful syntactical notions. A formula of $\mathcal{L}^O_{PS}$ is

- a *doxastic atom iff* it is a formula $\mathbf{B}_i \varphi$ where φ is objective.
- a *O-doxastic atom iff* it is a formula $\mathbf{O}_i \varphi$ where φ is objective.
- a *normal positive doxastic* (NPD) *formula iff* Φ is of the form $\mathbf{B}_\infty \varphi_\infty \wedge \mathbf{B}_n \varphi_n \wedge \cdots \wedge \mathbf{B}_1 \varphi_1$, where $\varphi_\infty, \varphi_1, \ldots, \varphi_n$ are objective formulas such that for all j and $i > j$ we have $\models \varphi_j \rightarrow \varphi_i$.
- a *normal O* (NO) *formula iff* it is of the form $\mathbf{O}_\infty \varphi_\infty \wedge \mathbf{O}_{n+1} \varphi_\infty \wedge \mathbf{O}_n \varphi_n \wedge \cdots \wedge \mathbf{O}_1 \varphi_1$, where $\varphi_\infty, \varphi_1, \ldots, \varphi_n$ are objective formulas such that for all j and $i > j$ we have $\models \varphi_j \rightarrow \varphi_i$.

EXAMPLE 3.

- $\mathbf{B}_3 \neg x$, $\mathbf{K}(\neg x \vee \neg y)$ are doxastic atoms;
- $\mathbf{O}_3 \neg x$ is a O-doxastic atom;
- $\mathbf{K} \top \wedge \mathbf{B}_4 \top \wedge \mathbf{B}_3 a \wedge \mathbf{B}_2 a \wedge \mathbf{B}_1 (a \wedge b)$ is a NPD formula;
- $\mathbf{O}_\infty \top \wedge \mathbf{O}_4 \top \wedge \mathbf{O}_3 a \wedge \mathbf{O}_2 a \wedge \mathbf{O}_1 (a \wedge b)$ is a NO formula.

When writing a normal positive doxastic formula $\mathbf{B}_\infty \varphi_\infty \wedge \mathbf{B}_n \varphi_n \wedge \cdots \wedge \mathbf{B}_1 \varphi_1$, we omit subformulas $\mathbf{B}_i \varphi_i$ such that $\varphi_{i+1} \equiv \varphi_i$, as well as tautological subformulas of the form $\mathbf{B}_i \top$: for instance,

$$\mathbf{B}_\infty \top \wedge \cdots \wedge \mathbf{B}_4 \top \wedge \mathbf{B}_3 a \wedge \mathbf{B}_2 a \wedge \mathbf{B}_1 (a \wedge b)$$

is simply denoted by its equivalent simplified form

$$\mathbf{B}_3 a \wedge \mathbf{B}_1 (a \wedge b)$$

Henceforth, formulas such as $\mathbf{B}_2 a, \mathbf{B}_\infty \neg a \wedge \mathbf{B}_1 (b \wedge \neg a)$ are considered as normal positive doxastic formulas. The limit case where all φ_i are $\top$ is simply denoted by $\top$ – which is therefore a NPD formula as well. Likewise, $\bot$ is also a NO formula.

Since $\mathbf{B}_i (\varphi \wedge \psi) \leftrightarrow \mathbf{B}_i \varphi \wedge \mathbf{B}_i \psi$ and $\mathbf{B}_i \varphi \rightarrow \mathbf{B}_j \varphi \, (i \geqslant j)$ are valid in $\mathsf{KD45}_G$, any conjunction of doxastic atoms can be equivalently rewritten in NPD form. For instance,

$$\mathbf{B}_3 a \wedge \mathbf{B}_1 (a \rightarrow b) \wedge \mathbf{B}_1 c$$

is equivalent to $\mathbf{B}_3 a \wedge \mathbf{B}_1 (a \wedge b \wedge c)$.

We also make use of the following syntactical shortcut: for any NPD formula $\Phi = \mathbf{B}_\infty \varphi_\infty \wedge \mathbf{B}_n \varphi_n \wedge \cdots \wedge \mathbf{B}_1 \varphi_1$, $\mathbf{Only}(\Phi)$ is the formula $\mathbf{O}_\infty \varphi_\infty \wedge \mathbf{O}_{n+1} \varphi_\infty \wedge \mathbf{O}_n \varphi_n \wedge \cdots \wedge \mathbf{O}_1 \varphi_1$. Such formulas completely express the agent's belief state; they are satisfied by a single OCF, namely $\kappa_\Phi = G(\Phi)$ defined in Section 2. For instance,

$$\mathbf{Only}(\mathbf{B}_3 a \wedge \mathbf{B}_1(a \wedge b))$$
$$= \mathbf{O}_\infty \top \wedge \cdots \wedge \mathbf{O}_4 \top \wedge \mathbf{O}_3 a \wedge \mathbf{O}_2 a \wedge \mathbf{O}_1(a \wedge b)$$

Any belief state κ corresponds to a NO formula Φ_κ, unique up to logical equivalence:

DEFINITION 4 (From belief states to NO formulas and vice versa).

1. for any belief structure κ, $H(\kappa) = \Phi_\kappa$ is the NO formula (unique up to logical equivalence) defined by

$$\Phi_\kappa = \mathbf{O}_\infty \varphi_\infty \wedge \mathbf{O}_{n+1} \varphi_\infty \wedge \mathbf{O}_n \varphi_n \wedge \cdots \wedge \mathbf{O}_1 \varphi_1$$

    where

    - $n = \max\{\kappa(s) \mid s \in S \text{ and } \kappa(s) < \infty\}$
    - for all $i \in \{1, \ldots, n, \infty\}$, $\varphi_i = \text{Form}(\{s \in S \mid \kappa(s) < i\})$.

2. given a NO formula $\Phi = \mathbf{O}_\infty \varphi_\infty \wedge \mathbf{O}_{n+1} \varphi_\infty \wedge \mathbf{O}_n \varphi_n \wedge \cdots \wedge \mathbf{O}_1 \varphi_1$, $G(\Phi) = \kappa_\Phi$ is the OCF defined by

$$\kappa_\Phi(s) = \begin{cases} 0 & \text{if } s \models \varphi_1 \\ i & \text{if } s \models \varphi_{i+1} \wedge \neg\varphi_i \text{ and } i = 1, \ldots, n-1 \\ n & \text{if } s \models \varphi_\infty \wedge \neg\varphi_n \\ +\infty & \text{if } s \not\models \varphi_\infty \end{cases}$$

EXAMPLE 4. Let κ defined by $\kappa([a, \neg b]) = 0$, $\kappa([a, b]) = 1$, $\kappa([\neg a, b]) = 1$ and $\kappa([\neg a, \neg b]) = \infty$. Then

$$H(\kappa) = \mathbf{O}_\infty(a \vee b) \wedge \mathbf{O}_2(a \vee b) \wedge \mathbf{O}_1(a \wedge \neg b)$$

The following property tells that there is a one-to-one correspondence between OCFs and equivalence classes (w.r.t. equivalence on $\mathrm{KD45}_G$) of NO formulas:

PROPOSITION 4. For any NO formula $\Phi = \mathbf{O}_\infty \varphi_\infty \wedge \mathbf{O}_{n+1}\varphi_\infty \wedge \mathbf{O}_n\varphi_n \wedge \cdots \wedge \mathbf{O}_1\varphi_1$, $\kappa \models \Phi$ iff $\kappa = \kappa_\Phi$.

Proof. Let $\Phi = \mathbf{O}_\infty \varphi_\infty \wedge \mathbf{O}_{n+1}\varphi_\infty \wedge \mathbf{O}_n\varphi_n \wedge \cdots \wedge \mathbf{O}_1\varphi_1$.

$\Rightarrow$ Suppose $\kappa \models \Phi$, which is equivalent to the following condition: for all $s \in S$ and every i, $s \models \varphi_i$ iff $\kappa(s) < i$. This helps us remarking that (a) for all $s \in S$, $\kappa(s) > n$ implies $\kappa(s) = +\infty$. Consider now the following four cases:

- $\kappa(s) = 0$. In this case, $s \models \varphi_1$ and by definition of κ_Φ, we have $\kappa_\Phi(s) = 0$;
- $\kappa(s) = i$ where $1 \leqslant i \leqslant n - 1$. In this case, $s \models \neg\varphi_i \wedge \varphi_{i+1}$ and by definition of κ_Φ, we have $\kappa_\Phi(s) = i$.
- $\kappa(s) = n$. In this case, $s \models \varphi_\infty \wedge \neg\varphi_n$, and by definition of κ_Φ, we have $\kappa_\Phi(s) = n$.
- $\kappa(s) = +\infty$. In this case, $s \models \neg\varphi_\infty$ and by definition of κ_Φ, we have $\kappa_\Phi(s) = \infty$.

Due to (a), these cover all possible cases, therefore $\kappa_\Phi = \kappa$.

$\Leftarrow$ We have to verify that $\kappa_\Phi \models \Phi$. First, we check that for all $s \in S, s \models \neg\varphi_\infty$ iff $\kappa_\Phi(s) = +\infty$, therefore $\kappa_\Phi \models \mathbf{O}_\infty \varphi_\infty$. Next, for all $s \in S$ and all $i \leqslant n$, $s \models \neg\varphi_i$ iff $\kappa_\Phi(s) \geqslant i$, therefore $\kappa_\Phi \models \mathbf{O}_i \varphi_i$. Hence, $\kappa_\Phi \models \Phi$. $\square$

COROLLARY 1. $\kappa_{\Phi_\kappa} = \kappa$ and $\Phi_{\kappa_\Phi} \equiv \Phi$.

Proof. Let κ be a belief state. It is easily checked that $\kappa \models \Phi_\kappa$. Now, letting $\Phi = \Phi_\kappa$ in Proposition 4 gives $\kappa \models \Phi_\kappa$ iff $\kappa = \kappa_{\Phi_\kappa}$, hence $\kappa_{\Phi_\kappa} = \kappa$. This shows that $H = G^{-1}$ (where NO formulas are identified, by a slight abuse of notation, with their equivalence class w.r.t. logical equivalence), therefore $\Phi_{\kappa_\Phi} \equiv \Phi$. $\square$

Notice that when writing $\Phi_\kappa = \mathbf{O}_\infty \varphi_\infty \wedge \mathbf{O}_{n+1}\varphi_\infty \wedge \mathbf{O}_n\varphi_n \wedge \cdots \wedge \mathbf{O}_1\varphi_1$, φ_i is the formula expressing all the agent believes to the degree i in the belief state κ.

2.4. *Related Work on Modal Logics of Graded Belief*

Although it is original, the construction given in this Section is not the primary goal of the paper. It is very similar to the work on stratified belief bases and possibilistic logic (e.g. (Dubois et al. 1994)) where the duality between (semantical) belief states and (syntactical) NPD formulas can be expressed as well. A multimodal system (with no account for only believing) for possibilistic logic is given in Fariñas del Cerro and Herzig (1991). As for gradual doxastic logics, van der Hock and ch Meyer 1991 define a gradual version of **KD45** as well. The interpretation of graded belief is, however, totally different from ours, since $\mathbf{B}_n \varphi$ expresses that φ is true in all worlds except n or less.

3. OBSERVATIONS AND REVISION

3.1. *Combination of Belief States*

We now define the *combination* of belief states, and by isomorphism, the combination of NO formulas. Calling it a "connective" is an abuse of language, since it only connects NO formulas and is therefore not a full-fledged connective.

DEFINITION 5 (OCF combination). Let κ_1 and κ_2 be two OCFs. If $\min_S(\kappa_1 + \kappa_2) = \infty$, then $\kappa_1 \oplus \kappa_2$ is undefined; otherwise, $\kappa \oplus \kappa_2$ is defined by

$$\forall s \in S, \quad (\kappa_1 \oplus \kappa_2)(s) = \kappa_1(s) + \kappa_2(s) - \min_S(\kappa_1 + \kappa_2)$$

When defined, we have $\min_S(\kappa_1 \oplus \kappa_2) = 0$, therefore $\kappa_1 \oplus \kappa_2$ is an OCF.

In the particular case of κ_φ defined by

$$\kappa_\varphi(s) = \begin{cases} 0 & \text{if } s \models \varphi \\ +\infty & \text{if } s \models \neg\varphi \end{cases}$$

then

$$(\kappa \oplus \kappa_\varphi)(s) = \begin{cases} \kappa(s) - \kappa(\varphi) & \text{if } s \models \varphi \\ +\infty & \text{if } s \models \neg\varphi \end{cases}$$

provided that $\kappa(\varphi) < \infty$. Therefore, $(\kappa \oplus \kappa_\varphi)(s) = \kappa(s|\varphi)$, where $\kappa(.|\varphi)$ is Spohn's conditioning (Spohn 1988).

The intuitive idea behind OCF combination is first illustrated when $\min_S(\kappa_1 + \kappa_2)$. When combining the beliefs coming from the sources 1 and 2 (corresponding respectively to κ_1 and κ_2), the combined exceptionality degree of a state s is the sum of the exceptionality of s according to 1 and of that according to 2.

[234]

EXAMPLE 5. Consider $\kappa_1 = \kappa_{\Phi_1}$ and $\kappa_2 = \kappa_{\Phi_1}$, where $\Phi_1 = \textbf{Only}$
$(\textbf{B}_\infty(a \vee b) \wedge \textbf{B}_2 a \wedge \textbf{B}_1(a \wedge b))$ and $\Phi_2 = \textbf{Only}(\textbf{B}_1 b)$.

		κ_1	κ_2	$\kappa_1 \oplus \kappa_2$
ab	:	0	0	0
$a\bar{b}$	:	1	1	2
$\bar{a}b$	:	2	0	2
$\bar{a}\bar{b}$	:	∞	1	∞

κ_1 and κ_2 do not conflict: there is a state, namely ab, considered normal by both; hence the identity $\kappa_1 \oplus \kappa_2 = \kappa_1 + \kappa_2$. Now, $(\kappa_1 \oplus \kappa_2)(ab) = 0$ intuitively means that the state ab, considered normal by both κ_1 and κ_2, is considered normal by their combination as well. Next, $a\bar{b}$ being considered simply exceptional by both κ_1 and κ_2, the combination of both considered it doubly exceptional $((\kappa_1 \oplus \kappa_2)(a\bar{b}) = 2.)$ This is justified by the fact that κ_1 and κ_2 are considered as two independent sources: intuitively, if $a\bar{b}$ is the actual state then both sources 1 and 2 have to be wrong. Considering now that source 1 (resp. 2) is wrong about $a\bar{b}$ with probability $o(\varepsilon)$ (because $\kappa_1(a\bar{b}) = \kappa_2(a\bar{b}) = 1$), the probability that both sources are wrong is in $o(\varepsilon^2)$.

When both sources κ_1 and κ_2 conflict, we end up with a $\kappa_1 + \kappa_2$ without any normal state. Renormalizing then just corresponds to making the least exceptional states normal.

EXAMPLE 6. Consider κ_1 as above and $\kappa_3 = \kappa_{\Phi_3}$, where

$$\Phi_3 = \textbf{Only}(\textbf{B}_\infty(\neg a \vee \neg b))$$

		κ_1	κ_3	$\kappa_1 + \kappa_3$	$\kappa_1 \oplus \kappa_3$
ab	:	0	∞	∞	∞
$a\bar{b}$	:	1	0	1	0
$\bar{a}b$	:	2	0	2	1
$\bar{a}\bar{b}$	:	∞	0	∞	∞

No state being considered normal by both sources, $a\bar{b}$, being the "closest to normality" when considering both sources, is made normal in their combination.

Up to an isomorphism, $\oplus$ corresponds to the "product combination" of possibility distributions (see Section 3.4. of (Benferhat et al.

2001)), as well as to an infinitesimal version of Dempster's rule of combination (Dempster 1967). The details are in Appendix.

By isomorphism, NO formulas can be combined as well:

DEFINITION 6. For Φ and Ψ two NO formulas we have:

$$\Phi \otimes \Psi = \begin{cases} H(\kappa_\Phi \oplus \kappa_\Psi) = H(G(\Phi) \oplus G(\Psi)) & \text{if defined} \\ \bot & \text{otherwise} \end{cases}$$

Since, due to Corollary 1, there is a one-to-one correspondence between NO formulas (modulo logical equivalence) and belief states, the following holds: let Φ, Ψ are two NO formulas such that $\Phi \otimes \Psi \not\equiv \bot$, then $\kappa \models \Phi \otimes \Psi$ iff $\kappa = \kappa_\Phi \oplus \kappa_\Psi$.

PROPOSITION 5. The following formulas are valid:

1. $\mathbf{Only}(\mathbf{B}_i \varphi) \otimes \mathbf{Only}(\mathbf{B}_j \varphi) \equiv \mathbf{Only}(\mathbf{B}_{i+j} \varphi)$;
2. $\mathbf{Only}(\mathbf{B}_i \varphi) \otimes \mathbf{Only}(\mathbf{B}_j \neg \varphi) \equiv \begin{cases} \mathbf{Only}(\mathbf{B}_{i-j} \varphi) & \text{if } i > j \\ \mathbf{Only}(\mathbf{B}_{j-i} \neg \varphi) & \text{if } j > i \\ \mathbf{Only}(\mathbf{K} \top) & \text{if } i = j \end{cases}$
3. $\Phi \otimes \Psi \equiv \Psi \otimes \Phi$;
4. $\Phi \otimes (\Psi \otimes \Xi) \equiv (\Phi \otimes \Psi) \otimes \Xi)$;
5. $\Phi \otimes \top \equiv \Phi$

Proof.

1. By definition,

$$\kappa_{\mathbf{Only}(\mathbf{B}_i \varphi)}(s) = \begin{cases} 0 & \text{if } s \models \varphi \\ i & \text{if } s \not\models \varphi \end{cases}$$

and

$$\kappa_{\mathbf{Only}(\mathbf{B}_j \varphi)}(s) = \begin{cases} 0 & \text{if } s \models \varphi \\ j & \text{if } s \not\models \varphi \end{cases}$$

Therefore

$$(\kappa_{\mathbf{Only}(\mathbf{B}_i \varphi)} \oplus \kappa_{\mathbf{Only}(\mathbf{B}_j \varphi)})(s) \begin{cases} 0 & \text{if } s \models \varphi \\ i+j & \text{if } s \not\models \varphi \end{cases}$$

2. We have

$$\kappa_{\mathbf{Only}(\mathbf{B}_i\varphi)}(s) = \begin{cases} 0 & \text{if } s \models \varphi \\ i & \text{if } s \not\models \varphi \end{cases}$$

and

$$\kappa_{\mathbf{Only}(\mathbf{B}_j\neg\varphi)}(s) \begin{cases} j & \text{if } s \models \varphi \\ 0 & \text{if } s \not\models \varphi \end{cases}$$

Assume $i > j$. Then $\min(\kappa_{\mathbf{Only}(\mathbf{B}_i\varphi)} + \kappa_{\mathbf{Only}(\mathbf{B}_j\neg\varphi)}) = j$; now, if $s \models \varphi$ then $(\kappa_{\mathbf{Only}(\mathbf{B}_i\varphi)} + \kappa_{\mathbf{Only}(\mathbf{B}_j\neg\varphi)})(s) = i - j$ and if $s \models \neg\varphi$ then $(\kappa_{\mathbf{Only}(\mathbf{B}_i\varphi)} + \kappa_{\mathbf{Only}(\mathbf{B}_j\neg\varphi)})(s) = 0$. The case $j > i$ is symmetric. Lastly, if $i = j$ then $\min(\kappa_{\mathbf{Only}(\mathbf{B}_i\varphi)} + \kappa_{\mathbf{Only}(\mathbf{B}_j\neg\varphi)}) = i$, and for every s, $(\kappa_{\mathbf{Only}(\mathbf{B}_i\varphi)} + \kappa_{\mathbf{Only}(\mathbf{B}_j\neg\varphi)})(s) = 0$, hence $\kappa_{\mathbf{Only}(\mathbf{B}_i\varphi \otimes \mathbf{B}_j\neg\varphi)}\kappa_{\text{void}}$.

3. obvious.

4. $((\kappa_1 \oplus \kappa_2) + \kappa_3)(s) = \kappa_1(s) + \kappa_2(s) - \min_S(\kappa_1 + \kappa_2) + \kappa_3(s) - \min_S((\kappa_1 \oplus \kappa_2) + \kappa_3)$. Now, $\min_S((\kappa_1 \oplus \kappa_2) + \kappa_3) = \min_{s \in S}(\kappa_1(s) + \kappa_2(s) - \min_S(\kappa_1 + \kappa_2) + \kappa_3(s)) = \min_{s \in S}(\kappa_1(s) + \kappa_2(s) + \kappa_3(s)) - \min_S(\kappa_1 + \kappa_2)$; therefore, $((\kappa_1 \oplus \kappa_2) \oplus \kappa_3)(s) = \kappa_1(s) \oplus \kappa_2(s) + \kappa_3(s) - \min_S(\kappa_1 + \kappa_2 + \kappa_3)$. This expression is symmetric in κ_1, κ_2 and κ_3, therefore, $(\kappa_1 \oplus \kappa_2) \oplus \kappa_3 = (\kappa_2 \oplus \kappa_3) \oplus \kappa_1)$; by commutativity, we then get $(\kappa_1 \oplus \kappa_2) \oplus \kappa_3 = \kappa_1 \oplus (\kappa_2 \oplus \kappa_3)$. Lastly, by isomorphism we get $\Phi \otimes (\Psi \otimes \Xi) \equiv (\Phi \otimes \Psi) \otimes \Xi)$;

5. obvious from $\kappa_\top = \kappa_{\text{void}}$ and $\kappa \oplus \kappa_{\text{void}} = \kappa$. $\qquad\square$

An important corollary of point 1 is that $\Phi \otimes \Phi$ is generally not equivalent to Φ.

As an example, we consider $\Phi_1 \otimes \Phi_2$ where $\Phi_1 = \mathbf{Only}(\mathbf{B}_\infty(a \vee b) \wedge \mathbf{B}_2 a \wedge \mathbf{B}_1(a \wedge b))$ et $\Phi_2 = \mathbf{Only}(\mathbf{B}_1 b)$. We show with the array above that $\Phi_1 \otimes \Phi_2 \equiv \Phi_3$ where $\Phi_3 = \mathbf{Only}(\mathbf{B}_2(a \wedge b) \wedge \mathbf{B}_\infty(a \vee b))$.

	κ_{Φ_1}	κ_{Φ_2}	κ_{Φ_3}
ab :	0	0	0
$a\bar{b}$:	1	1	2
$\bar{a}b$:	2	0	2
$\bar{a}\bar{b}$:	∞	1	∞

3.2. Observations

Let us now introduce observations and revision of a belief state by an observation. The feedback of a sensing action is an *observation*. The simplest sensing actions are *basic tests*, whose feedback consists of the truth value of a given objective formula. Unlike most approaches to sensing in reasoning about action and planning, assuming that all sensing actions are basic tests such as in (Scherl and Levesque 1993; van Linder et al. 1994; Levesque 1996; Herzig et al. 2001) becomes a loss of generality when considering belief instead of knowledge: we want to allow for more general sensing actions, whose feedback might be imprecise and/or unreliable.

DEFINITION 7. An Observational believe state, or, for short, an observation, is a belief state κ_{obs}, corresponding to a NO formula obs $= H(\kappa_{\mathrm{obs}}) = \mathbf{Only}(\mathbf{B}_\infty o \wedge \mathbf{B}_n o_n \wedge \cdots \wedge \mathbf{B}_1 o_1)$ (by convention we write $o_\infty = o$).

An observation is therefore defined by the belief state it conveys (which, in practice, may be a function of the belief state of the source and the belief that the agent has on the reliability of the source): κ_{obs} is *all we observe* when getting the observation obs. κ_{obs} can also be viewed as the belief state the agent gets into when obtaining obs in the void belief state κ_{void}. The *void observation* $\mathrm{obs}_{\mathrm{void}}$ is defined by $\mathrm{obs}_{\mathrm{void}} = \mathbf{Only}(\mathbf{K}\top)$ – i.e., $\kappa_{\mathrm{obs}_{\mathrm{void}}} = \kappa_{\mathrm{void}}$.

The outcome of a reliable truth test for a given variable x is an observation of the form $obs \equiv \mathbf{Only}(\mathbf{B}_\infty o)$, where $o = x$ or $o = \neg x$. In this case, obs is a *reliable and fully informative observation about* x. If $obs \equiv \mathbf{Only}(\mathbf{B}_\infty o)$ where o is a more general formula (such as, for instance, $x \vee y$), then obs is reliable but incomplete; a degenerate case is when $o = \top$: the tautology is observed – obviously with full reliability. Now, a simple observation $\mathrm{obs} \equiv \mathbf{Only}(\mathbf{B}_k o_k)$, where $k < \infty$, is only *partially reliable*. A complex observation is composed of a reliable part (possibly conveying little information, sometimes none at all) and some partially reliable parts – the amount of information obviously decreasing with the reliability level. This rather complex definition is due to the fact that a single observation generally relates to the real state of the world in several ways, with various degrees of uncertainty (exactly as in the Bayesian case). Consider for instance reading the value θ on a temperature sensor, which may for instance correspond to the observation $\mathrm{obs} = \mathbf{Only}(\mathbf{B}_1(t - 1 \leqslant \theta \leqslant t + 1) \wedge \mathbf{B}_2(t - 2 \leqslant \theta \leqslant t + 2) \wedge \mathbf{B}_\infty(t - 5 \leqslant \theta \leqslant t + 5))$.

Here is another example. *At 8 in the morning, the agent hears on the radio "due to a strike of a part of the airport staff, today the air traffic will be subject to strong perturbations; as for now, no flight has been scheduled yet". The agent, who namely has a ticket for a 11.00 flight to destination D, views this as an observation that: (a) for sure, perturbations will occur; (b) there is a strong (but not total) evidence that he will not leave at 11.00 as initially planned; (c) there is a weaker evidence that he won't be able to leave today at all, Therefore, using the variables p (perturbations), m (the agent gets a flight in the morning as planned) and l (the agents gets a flight later in the day), the complex observation brought by the radio information may be* $\mathbf{Only}$ $(\mathbf{B}_\infty p \wedge \mathbf{B}_2(p \wedge \neg m) \wedge \mathbf{B}_1(p \wedge \neg m \wedge \neg l))$.

3.3. *Revision*

Now, the agent revises her current belief state by an observation *simply by combining both.*

DEFINITION 8. Let κ be a belief state and κ_{obs} an observational belief state. The revision of κ by κ_{obs} is the combination of κ and κ_{obs}, i.e., $\mathrm{rev}(\kappa, o) = \kappa \oplus \kappa_{obs}$.

One may be somewhat surprised by the fact that revision is defined by a symmetric operator, while most standard approaches to belief revision are definitely non-commutative. The latter (apparent) non-commutativity comes from the status of the observation, which is considered as definitely true and must be accepted in any case.[2] However, belief revision with fully reliable observations is a particular case of our general revision, which argues that standard (AGM) belief revision can also be considered as commutative, provided that each piece of information is labeled by its status (reliable or not).

Now, by isomorphism, revision can be performed syntactically: $\Phi = \mathbf{Only}(\mathbf{B}_\infty \varphi \wedge \mathbf{B}_n \varphi_n \wedge \cdots \wedge \mathbf{B}_1 \varphi_1)$ being a NO formula and $\mathrm{obs} = \mathbf{Only}(\mathbf{B}_\infty o \wedge \mathbf{B}_p o_p \wedge \cdots \wedge \mathbf{B}_1 o_1)$ an observation, the revision Φ by *obs* is $\Phi \otimes \mathrm{obs}$. The following result shows how the latter expression can be computed syntactically in a compact way, *without performing revision state by state:*

PROPOSITION 6. Given Φ and obs two NO formulas,

$$\Phi \otimes \mathrm{obs} \equiv \mathbf{Only}(\mathbf{B}_1 \psi_p \wedge \cdots \wedge \mathbf{B}_m \psi_{p+m-1} \wedge \mathbf{B}_\infty \psi)$$

where

- $\psi = \varphi \wedge o$;
- $\forall i \in \mathbb{N}, \psi_i = (\varphi_1 \wedge o_i) \vee (\varphi_2 \wedge o_{i-1}) \vee \cdots \vee (\varphi_i \wedge o_1)$;
- $p = \min\{j, \psi_j \not\equiv \perp\}$;
- $m = \max\{j, \psi_{p+j-1} \not\equiv \psi\}$.

Proof. First we show that $\forall i \in \mathbb{N}$, $\mathrm{Mod}(\psi_j) = \{s | \kappa_\Phi(s) + \kappa_{\mathrm{obs}}(s) < j\}$. Let $s \in S$ such that $\kappa_\Phi(s) + \kappa_{\mathrm{obs}}(s) < j$, then $\kappa_\Phi(s) < j - \kappa_{\mathrm{obs}}(s)$, hence $s \models \varphi_{j-\kappa_{\mathrm{obs}}(s)}$ (cf. Definition 4). Furthermore, the same definition implies $s \models o_{\kappa_{\mathrm{obs}}(s)+1}$. Therefore, $s \models \psi_j$. Conversely, let $s \models \psi_j$. Then, by construction of ψ_j, there exist u and v such that $u + v = j + 1$ and $s \models \varphi_u \wedge o_v$. Using definition 4, this implies that $\kappa_\Phi(s) < u$ and $\kappa_{\mathrm{obs}}(s) < v$, *i.e.*, $\kappa_\Phi(s) + \kappa_{\mathrm{obs}}(s) < j$.

This property shows first that $\min_S(\kappa + \kappa_{\mathrm{obs}}) = p - 1$, and then that $\mathrm{Mod}(\psi_{p+i-1}) = \{s | \kappa(s) + \kappa_{\mathrm{obs}}(s) < p + i - 1\}$, *i.e.*, $\mathrm{Mod}(\psi_{p+i-1}) = \{s | \kappa(s) + \kappa_{\mathrm{obs}}(s) - \min_S(\kappa + \kappa_{\mathrm{obs}}) < i\} = \{s | \kappa_\Phi \oplus \kappa_{\mathrm{obs}}(s) < i\}$. Furthermore, it obviously holds that $\mathrm{Mod}(\psi) = \{s \in S | \kappa_\Phi(s) < \infty$ and $\kappa_{\mathrm{obs}}(s) < \infty\} = \{s \in S | (\kappa_\Phi \oplus \kappa_{\mathrm{obs}})(s) < \infty\}$. This shows that $\kappa_{\mathbf{O}_\infty \psi} \wedge \mathbf{O}_m \psi_{p+m-1} \wedge \cdots \wedge \mathbf{O}_1 \psi_p = \kappa_\Phi \oplus \kappa_{\mathrm{obs}}$. Hence, by isomorphism, $\mathbf{O}_\infty \psi \wedge \mathbf{O}_m \psi_{p+m-1} \wedge \cdots \wedge \mathbf{O}_1 \psi_p \equiv \Phi \otimes (\mathbf{O}_\infty o \wedge \mathbf{O}_r o_r \wedge \cdots \wedge \mathbf{O}_1 o_1)$. $\square$

The semantical expression (immediate from Section 3) of the combination of Φ (corresponding to, κ_Φ) by *obs* (corresponding to κ_{obs}) is simply $\kappa(s | obs) = \kappa(s) + \kappa_{\mathrm{obs}}(s) - \min_S(\kappa + \kappa_{\mathrm{obs}})$, i.e., $\kappa(.|obs) = \kappa \oplus \kappa_{\mathrm{obs}}$.

Applying Proposition 6 to the specific case of simple observations – of the form obs = $\mathbf{Only}(\mathbf{B}_k o_k)$ – gives a rather long formula that we will not write down here, except in two cases: $k = +\infty$ and $k = 1$. First, when $k = 1$:

COROLLARY 2. Let $\Phi = \mathbf{Only}(\mathbf{B}_\infty \varphi_\infty \wedge \mathbf{B}_n \varphi_n \wedge \cdots \wedge \mathbf{B}_1 \varphi_1)$ and obs = $\mathbf{Only}(\mathbf{B}_1 o_1)$. Then $\Phi \otimes obs \equiv \mathbf{Only}(\Psi)$ where Ψ is as follows:

Case 1: $\varphi_1 \wedge o_1 \not\equiv \perp$

$$\Psi = \mathbf{B}_1(\varphi_1 \wedge o_1) \wedge \mathbf{B}_2(\varphi_1 \vee (\varphi_2 \wedge o_1)) \wedge \cdots$$
$$\wedge \mathbf{B}_n(\varphi_{n-1} \vee (\varphi_n \wedge o_1)) \wedge \mathbf{B}_{n+1}(\varphi_n \vee (\varphi_\infty \wedge o_1))$$
$$\wedge \mathbf{B}_\infty \varphi_\infty$$

Case 2: $\varphi_1 \wedge o_1 \equiv \perp$

$$\Psi = \mathbf{B}_1(\varphi_1 \vee (\varphi_2 \wedge o_1)) \wedge \cdots \wedge \mathbf{B}_{n-1}(\varphi_{n-1} \vee (\varphi_n \wedge o_1))$$
$$\wedge \mathbf{B}_n(\varphi_n \vee (\varphi_\infty \wedge o_1)) \wedge \mathbf{B}_\infty \varphi_\infty$$

Then, when $obs = \mathbf{Only}(\mathbf{B}_\infty o)$ is a reliable observation, applying Proposition 6 gives

COROLLARY 3. Let $\Phi = \mathbf{Only}(\mathbf{B}_\infty \varphi_\infty \wedge \mathbf{B}_n \varphi_n \wedge \cdots \wedge \mathbf{B}_1 \varphi_1)$ and $obs = \mathbf{Only}(\mathbf{B}_\infty o)$. Assume $\varphi_\infty \wedge o \not\equiv \bot$ and let $p = \min\{j, \varphi_j \wedge o \not\equiv \bot\}$; Then (if $p \leqslant n$)

$$\Phi \otimes obs \equiv \mathbf{Only}(\mathbf{B}_\infty(\varphi_\infty \wedge o)$$
$$\wedge \mathbf{B}_{n-p+1}(\varphi_n \wedge o) \wedge \cdots \wedge \mathbf{B}_1(\varphi_p \wedge o))$$

Here is a more intuitive example.

EXAMPLE 7. Consider an agent asking pedestrians about the way to the railway station. Assume there are only two directions, r (right) and $\neg r$ (left). The agent's initial belief state is void ($\kappa_0 = \kappa_{\text{void}}$). When asking a pedestrian, five observations are possible:

- $\Phi_{\text{obs}_1} = \mathbf{Only}(\mathbf{B}_2 r)$, corresponding to a pedestrian answering "the station is on the right" without hesitation (however, the observation is considered as not fully reliable – since it is known that pedestrians sometimes give wrong indications even when seem to be sure);
- $\Phi_{\text{obs}_2} = \mathbf{Only}(\mathbf{B}_1 r)$, corresponding to a pedestrian answering "I believe it's on the right but I might be wrong");
- $\Phi_{\text{obs}_3} = \mathbf{Only}(\mathbf{B}_2 \neg r)$;
- $\Phi_{\text{obs}_4} = \mathbf{Only}(\mathbf{B}_1 \neg r)$;
- $\Phi_{\text{obs}_5} = \mathbf{Only}(\mathbf{B}_\infty \top)$ (the pedestrian answers "I have no clue").

We have for instant $\kappa_{\text{obs}_1} = \{(r, 0); (\neg r, 2)\}$ and $\kappa_{\text{obs}_4} = \{(r, 1); (\neg r, 0)\}$. Obviously, $\kappa_0 \oplus \kappa_{\text{obs}_i} = \kappa_{\text{obs}_i}$ for any i.

- After observing obs$_2$, we have $\kappa_1 = \kappa_0 \oplus \kappa_{\text{obs}_2} = \kappa_{\text{obs}_2}$ and $\Phi_1 = \Phi_0 \otimes \Phi_{\text{obs}_2} = \mathbf{Only}(\mathbf{B}_1 r)$.
- Assume now that the second pedestrian gives obs$_2$ too, Using Proposition 6, we get:

    - $\psi = \top \wedge \top = \top$;
    - $\psi_1 = r \wedge r = r$;
    - $\psi_2 = (r \wedge \top) \vee (\top \wedge r) = r$;
    - $\psi_3 = (r \wedge \top) \vee (\top \wedge \top) \vee (\top \wedge r) = \top$

    therefore $p = 1$ and $p + m = 3$, hence $\Phi_1 \otimes \Phi_{\text{obs}_1} = \mathbf{Only}(\mathbf{B}_2 r \wedge \mathbf{B}_\infty \top) = \mathbf{Only}(\mathbf{B}_2 r)$.

If the second observation had been obs_4 instead of obs_2 we would have had $\Phi_2 = \Phi_1 \otimes \Phi_{obs_4} = \mathbf{Only}(\mathbf{B}_\infty \top)$ (the agent comes back to his initial belief state). Indeed,

- $\psi = \top \wedge \top = \top$;
- $\psi_1 = r \wedge \neg r = \bot$;
- $\psi_2 = (r \wedge \top) \vee (\top \wedge \neg r) = \top$

therefore $p = 2$ and $p + m = 2$, hence $\Phi_1 \otimes \Phi_{obs_2} = \mathbf{Only}(\mathbf{B}_1 \top \wedge \mathbf{B}_\infty \top) = \mathbf{Only}(\mathbf{B}_\infty \top)$.

It can be shown by induction that after p_1 occurrences of obs_1, p_2 of obs_2, p_3 of obs_3, p_4 of obs_4 and p_5 of obs_5 (in any order), iterated combination leads to

- $\mathbf{Only}(\mathbf{B}_q r)$ if $2p_1 + p_2 > 2p_3 + p_4$ and $q = (2p_1 + p_2) - (2p_3 + p_4)$;
- $\mathbf{Only}(\mathbf{B}_q \neg r)$ if $2p_1 + p_2 < 2p_3 + p_4$ and $q = (2p_3 + p_4) - (2p_1 + p_2)$;
- $\mathbf{Only}(\mathbf{B}_\infty, \top)$ if $2p_1 + p_2 = 2p_3 + p_4$.

This example shows how observations *reinforce* prior beliefs when they are consistent with them[3]. It clearly appears that the crucial hypothesis underlying the combination rule is *independence* between the successive observations. Thus, on Example 7, the successive answers are independent (pedestrians do not listen to the answers given by their predecessors). If, on the other hand, we want to express that successive actions are dependent of each other, then we just have to add one or several hidden variables (as commonly done in Markov processes) which would have the effect of blocking (or limiting) the reinforcement[4].

We conclude this section by a discussion on the recent paper (van Ditmarsch 2004), which defines 5 revision operators, four of which appear to be instances of our revision operator:

- *minimal revision* (*1 in (van Ditmarsch 2004)) is the weakest form of revision (in the weak sense, that is, without the so-called "success postulate" telling that when revising by φ, then φ should be believed afterward; it coincides with revision by $\mathbf{Only}(\mathbf{B}_1 \varphi)$– however, the 'eventually successful' property does not hold in our framework because we also consider worlds with an infinite rank, so that if initially all φ-states have an infinite rank, then after any number of revisions by $\mathbf{Only}(\mathbf{B}_1 \varphi)$, the agent still believes $\neg \varphi$.

- *maximal revision* (*2 in (van Ditmarsch 2004)) corresponds to a revision by $\mathbf{Only}(\mathbf{B}_\infty\varphi)$ and therefore to the usual Spohnian revision, used as well in iterated revision frameworks such as (Darwiche and Pearl 1997).
- *"focus on φ" revision* (*4 in (van Ditmarsch 2004)) corresponds to a revision by $\mathbf{Only}(\mathbf{B}_k\varphi)$ such that $k = \max\{\kappa(s)|\kappa(s) < \infty\}$. The effect of such a revision is to make all φ-states that are initially possible more plausible than all $\neg\varphi$-states.
- *"successful minimal"* revision (*5 in (van Ditmarsch 2004)) corresponds to a revision by $\mathbf{Only}(\mathbf{B}_k\varphi)$ with $k = \kappa(\varphi) + 1$ – intuitively, k is the smallest integer such that the revision by $\mathbf{Only}(\mathbf{B}_k\varphi)$ ensures that φ is more believed than $\neg\varphi$.

3.4. *Related Work on Revision by Uncertain Observations*

In addition to (van Ditmarsch 2004) (discussed in Section 3), a close work to ours is (Boutilier et al. 1998), where observational systems allowing for unreliable observations are modeled using OCFs, Their work is less specific than ours (notice that in the absence of ontic actions, our revision process falls in the the class of Markovian observation systems). The main difference between (Boutilier et al. 1998) and our Section 3 is that the revision functions in (Boutilier et al. 1998) remain defined at the semantical level, which, if computed state by state following the definition, needs an exponentially large data structure. Our approach can therefore be viewed as providing a compact representation for a specific class of observation systems. In another line of work, namely (Bacchus et al. 1999), models noisy observations in a probabilistic version of the situation calculus (again, compact representation issues are not considered). (Thielscher 2001) considers noisy sensors as well in a logical framework, but with no graded uncertainty.

Belief transmutations and adjustments (Willams 1994) are based on OCFs too; however, they are based on Spohn's notion of α-conditionalization which, similarly to Jeffrey's rule in probability theory, consist in changing minimally a belief state so as to force a given formula to have the exceptionality degree α; this totally differs from a revision rule enabling an *implicit* reinforcement of belief when the observation is consistent with the initial belief state, as seen in Example 7. Likewise, the work of (Aucher 2004), which defines a logic for public and private announcements with graded plausibility, is based on Spohn's conditionalization as well. The difference between both revision is salient in probability theory as well: Jeffrey's rule has no implicit reinforcing behavior, while Pearls'rule does – see a discussion on both in (Chan and Darwiche 2003). See also (Dubois and Prade 1997) for a panorama of revision rules in numerical formalisms, including OCFs.

4. PROGRESSION

We now consider the case of ontic (or physical) actions. Progressing a belief state by an ontic action is the process consisting of projecting the expected changes implies by the action on the current belief state so as to produce a new belief state, representing the agent's beliefs after the action is performed.

Purely ontic actions may change the state of the world but do not give any feedback. Therefore, given an initial belief state κ and an ontic action α, it is possible to determine the future belief state (after the action is performed) by projecting the possible outcomes of α on the current belief state. This operation is usually called *progression*: $\text{prog}(\kappa, \alpha)$ is the belief state obtained after α is performed in belief state κ. By isomorphism, if Φ is a NO formula, we also define $\text{Prog}(\Phi, \alpha) = H(\text{prog}(G(\Phi), \alpha))$.

4.1. *Semantical Characterization of Progression*

The semantics of progression is defined as in [Boutilier 1998] by means of *OCF transition models*.

DEFINITION 9. An OCF transition model for action α is a collection of OCFs $\{\kappa_{\alpha(.|s), s \in S}\}$.

$\kappa_\alpha(s'|s)$ is the exceptionality degree of the outcome s' when performing action α in state s. Notice that for all $s \in S$, $\min_{s' \in S} \kappa_\alpha(s'|s) = 0$ holds, therefore $\kappa_\alpha(.|s)$ is an OCF. κ_α can be seen as the ordinal counterpart of stochastic transition functions.

DEFINITION 10 (Progression of κ by an ontic action). Given an initial belief state κ and an ontic action α whose dynamics is expressed by the OCF transition model κ_α, the progression of κ by α is the belief state $\kappa' = \text{prog}(\kappa, \alpha)$ defined by

$$\forall s' \in S \quad \kappa'(s') = \min_{s \in S}\{\kappa(s) + \kappa_\alpha(s'|s)\}$$

This definition appears in several places, including (Goldszmidt and Pearl 1992; Boutilier 1998, Boutilier et al. 1998). It is the ordinal counterpart of $p'(s') = \sum_{s \in S} p(s)p(s'|s, \alpha)$. Notice that κ' is a belief state, because the normalization of *both* κ and $\kappa_\alpha(.|s)$ imply

$$\min_{s \in S}\left\{\min_{s \in S}\{\kappa(s) + \kappa_\alpha(s'|s)\}\right\} = 0$$

i.e., $\min_S \kappa' = 0$

EXAMPLE 8. Consider two blocks A and B lying down on a table; the propositional variable x is true if A is on top of B, false otherwise. A robot can perform the action α consisting in try to put A on B. If A is on B in the initial state, the action has no effect; otherwise, it normally succeeds (*i.e.*, x becomes true), and exceptionally fails (in that case, x remains false). The OCF transition model for α is: $\kappa_\alpha(x|x) = 0$; $\kappa_\alpha(\neg x|x) = \infty$; $\kappa_\alpha(x|\neg x) = 0$; $\kappa_\alpha(\neg x|\neg x) = 1$.

Assume the initial state is κ_{void}, then $\kappa' = \text{prog}(\kappa_{\text{void}}, \alpha) = \{(x, 0), (\neg x, 1)\}$; now, $\text{prog}(\kappa', \alpha) = \kappa'' = \{(x, 0), (\neg x, 2)\}$. More generally, after performing α n times without performing any sensing action (starting from κ_{void}), we get $\text{prog}(\kappa_{\text{void}}, \alpha^n) = \{(x, 0), (\neg x, n)\}$, whose associated NO formula is $\mathbf{O_n}x$: after performing action α n times (without sensing), the agent believes to the degree n that A is on B.

Example 8 shows that once again, the underlying hypothesis is the independence between the outcomes of the different occurrences of actions. Indeed, the intuitive explanation of the result of previous example is that after these n executions of α, A is still not on B if and only if all n occurrences of α failed; each of the failures has an exceptionality degree of 1 and failures are independent, henceforth, n successive failures occur with an exceptionality degree of n. Notice that this reinforcement effect is a consequence of the use of $\oplus$ (if conjunction were used instead, we would still get $\mathbf{O_1}x$ after performing α n times). Again (see Section 3), this reinforcement can be limited or blocked using hidden variables expressing some correlations between the outcomes of the different action occurrences.

In the rest of this section we now show how progression can be computed syntactically, which avoids explicitly computing progression state by state consisting of a straightforward application of the definition.

4.2. *Action Theories with Exceptional Effects*

The first thing we need is a syntactical description of action effects. Therefore, we show that action effects can be described by *graded action theories*, generalizing action theories so as to allow for more or less exceptional action effects.

We first recall briefly that an action theory is a logical theory describing the effects of a given action on a set of variables (or fluents), in a language equipped with a syntactical way of distinguishing between the states of the world *before* and *after* the action is performed. Propositional action theories are usually written by duplicating each variable x of PS in x_t et x_{t+l} (representing x respectively before and after the execution of the action)[5]; this is the way we use for representing graded action theories.

Thus, let $PS_t = \{x_t \mid x \in PS\}$, $PS_{t+1}\{x_{t+1} \mid x \in PS\}$, $S_t = 2^{PS_t}$ and $S_{t+1} = 2^{PS_{t+1}}$. For any formula Φ, let Φ_t (resp. Φ_{t+1}) be the formula obtained from Φ by replacing each occurrence of x by x_t (resp. x_{t+1}). A *graded action theory* is a NO formula of this extended language: $\Sigma_\alpha = \mathbf{Only}(\mathbf{B}_\infty r \wedge \mathbf{B}_n r_n \wedge \cdots \mathbf{B}_1 r_1)$. We just give the graded action theory corresponding to Example 8:

$$\Sigma_\alpha = \mathbf{Only}(\mathbf{B}_\infty (x_t \rightarrow x_{t+1}) \wedge \mathbf{B}_1 x_{t+1})$$

The graded action theory can be obtained from a set of causal (dynamic or static) rules through a completion process whose technical details are omitted because they are only little relevant to the subject of this paper. This completion does not present any particular difficulty: it is an easy extension of completion for nondeterministic action theories such as in (Lin 1996; Giunchiglia et al. 2003).

4.3. *Syntactical Characterization of Progression*

We now show how progression can be computed syntactically, which avoids explicitly computing progression state by state consisting of a straightforward application of the definition.

Like for the static case, any OCF transition models correspond to graded action theories and *vice versa*: $\{\kappa_\alpha(.|s), s \in S\}$ induces $\Sigma_\alpha = \mathbf{Only}(\mathbf{B}_\infty r \wedge \mathbf{B}_n r_n \wedge \cdots \wedge \mathbf{B}_1 r_1)$ where $r_i = \mathrm{Form}\{(s'_{t+1}, s_t) \mid \kappa_\alpha(s'_{t+1} \mid s_t) < i\}$.

Now, we recall the definition of *forgetting* a subset of propositional variables X from an objective propositional formula ψ (Lin and Reiter 1994):

1. $\text{forget}(\{x\}, \psi) = \psi_{x \leftarrow \top} \vee \psi_{x \leftarrow \bot}$;
2. $\text{forget}(X \cup \{x\}, \psi) = \text{forget}(\{x\}, \text{forget}(X, \psi))$.

Forgetting is extended to S5 formulas in (Herzig et al. 2003) and is here extended to NO formulas in the following way: if $\Phi = \mathbf{Only}(\mathbf{B}_\infty \varphi \wedge \mathbf{B}_n \varphi_m \wedge \cdots \wedge \mathbf{B}_1 \varphi_1)$ and $X \subset \text{Var}(\Phi)$, then $\text{Forget}(X, \Phi) = \mathbf{Only}(\mathbf{B}_\infty \text{forget}(X, \varphi) \wedge \mathbf{B}_n \text{forget}(X, \varphi_n) \wedge \cdots \wedge \mathbf{B}_1 \text{forget}(X, \varphi_1))$.

Now we have the following syntactical characterization of progression:

PROPOSITION 7. Let Φ be the NO formula corresponding to the initial belief state κ, and α an ontic action described by an action theory as previously defined. Then

$$\text{Prog}(\Phi, \alpha) \equiv \text{Forget}(PS_t, \Phi_t \otimes \Sigma_\alpha)$$

We start by proving the following Lemma.

LEMMA 1. *Let $\{X, Y\}$ be a partition of PS and κ an OCF on 2^{PS}. Define $\kappa_X : 2^X \times \overline{\mathbb{N}}$ by: for all $s_X \in 2^X$, $\kappa_X(s_X) = \min\{\kappa(s_X, s_Y)$ s.t. $s_Y \in 2^Y\}$. Then $\Phi_{\kappa_X} = \text{Forget}(\Phi_\kappa, Y)$.*

Proof. Notice first that $\min \kappa_X = 0$, therefore κ_X is an OCF. Now, let $i \in \{1, \ldots, n, \infty\}$ and $s_X \in 2^X$. Assume $s_X \not\models \text{forget}(Y, \varphi_i)$. Then, by Corollary 5 of Proposition 20 in (Lang et al. 2003), there is no $s_Y \in 2^Y$ such that $(s_X, s_Y) \models \varphi_i$; therefore, for all $s_Y \in 2^Y$ we have $\kappa(s_X, s_Y) \geqslant i$ and $\min_{s_Y \in 2^Y} \kappa(s_X, s_Y) \geqslant i$, i.e., $\kappa_X(s_X) \geqslant i$. Conversely, assume $s_X \models \text{forget}(Y, \varphi_i)$. Then, again from Corollary 5 of Proposition 20 in (Lang et al. 2003), there exists a $s_Y \in 2^Y$ such that $(s_X, s_Y) \models \varphi_i$, therefore $\min_{s_Y \in 2^Y} \kappa(s_X, s_Y) < i$, i.e., $\kappa_X(s_X) < i$. In summary, for every i and every s_X, $\kappa_X(s_X) < i$ iff $s_X \models \text{forget}(Y, \varphi_i)$, which enables us to conclude that $\Phi_{\kappa_X} = \text{Forget}(\Phi_\kappa, Y)$. $\square$
We now prove Proposition 7.

Proof. We start by defining the cylindrical extension $\tilde{\kappa}$ of κ to $2^{S_t \times S_{t+1}}$ by: for all $s_t \in S_t$, $\tilde{\kappa}(s_t, s_{t+1}) = \kappa(s_t)$. Then, by definition 10, and using $\min_{(s_t, s_{t+1}) \in S_t \times S_{t+1}} \{\kappa(s_t) + \kappa_\alpha(s_{t+1} \mid s_t)\} = 0$ we get $\kappa'(s_{t+1}) = \min_{s_t \in S_t} \{(\tilde{\kappa} \oplus \kappa_\alpha)(s_t, s_{t+1})\}$. Now, the definition of r_i ($i = 1, \ldots, n, +\infty$) implies $\kappa_\alpha = \kappa_{\Sigma_\alpha}$ and the definition of $\tilde{\kappa}$ implies $\tilde{\kappa} = \kappa_{\Phi_t}$. Therefore, by Definition 5, we get: $\tilde{\kappa} \oplus \kappa_\alpha = G(\Phi_t \otimes \Sigma_\alpha)$. Now, $\kappa'(s_{t+1}) = \min_{s_t \in S_t} \kappa(s_t) + \kappa_\alpha(s_{t+1}|s_t) = \min_{s_t \in S_t} (\tilde{\kappa} \oplus \kappa_\alpha)(s_t, s_{t+1})$. Now, using Lemma 1, $\kappa'(s_{t+1}) = \kappa_{\mathrm{Forget}(\mathrm{Var}_t, \Phi_t \otimes \Sigma_\alpha)}$, which, by isomorphism, is equivalent to $\mathrm{Prog}(\Phi, \alpha) = \mathrm{Forget}(PS_t, \Phi_t \otimes \Sigma_\alpha)$.   □

Thus, progression amounts to a combination followed by a forgetting. For the first step, Proposition 6 can be applied again, as shown on the following example. The second step amounts to a sequence of classical forgetting operations.

EXAMPLE 8 (Continued). We have

$$\Sigma_\alpha = \mathbf{Only}(\mathbf{B}_\infty(x_t \to x_{t+1}) \wedge \mathbf{B}_1 x_{t+1})$$

The initial belief state corresponds to

$$\Phi = \mathbf{Only}(\mathbf{B}_1 x)$$

Then,

$$\Phi_t \otimes \Sigma_\alpha = \mathbf{Only}(\mathbf{B}_\infty \psi \wedge \mathbf{B}_n \psi_n \wedge \cdots \wedge \mathbf{B}_1 \psi_1)$$

where

$$\psi = \top \wedge (x_t \to x_{t+1});$$
$$\psi_1 = x_t \wedge x_{t+1};$$
$$\psi_2 = (x_t \wedge (x_t \to x_{t+1})) \vee (\top \wedge x_{t+1});$$
$$\psi_3 = (x_t \wedge (x_t \to x_{t+1})) \vee (\top \wedge (x_t \to x_{t+1}) \vee (\top \wedge x_{t+1}))$$

After simplifying the expression we get $\psi = x_t \to x_{t+1}$; $\psi_1 = x_t \wedge x_{t+1}$; $\psi_2 = x_{t+1}$; $\psi_3 = x_t \to x_{t+1} = \psi$. Next, we get $\Phi_t \otimes \Sigma_\alpha \equiv \mathbf{Only}(\mathbf{B}_\infty(x_t \to x_{t+1}) \wedge \mathbf{B}_1(x_t \wedge x_{t+1}) \wedge \mathbf{B}_2 x_{t+1})$ and $\mathrm{Forget}(PS_t, \Phi_t \otimes \Sigma_\alpha) = \mathbf{Only}(\mathbf{B}_\infty \top \wedge \mathbf{B}_1 x_{t+1} \wedge \mathbf{B}_2 x_{t+1}) = \mathbf{Only}(\mathbf{B}_2 x_{t+1})$, and finally $\mathrm{Prog}(\Phi, \alpha) = \mathbf{Only}(\mathbf{B}_2 x)$.

Note the importance of combination, which explains the reinforcement obtained when chaining several actions. Such a reinforcement would not be obtained if conjunction were used instead of combination: doing α many times would give $\mathbf{B}_1 x$ again and again.

4.4. Related Work on Actions with Exceptional Effects

Goldszmidt and Pearl 1992 and Boutilier 1999 study belief update operators with belief states modeled by OCFs, so as to model exceptional effects of actions. These operators are very similar to our progression for ontic actions from a semantical point of view – but they do not give any syntactical characterization of progression. Shapiro et al. 2000 considers physical and sensing actions in a situation calculus setting, where states are mapped to a plausibility values; these plausibility values are simply inherited from plausibility values in the initial belief state (noisy observations and exceptional effects actions are not considered). See also Baral and Lobo (1997) for a language for describing normal effects in action theories. Lang et al. 2001 define also an update operator for belief states modeled by OCFs, but this operator, which plays more or less for belief update the role played by transmutations for belief revision, is very different from the one given in this article and could not even handle our simple Example 8.

5. ON-LINE EXECUTION OF BBPs

5.1. *Execution and Progression*

BBPs have been defined in Section 2 from a set of propositional symbols PS and a set of primitive actions ACT. For the sake of simplicity, primitive actions are assumed to be either purely physical (or ontic) or purely informative actions: $ACT = ACT_P \cup ACT_I$ (where actions in ACT_P are physical and actions in ACT_I are purely informative, that is, pure sensing actions). This simplification is usual (see Scherl and Levesque 1993; Herzig et al. 2000; Reiter 2001a) and does not induce any loss of generality, as any complex action with both physical and informative effects can be decomposed in two actions performed in sequence, the first one being purely physical and the second one purely informative.

The *on-line execution* of a belief program is a function mapping a pair consisting of an initial belief state and a program to a set of *traces* of the program.

DEFINITION 11 (Traces). A trace is a sequence $\tau = \langle \langle \kappa_t, \alpha_t, \text{obs}_t \rangle_{0 \leq t \leq T-1}, \kappa_T \rangle$ where $T \geq 0$ and for all t, κ_t is a belief state, α_t an action and obs_t an observation. (*If $T = 0$ then $\tau = \langle \kappa_0 \rangle$.*)

We make use of the following notations:

- if $T \neq 0$ then we write $\tau = \langle \kappa_0, \alpha_0, \text{obs}_0 \rangle.\tau'$, where $\tau' = \langle \langle \kappa_t, \alpha_t, \text{obs}_t \rangle_{1 \leq t \leq T-1}, \kappa_T \rangle$.
- $\text{tail}(\tau) = \kappa_T$.

[249]

As in Reiter (2001b), each time a program interpreter adds a new action α_t to its action history, the robot (or whatever entity executing the program) also physically performs this action. Since some of these actions are informative actions, we cannot predict off-line the outcome of the program, therefore we must consider a set of possible executions of the program, i.e., a set of traces.

We first have to define an informative action formally. As we said previously, the notion of informative action we need is more complex that actions of the type sense(ψ) used e.g. in Reiter(2001b). These reliable and precise test actions sense(ψ), that send back obs($\mathbf{K}\psi$) if ψ is true in the actual state and obs($\mathbf{K}\neg\psi$) otherwise, are generally assumed to be deterministic, that is, the observation they send back is a function of the actual state of the world. Because we want to allow for possibly unreliable observations, we cannot assume informative actions to be deterministic: the possibility of gathering unreliable pieces of information must come together with the possibility of having several possible observations *even if the state of the world is given*: for instance, in Example 7, given that the station is on the right ($s = r$), we may, for instance, observe either $\mathbf{B}_1 r$, $\mathbf{B}_2 r$, $\mathbf{K}r$, $\top$, $\mathbf{B}_1 \neg r$ and $\mathbf{B}_2 \neg r$, $\mathbf{K}\neg r$ cannot occur as an observation in that state.

DEFINITION 12 (Feedback function). A feedback function

$$\text{feedback: } ACT \times S \rightarrow 2^{OBS}$$

maps each action and each state to a set of observations, satisfying the following requirements:

1. feedback$(\alpha, s) \neq \emptyset$;
2. if α is ontic then feedback$(\alpha, s) = \{\text{obs}_{\text{void}}\}$
3. if obs $\in$ feedback(α, s) then obs$(s) < \infty$.

feedback(α, s) is the set of possible observation obtained after performing the sensing action α in state s. Condition 1 requires each action to send back a feedback (possibly void). Ontic actions cannot send any non-void feedback (Condition 2) (alternatively, we could have restricted the definition of the feedback function to informative actions only, but not doing this allows for simpler and shorter definitions further on). Condition 3 ensures a minimum level of consistency between the feedback and the current state, since we exclude that an observation occurs in a state that it totally excludes. The

[250]

reason for requirement (3) is that revision of κ by obs($\mathbf{Only(K\varphi)}$) is not defined when $\kappa(\varphi) = \infty$ (cf. Section 3); thus, (3) excludes fully contradicting sequences of observations such as obs($\mathbf{K\varphi}$) followed by obs($\mathbf{K\neg\varphi}$) without any ontic action being executed inbetween. In particular, a fully reliable test action such as sense(ψ) as in (Scherl and Levesque 1993; Reiter 2001b) is modeled by the following feedback function:

$$\text{feedback}(\alpha, s) = \begin{cases} \{\text{obs}(\mathbf{Only(K\psi)})\} & \text{if } s \models \psi \\ \{\text{obs}(\mathbf{Only(K\neg\psi)})\} & \text{if } s \models \neg\psi \end{cases}$$

But generally, there may be any number of possible outcomes for a given sensing action, including possible void observations (obs$_{\text{void}}$ = $\mathbf{Only(K\top)}$).

Now, the agent generally does not know the actual state of the world with precision, which calls for extending the feedback function from states to belief states:

DEFINITION 13 (Subjective feedback). Let feedback be a feedback function. Then the subjective feedback function feedback$_S$ induced by feedback is the function

$$\text{feedback}_S : \text{ACT} \times \text{BS} \to 2^{\text{OBS}}$$

mapping each action and each belief state to a set of observations, defined by

$$\text{feedback}_S(\alpha, \kappa) = \bigcup \{\text{feedback}(\alpha, s) \mid \kappa(s) < \infty\}$$

It is easily checked that the following properties follow immediately from Definitions 12 and 13.

1. feedback$_S(\alpha, \kappa) \neq \emptyset$;
2. if α is ontic then feedback$_S(\alpha, \kappa) = \{\text{obs}_{\text{void}}\}$;
3. if obs $\in$ feedback$_S(\alpha, \kappa)$ then rev(κ, obs) is defined.

In the specific case where α is a fully reliable truth test that sends back obs($\mathbf{K\varphi}$) or obs($\mathbf{K\neg\varphi}$) then

$$\text{feedback}_S(\alpha, k) = \begin{cases} \{\text{obs}(\mathbf{K\varphi})\} & \text{if } \kappa \models \mathbf{K\varphi} \\ \{\text{obs}(\mathbf{K\neg\varphi})\} & \text{if } \kappa \models \mathbf{K\neg\varphi} \\ \{\text{obs}(\mathbf{K\varphi})\}, \{\text{obs}(\mathbf{K\neg\varphi})\} & \text{otherwise} \end{cases}$$

In other words, if the agent already *knows* that φ is true, then the only possible feedback is observing that φ is true (otherwise the agent would have had an incorrect infinite belief, that is, incorrect knowledge).

We now define the set of possible executions of a BBP π in an initial belief state κ.

DEFINITION 14 (possible executions of a BBP). A trace $\tau = \langle\langle \kappa_t, \alpha_t, \text{obs}_t\rangle_{0 \leqslant t \leqslant T-1}, \kappa_T \rangle$ is a possible execution of the BBP π in the belief state κ iff one of the following conditions is satisfied:

1. $\tau = \langle \kappa \rangle$ and $\pi = \lambda$;
2. (a) $\tau = \langle \kappa, \alpha, \text{obs}\rangle.\tau'$ where $\alpha \in \text{ACT}_P$ (b) $\pi = \alpha; \pi'$, (c) obs $=$ obs$_{\text{void}}$ and (d) τ' is a possible execution of ϕ' in prog (κ, α);
3. (a) $\tau = \langle \kappa, \alpha, \text{obs}\rangle.\tau'$ where $\alpha \in \text{ACT}_I$, (b) $\pi = \alpha; \pi'$, (c) obs $\in$ feedback$_S(\alpha, \kappa)$ and (d) τ' is a possible execution of π' in rev (κ, obs);
4. (a) $\tau = \langle \kappa, \alpha, \text{obs}\rangle.\tau'$, (b) $\pi = $ if Φ then π_1 else $\pi_2; \pi_3$, and (c) either $\kappa \models \Phi$ and τ is a possible execution of $(\pi_1; \pi_3)$ in κ, or $\kappa \models \neg\Phi$ and τ is a possible execution of $(\pi_2; \pi_3)$ in κ.
5. (a) $\tau = \langle \kappa, \alpha, \text{obs}\rangle.\tau'$, (b) $\pi = $ while Φ do $\pi_1; \pi_2$, and (c) either $\kappa \models \Phi$ and τ is a possible execution of $(\pi_1; \pi)$ in κ, or $\kappa \models \neg\Phi$ and τ is a possible execution of π_2 in κ.

We denote by exec(π, κ) the set of possible executions of π in κ.

It is easily shown that any BBP has at least one possible execution in any belief state. There may be infinitely many such possible executions, as shown in the following example.

EXAMPLE 9. Consider Example 7 again. Here are some possible executions of $\pi = $ while $\neg(\mathbf{B}_2 r \vee \mathbf{B}_2 \neg r)$ do ask in $\kappa_{\mathbf{Only}(\top)}$, where $\kappa_0 = \kappa_{\mathbf{Only}(\top)}$, $\kappa_1 = \kappa_{\mathbf{Only}(\mathbf{B}_1 r)}$, $\kappa_2 = \kappa_{\mathbf{Only}(\mathbf{B}_1 \neg r)}$, $\kappa_3 = \kappa_{\mathbf{Only}(\mathbf{B}_2 r)}$, $\kappa_4 = \kappa_{\mathbf{Only}(\mathbf{B}_2 \neg r)}$, obs$_1 = obs(\mathbf{Only}(\mathbf{B}_1 r))$ and obs$_2 = \mathbf{Only}(\mathbf{B}_1 \neg r)$.

- $\langle\langle \kappa_0, \text{ask}, \text{obs}_1\rangle, \langle \kappa_1, \text{ask}, \text{obs}_1\rangle, \kappa_2\rangle$;
- $\langle\langle \kappa_0, \text{ask}, \text{obs}_2\rangle, \langle \kappa_3, \text{ask}, \text{obs}_1\rangle, \kappa_4\rangle$;
- $\langle\langle \kappa_0, \text{ask}, \text{obs}_1\rangle, \langle \kappa_1, \text{ask}, \text{obs}_2\rangle, \langle \kappa_0, \text{ask}, \text{obs}_1\rangle,$
  $\langle \kappa_1, \text{ask}, \text{obs}_1\rangle, \kappa_2\rangle$;
- $\langle\langle \kappa_0, \text{ask}, \text{obs}_1\rangle, \langle \kappa_1, \text{ask}, \text{obs}_2\rangle, \langle \kappa_0, \text{ask}, \text{obs}_2\rangle,$
  $\langle \kappa_3, \text{ask}, \text{obs}_2\rangle, \kappa_2\rangle$; etc.

There are infinitely many possible executions of π. They can be finitely described by the regular expression $[(ab \cup cd)^*; (ae \cup cf)]$, where $a = \langle \kappa_0, \text{ask}, \text{obs}_1 \rangle, b = \langle \kappa_1, \text{ask}, \text{obs}_2 \rangle, c = \langle \kappa_0, \text{ask}, \text{obs}_2 \rangle, d = \langle \kappa_3, \text{ask}, \text{obs}_1 \rangle, e = \langle \kappa_1, \text{ask}, \text{obs}_1 \rangle$ and $f = \langle \kappa_3, \text{ask}, \text{obs}_2 \rangle)$.

We now show how progression can be extended from single actions to belief-based programs, and then show the correspondance with the set of possible executions of the program.

DEFINITION 15 (Progression of an initial belief state by a BBP). Given a BPP π and an *NO* formula Φ, the progression of Φ by π is the set of *NO* formulas $\text{Prog}(\Phi, \pi)$ defined inductively by

- $\text{Prog}(\Phi, \lambda) = \{\Phi\}$;
- if $\pi = \alpha; \pi'$ with $\alpha \in \text{ACT}_P$ then $\text{Prog}(\Phi, \pi) = \text{Prog}(\text{Prog}(\Phi, \alpha), \pi')$
- if $\pi = \alpha; \pi'$ with $\alpha \in \text{ACT}_I$ then

$$\text{Prog}(\Phi, \pi) = \bigcup_{\text{obs} \in \text{feedback}(\alpha, \kappa_\Phi)} \text{Prog}(\Phi \otimes \text{obs}, \pi')$$

- if $\pi = (\text{if } \Psi \text{ then } \pi_1 \text{ else } \pi_2); \pi_3$ then

$$\text{Prog}(\Phi, \pi) = \begin{cases} \text{Prog}(\Phi, (\pi_1; \pi_3)) & \text{if } \Phi \models \Psi \\ \text{Prog}(\Phi, (\pi_2; \pi_3)) & \text{otherwise} \end{cases}$$

- if $\pi = (\text{while } \Psi \text{ do } \pi_1); \pi_2$ then

$$\text{Prog}(\Phi, \pi) = \begin{cases} \text{Prog}(\Phi, (\pi_1; \pi)) & \text{if } \Phi \models \Psi \\ \text{Prog}(\Phi; \pi_2) & \text{otherwise} \end{cases}$$

The following result guarantees that the syntactical way of computing progression is correct.

PROPOSITION 8. $\Psi \in \text{Prog}(\pi, \Phi)$ iff there is an execution $\tau \in \text{exec}(\pi, \kappa_\Phi)$ such that $\text{tail}(\tau) = \kappa_\Psi$.
Remark that an equivalent formulation of the above identity is

$$\text{tail}(\text{exec}(\pi, \kappa)) = \{\kappa_\Psi, \Psi \in \text{Prog}(\pi, \Phi_\kappa)\}$$

where $\text{tail}(X) = \{\text{tail}(\tau) \text{ s.t. } \tau \in X\}$.

Proof. By induction on the size of π. We first define the size of a BBP inductively by: $\text{size}(\lambda) = 0$; $\text{size}(\alpha) = 1$ for $\alpha \neq \lambda$; $\text{size}(\pi;\pi')$ $= \text{size}(\pi) + \text{size}(\pi')$; $\text{size}(\text{if } \Phi \text{ then } \pi_1 \text{ else } \pi_2) = \max(\text{size}(\pi_1), \text{size}(\pi_2)) + 1$; $\text{size}(\text{while } \Phi \text{ do } \pi') = \text{size}(\pi') + 1$. Let us consider the induction hypothesis

$$I(m): \text{ for all } \pi \text{ such that } \text{size}(\pi) \leqslant m \text{ and for all } \kappa,$$
$$\text{tail}(\text{exec}(\pi, \kappa)) = \{\kappa_\Psi \text{ s.t. } \Psi \in \text{Prog}(\Phi_{\kappa,\pi})\}$$

If $\pi = \lambda$ then $\text{exec}(\pi, \Phi) = \{\Phi\} = \{\Phi\}_{\kappa_\Phi}$ by Corollary 1. Therefore $I(0)$ is verified. Assume now that $I(m)$ is verified and let π be a BBP such that $\text{size}(\pi) = m + 1$.

- if $\pi = \alpha; \pi'$ and $\alpha \in \text{ACT}_P$. Since $\text{size}(\pi') = m$, $I(m)$ implies that $\text{exec}(\pi', \kappa) = \{\kappa_\Psi, \Psi \in \text{Prog}(\Phi_\kappa, \pi')\}$. Then we have the following chains of equivalences:
  $\Psi \in \text{Prog}(\Phi, \pi)$
  iff $\Psi \in \text{Prog}(\Phi, (\alpha; \pi'))$
  iff $\Psi \in \text{Prog}(\text{Prog}(\Phi, \alpha)\pi')$
  iff there is a $\tau' \in \text{exec}(\pi', \kappa_{\text{Prog}(\Phi,\alpha)})$ such that $\text{tail}(\tau') = \kappa_\Psi$
  iff there is a $\tau' \in \text{exec}(\pi', \text{Prog}(\kappa_\Phi, \alpha))$ such that $\text{tail}(\tau') = \kappa_\Psi$
  iff there is a $\tau \in \text{exec}((\alpha; \pi'), \kappa_\Phi)$ such that $\text{tail}(\tau') = \kappa_\Psi$
  iff there is a $\tau \in \text{exec}(\pi, \kappa_\Phi)$ such that $\text{tail}(\tau) = \kappa_\Psi$.
  iff $\kappa_\Psi \in \text{tail}(\text{exec}(\pi, \kappa_\Phi))$.

- let $\pi = \alpha; \pi'$ and $\alpha \in \text{ACT}_I$. Again, $\text{exec}(\pi', \kappa) = \{\kappa_\Psi, \Psi \in \text{Prog}(\Phi_\kappa, \pi')\}$ holds the induction hypothesis. Then we have the following chains of equivalences:
  $\Psi \in \text{Prog}(\Phi, \pi)$
  iff $\Psi \in \text{Prog}(\Phi, (\alpha; \pi'))$
  iff $\Psi \in \bigcup \{\text{Prog}(\Phi \otimes \text{obs}) \mid \text{obs} \in \text{feedback}(\alpha, \Phi)\}$
  iff there is a $\text{obs} \in \text{feedback}(\alpha, \Phi)$ and a $\tau \in \text{exec}(\pi', \kappa_{\Phi \otimes \text{obs}})$ such that $\text{tail}(\tau) = \kappa_\Psi$
  iff there is a $\text{obs} \in \text{feedback}(\alpha, \Phi)$ and a $\tau \in \text{exec}(\pi', \text{rev}(\kappa_\Phi, \kappa_{\text{obs}}))$ such that $\text{tail}(\tau) = \kappa_\Psi$
  iff there is a $\tau \in \text{exec}((\alpha; \pi'), \kappa_\Phi)$ such that $\text{tail}(\tau) = \kappa_\Psi$
  iff $\kappa_\Psi \in \text{tail}(\text{exec}(\pi, \kappa_\Psi))$

- let $\pi = (\text{if } \Gamma \text{ then } \pi_1 \text{ else } \pi_2); \pi_3$. Note that $\text{size}(\pi_1; \pi_3) \leqslant m$ and $\text{size}(\pi_2; \pi_3) \leqslant m$, therefore the induction hypothesis can be applied to $\pi_1; \pi_3$ and to $\pi_2; \pi_3$. Then we have $\Psi \in \text{Prog}(\Phi, \pi)$

iff either $\Phi \models \Gamma$ and $\Psi \in \text{Prog}(\Phi, (\pi_1; \pi_3))$ or $\Phi \nvDash \Gamma$ and $\Psi \in \text{Prog}(\Phi, (\pi_2; \pi_3))$

iff either $\kappa_\Phi \models \Gamma$ and there is a $\tau \in \text{exec}$ $((\pi_1; \pi_3), \kappa_\Phi)$ such that $\text{tail}(\tau) = \kappa_\Psi$ or $\kappa_\Phi \nvDash \Gamma$ and there is a $\tau \in \text{exec}((\pi_2; \pi_3), \kappa_\Phi)$ such that $\text{tail}(\tau) = \kappa_\Psi$

iff $\kappa_\Psi \in \text{tail}(\text{exec}(\pi, \kappa_\Phi))$.

- the case $\pi = \text{while } \Gamma \text{ do } \pi_1; \pi_2$ is similar to the latter.

Therefore, in all cases we have $\Psi \in \text{Prog}(\Phi, \pi)$ *iff* $\kappa_\Psi \in \text{tail}(\text{exec}(\pi, \kappa_\Phi))$, or equivalently, $\text{tail}(\text{exec}(\pi, \kappa)) = \{\kappa_\Psi \text{ s.t. } \Psi \in \text{Prog}(\Phi_\kappa, \pi)\}$. This being true for any π of size $m + 1$, we have shown that the induction hypothesis carries on from m to $m + 1$, which completes the proof.

$\square$

EXAMPLE 10. Consider Example 7 again. We have the following:

$$\text{Prog}(\pi, \text{Only}(\top)) = \{\text{Only}(\mathbf{B}_2 r), \text{Only}(\mathbf{B}_2 \neg r)\}$$

Remark here that $\text{Prog}(\pi, \textbf{Only}(\top))$ is finite although exec $(\pi, \kappa_{\textbf{Only}(\top)})$ is infinite.

5.2. *BBP as Implicit and Compact Representations of POMDP Policies*

POMDPs are the dominant approach for planning under partial observability (including nondeterministic actions and unreliable observations) – see for instance (Kaelbling et al. 1998; Bonet and Geffner 2001) for two of the most relevant references on planning with POMDPs). The relative plausibility of observations given states, as well as the notion of progressing a belief state by an action, has its counterparts in POMDPs. Now, there are two important differences between POMDPs and our work.

A POMDP policy σ is a labeled automaton, that is, a graph whose vertices are labeled by actions and edges by observations, and the outcoming edges from a vertex v labeled by α are labeled by a possible feedback obs of α (in particular, if α is ontic then there is a unique outcoming edge from v, labeled by obs_{void}).

Unlike a BBP, a policy can be followed without needing to perform a deduction task for evaluating a branching condition: a policy is executed just by following the observation flow and executing the indicated actions.

[255]

Given a **BBP** π and an initial belief state, it is possible to "compile" π into a policy σ, by simulating its execution and evaluating the branching conditions for each possible observation sequence. For the sake of simplicity we define this induced policy as a tree; it is then possible to reduce the tree into a smaller graph by a standard automaton minimization process.

In the following definition we denote by Tree(α, $\langle$obs$_1$, $\tau_1\rangle$, ..., $\langle$obs$_p$, $\tau_p\rangle$) the tree whose root is labelled by α and containing p subtrees $\tau_1, \ldots, \tau_p$, labeled respectively by obs$_1, \ldots,$ obs$_p$.

DEFINITION 16. Let π be a **BBP** and κ a belief state. Then the policy $\sigma =$ policy (π, κ) induced by π and κ is defined inductively by

- policy(λ, κ) is the tree composed of a single vertex labeled by λ;
- if $\pi = \alpha; \pi'$ with $\alpha \in$ ACT$_P$ then

$$\text{policy}(\pi, \kappa) = \text{Tree}(\alpha, \langle\text{obs}_{\text{void}}, \text{policy}(\pi', \text{prog}(\kappa, \alpha))\rangle)$$

- if $\pi = \alpha; \pi'$ with $\alpha \in$ ACT$_I$ then

$$\text{policy}(\pi, \kappa) = \text{Tree}(\alpha, \langle\text{obs}_1, \text{policy}(\pi', \text{rev}(\kappa,$$
$$\text{obs}_1)\rangle), \ldots, \langle\text{obs}_p, \text{policy}(\pi', \text{rev}(\kappa, \text{obs}_p)\rangle)$$

where $\{\text{obs}_1, \ldots, \text{obs}_p\} = \text{feedback}_S(\alpha, \kappa)$.
- if $\pi = (\text{if } \Phi \text{ then } \pi_1 \text{ else } \pi_2); \pi_3$ then

$$\text{policy}(\pi, \kappa) = \begin{cases} \text{policy}((\pi_1; \pi_3), \kappa) & \textit{if } \kappa \models \Phi \\ \text{policy}((\pi_2; \pi_3), \kappa) & \text{otherwise} \end{cases}$$

- if $\pi = (\text{while } \Phi \text{ do } \pi'); \pi''$ then

$$\text{policy}(\pi, \kappa) = \begin{cases} \text{policy}((\pi'; \pi), \kappa) & \textit{if } \kappa \models \Phi \\ \text{policy}(\pi'', \kappa) & \text{otherwise} \end{cases}$$

The crucial difference between a **BBP** and the policy implementing it is in the expression of branching conditions :

- in a **BBP** branching conditions are *subjective*, since they refer to the current belief state of the agent.
- in a **POMDP** policy, branching conditions are *objective*: the next action is dictated by the last observation made.

This difference in the nature of branching conditions has two important practical consequences:

[256]

1. a policy is directly implementable (at each point in the policy execution, the next action to be performed is specified directly from the feedback and is therefore determined in linear time, just by following the edge corresponding to the observation made in the policy graph). Contrariwise, a BBP is not directly implementable, since branching conditions have first to be evaluated. Evaluating a branching condition is a coNP-hard problem that has to be solved on-line: thus, BBPs need a *deliberation phase* when being executed, while policies do not.
2. a BBP is a much more compact description of the policy than the explicit specification of the policy itself. Indeed, policies induced by BBPs without while statements are, in the worst case, exponentially larger than that the BBP they implement.

A policy is a particular case of a *protocol* in the sense of Fagin et al. (1995). A single-agent protocol maps the *local state* of the agent to an action; here, a local state is defined by the sequence of observations and actions performed so far (and thus corresponds to a vertex in the policy tree). A more extensive discussion on the differences between protocols and KBPs can be found in Fagin et al. (1995). We end up this discussion by giving two examples.

EXAMPLE 11. Consider the BBP

$$\text{while} \neg (\mathbf{B}_2 x \vee \mathbf{B}_2 \neg x) \text{ do ask}$$

applied in the void initial belief state κ_{void}.

Then the policy σ implementing π, as it is defined above, is an infinite tree, which can easily be shown to be reducible to the following finite graph $G = \langle V, E \rangle$ defined by:

- $V = \{v_\top, v_{\mathbf{B}_1 x}, v_{\mathbf{B}_2 x}, v_{\mathbf{B}_1 \neg x}, v_{\mathbf{B}_2 \neg x}\}$;
- $v_\top$, $v_{\mathbf{B}_1 x}$ and $v_{\mathbf{B}_1 \neg x}$ are labeled by ask whereas $v_{\mathbf{B}_2 x}$ and $v_{\mathbf{B}_2 \neg x}$ are labeled by λ.
- $E = \{(v_\top, \text{obs}(\mathbf{B}_1 x), v_{\mathbf{B}_1 x}), (v_\top, \text{obs}(\mathbf{B}_1 \neg x), v_{\mathbf{B}_1 \neg x}), (v_{\mathbf{B}_1 x}, \text{obs}(\mathbf{B}_1 x), v_{\mathbf{B}_2 x}), (v_{\mathbf{B}_1 x}, \text{obs}(\mathbf{B}_1 \neg x), v_\top), (v_{\mathbf{B}_1 \neg x}, \text{obs}(\mathbf{B}_1 \neg x), v_{\mathbf{B}_2 \neg x}), (v_{\mathbf{B}_1 \neg x}, \text{obs}(\mathbf{B}_1 x), v_\top)\}$, where $(v_\top, \text{obs}(\mathbf{B}_1 x), v_{\mathbf{B}_1 x})$ denotes an edge from $v_\top$ to $v_{\mathbf{B}_1 x}$ labeled by $\text{obs}(\mathbf{B}_1)x)$, etc.

EXAMPLE 12. Consider a model-based diagnosis problem, with n components $1, \ldots, n$. For each component i, the propositional variable $\text{ok}(i)$ represents the status of component i (working state if ok(i) is true, or failure state otherwise). Σ is a propositional formula expressing links between the components, given some background knowledge about the system plus possibly some initial measurements: for instance, $\Sigma = (\neg ok(1) \vee \neg ok(2)) \wedge (\neg ok(1) \vee \neg ok(3)) \wedge \neg ok(4)$ means that one of the components 1 and 2 is faulty, one of the components 1 and 3 is, and component 4 is faulty. Each component can be inspected by means of a purely informative action inspect(i) whose feedback is either $\mathbf{K}ok(i)$ or $\mathbf{K}\neg ok(i)$, and repaired by means of an ontic action repair(i) whose effect is ok(i). For the sake of simplicity we assume that beliefs are nongraded. Consider the following BBP π:

```
while ¬K(ok(1) ∧ ⋯ ∧ ok(n))
do
        pick a i such that ¬Kok(i)
        if K¬ok(i)
        then repair(i)
        else if ¬Kok(i)
             then inspect(i)
             end if
        end if
end while
```

It can be shown that this program is guaranteed to stop after less than $2n$ actions. The size of the policy σ induced by π is, in the worst case, exponential in n.

To sum up, BBPs are a smart and compact way of specifying policies, which, on the other hand, requires much more computational tasks at execution time than the explicit policy.

Our work can thus be seen as a first step towards bridging KBPs and POMDPs. It would be interesting to go further and to build a language for BBPs describing "real" POMDP (with probabilistic belief states). This would require a rather deep modification of our framework, since probabilistic modalities are more complex than our graded belief modalities. This issue of designing probabilistic programs as compact description of POMDP policies is a promising topic that we leave for further research.

5.3. Detailed Example

Let us consider a last example, inspired from Levesque (1996).

EXAMPLE 13. The agent has a bowl, initially empty, and a box of 3 eggs; each egg is either good or rotten. There are three actions:

- `takeNewEgg` is a pure ontic action resulting in the agent having in his hand a new egg from the box; since this new egg may be good or rotten, `takeNewEgg` is nondeterministic; however, its normal result is the agent having a good egg in his hand; getting a rotten egg in hand is 1-exceptional.
- `testEgg` is a pure sensing action, consisting of smelling the egg; its feedback contains two possible observations: **Only(B_1g)** and **Only($B_1\neg g$)**. (Note that smelling is here considered as not fully reliable).
- `putIntoBowl` is a pure ontic action consisting in breaking the egg into the bowl; it results in the content of the bowl being spoiled if the egg is rotten, and in the bowl containing one more egg if the egg is good.

This domain can be modeled using the following set of variables:

- `egg` (the agent holds an egg in his hand);
- `g` (the last egg taken from the box is a good one);
- `in`(i) for $i \in \{0, \dots, 3\}$ (the bowl contain exactly i eggs);
- `spoiled` (the bowl contain at least one rotten egg);
- and the derived fluents `om`(i), $i = 0, \dots, 3$, defined from the other fluents by: `om(0)` $\equiv$ `in(0)` $\vee$ `spoiled` and for all $i > 0$, `om`(i) $\equiv$ `in`(i) $\wedge \neg$`spoiled`.

The ontic action `takeNewEgg` is modeled by the following transition system: for any state s, let $s + (egg, g)$ (resp. $s + (egg, \neg g)$) the state obtained from s by (a) assigning g to true (resp. false) and (b) assigning egg to true, Then, for any s, $\kappa(s + (egg, g)|s) = 0$ and $\kappa(s + (egg, \neg g)|s) = 1$. The action theory corresponding to take-NewEgg is

$$\Sigma_{\texttt{takeNewEgg}} = \mathbf{K}\left(egg_{t+1} \wedge \left(\bigwedge_i in(i)_{t+1} \leftrightarrow in(i)_t\right)\right.$$
$$\left. \wedge (spoiled_{t+1} \leftrightarrow spoiled_t)\right) \wedge \mathbf{B}_1 g_{t+1}$$

[259]

The ontic action putIntoBowl is modeled by the following transition system: for any state s, let i_s be the number of eggs in the bowl in s (that is, $s \models \text{in}(i_s)$ and $s \models \neg\text{in}(j)$ for all $j \neq i(s)$). Let next(putIntoBowl, s) be the state defined by:

- if $s \models \text{egg} \wedge g$ then next(putIntoBowl, s) is the state obtained from s by (a) assigning egg to false; (b) assigning in(i_s) to false and in($i_s + 1$) to true (the rest being unchanged);
- if $s \models \text{egg} \wedge \neg g$ then next(putIntoBowl, s) is the state obtained from s by (a) assigning egg to false; (b) assigning in(i_s) to false and in($i_s + 1$) to true; (c) assigning spoiled to true (the rest being unchanged);
- if $s \models \neg\text{egg}$ then next(putIntoBowl, s) = s.

Then $\kappa_{\text{putIntoBowl}}$ (next(putIntoBowl, s)|s) = 0 and for all $s' \neq s$, $\kappa_{\text{putIntoBowl}}(s'|s) = +\infty$. The action theory corresponding to putIntoBowl is

$$\Sigma_{\text{putIntoBowl}} = \mathbf{K}\left(\neg\text{egg}_{t+1} \wedge \left(\bigwedge_i \text{in}(i+1)_{t+1} \leftrightarrow \text{in}(i)_t\right)\right.$$
$$\left.\wedge(\text{spoiled}_{t+1} \leftrightarrow (\text{spoiled}_t \vee g_t)) \wedge (g_{t+1} \leftrightarrow g_t)\right)$$

Let us now consider the BBP

$$\pi = (\text{takeNewEgg}; \text{testEgg}; \mathbf{if} \ \mathbf{B}_1 g \ \mathbf{then} \ \text{putIntoBowl})^3$$

(where $(\pi')^3$ means that the subplan π' is repeated three times) and the initial belief state Init = **Only**($\mathbf{K}$in(0)). Figure 1 shows the progression of Init by π.

Let us give some intuitive explanations about why these 4 belief states are obtained as possible outcomes of the program:

Case 1 The three tests came out to be negative, and therefore no egg has been put into the bowl: the final belief state is $\mathbf{K}(\text{in}(0))$.

Case 2 Only one of the three tests came out to be positive, and therefore one egg has been put into the bowl. In the final belief state, the agent knows for sure that there is one egg on the bowl ($\mathbf{K}(\text{in}(1))$; moreover the agent believes to the degree two that this egg is a good one ($\mathbf{B}_2\text{om}(1)$): indeed, when taking an egg out of the box, the agent has a prior belief (to the degree 1) that it is good ($\mathbf{B}_1 g$), and after testing it, a positive result reinforces this belief up to the

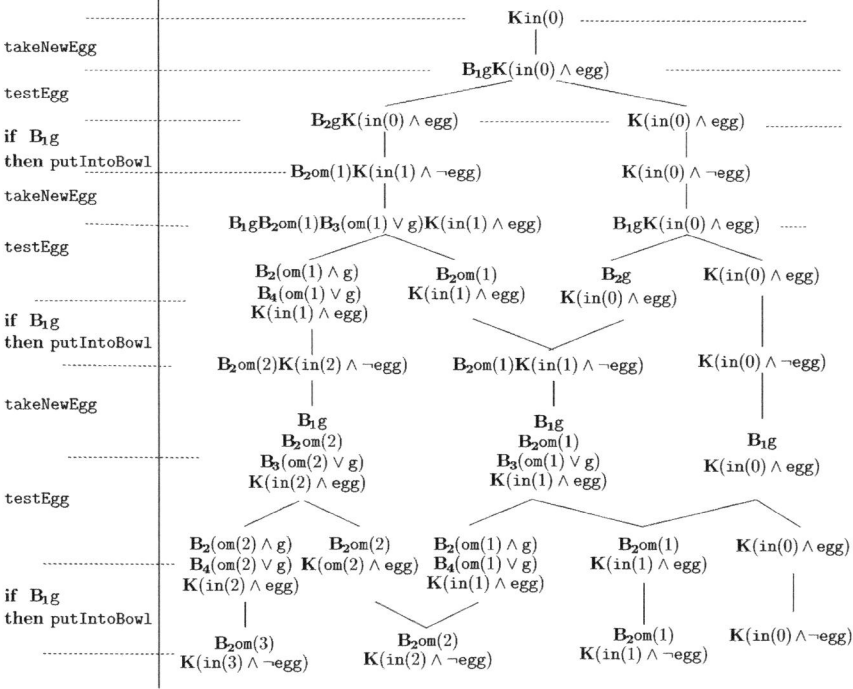

Figure 1.

degree 2, due to the reinforcement effect of combination (Section 3).

Case 3 Two out of the three tests came out to be positive: in the final belief state the agent knows that there are two eggs in the bowl, and for the same reasons as in Case 2, he believes to the degree 2 that both are good ($\mathbf{B}_2\mathrm{om}(2)$).

Case 4 All three tests being positive, the eggs have all been put into the bowl. For the same reasons as above the agent believes they are all good.

6. CONCLUSION

This paper has paved the way towards building a language for programming autonomous agents with actions, sensing (observations), and graded beliefs. Beliefs are expressed in a high-level language with graded modalities. Progression (by ontic actions and sensing) can be computed directly in this high-level language. We have shown how to compute a precompiled policy from a belief-based plan, that more or less corresponds to a policy in the POMDP meaning.

At least two issues for further research are expected in a near future.

6.1. *Integrating Belief-Based Programming and Golog*

A fairly close area is that of cognitive robotics, especially the work around *Golog* and the situation calculus (e.g., Reiter 2001a), which are concerned with logical specifications of actions and programs, including probabilistic extensions and partial observability.

First, we consider extending our belief-based plans towards a full belief-based programming language that could be an extension of *Golog* (e.g., Reiter 2001a). Golog allows for logical specifications of actions and programs, including sensing, in a high-level language which, on one hand, is far more expressive than ours, since it allows for many features (quantification over actions and states, nondeterministic choice, etc.) that are absent from our purely propositional language. Knowledge-based programming is also implementable in Golog (Reiter 2001b). On the other hand, knowledge-based programming in Golog (Reiter 2001b) does not allow for graded uncertainty.

Note that there have been several probabilistic extensions of the situation calculus and Golog: (Bacchus et al. (1999)) gives an account for the dynamics of probabilistic belief states after perceiving noisy observations and performing physical actions with noisy effectors, and (Grosskreutz and Lakemeyer (2000)) consider probabilistic Golog programs with partial observability, with the aim of turning off-line nondeterministic plans into programs that are guaranteed to reach the goal with some given probability. However, both lines of work consider simple branching conditions involving objective formulas, which is not suited to knowledge-based programming. As knowledge-based programming in Golog calls for an explicit knowledge modality as in (Scherl and Levesque 1993; Reiter 2001b), graded belief-based programming needs a collection of belief modalities, together with a syntactical way of making the agent's beliefs evolve after performing an action or an observation.

Thus, enriching BBPs with the highly expressive features of Golog and the situation calculus will ultimately results in a sophisticated language for belief-based programming, which is definitely an objective we want to pursue.

6.2. *Off-line Reasoning: Introducing Second-Order Uncertainty*

An issue that has not been considered here is the off-line evaluation of belief based plans. Many problems are raised by such an issue. First, in order to represent complex sensing actions, observations should be attached with their likelihood of occurrence (as in Boutilier et al. 1998). Given this, the projection of an initial belief state by a plan needs to introduce *second-order uncertainty*: for instance, on Example 7, given that accurate observations are more frequent than inaccurate, one would like to obtain that after asking two persons, then normally the agent is in the belief state O_2r or in the belief state $O_2\neg r$, and exceptionally he is in the void belief state. This calls for introducing belief states over belief states and a second family of modalities. These issues are investigated in the companion paper (Laverny and Lang 2005).

ACKNOWLEDGEMENTS

We are indebted to the two anonymous referees for their fruitful remarks, which helped us a lot rewriting this paper.

APPENDIX: OCF COMBINATION AND DEMPSTER'S RULE OF COMBINATION

A *mass assignment* is a mapping from $2^S \setminus \{\emptyset\}$ to $[0, 1]$ such that $\sum_{X \subseteq S} m(X) = 1$. Let m_1 and m_2 be two mass assignments, then Dempster's rule combines m_1 and m_2 into the mass assignment $m_1 \oplus m_2$ defined by

$$(m_1 \oplus m_2)(X) = \frac{\sum\{m_1(Y).m_2(Z) \text{ s.t. } Y, Z \subseteq S, X = Y \cap Z\}}{1 - \sum\{m_1(Y).m_2(Z) \text{ s.t. } Y, Z \subseteq S, Y \cap Z = \emptyset\}}$$

(and undefined when there is no pair (Y, Z) such that $Y \cap Z \neq \emptyset$ and $m_1(Y).m_2(Z) \neq 0$).

Now, let ε be an infinitesimal. To every OCF κ we associate the following family $MA(\kappa)$ of infinitesimal mass assignments, defined by: for all $s \in S$ such that $\kappa(s) \neq +\infty, m(\{s\}) = o(\varepsilon^{\kappa(s)})$; for all $s \in S$ such that $\kappa(s) = +\infty, m(\{s\}) = 0$; and for any non-singleton subset X of S, $m(X) = 0$. Clearly, for any infinitesimal mass assignment m giving a zero mass to all non-singleton subsets there is exactly one

κ such that $m \in MA(\kappa)$, and we then note $OM(m) = \kappa$ (where OM stands for *order of magnitude*). Then we have the following:

PROPOSITION 9. If $OM(m_1) = \kappa_1$ and $OM(m_2) = \kappa_2$ then $OM(m_1 \oplus m_2) = \kappa_1 \oplus \kappa_2$ (and $m_1 \oplus m_2$ is undefined iff $\kappa_1 \oplus \kappa_2$ is undefined).

Proof. Let m_1 and m_2 such that $OM(m_1) = \kappa_1$ and $OM(m_2) = \kappa_2$, with $\min(\kappa_1 + \kappa_2) \neq +\infty$. Since $m_1(X) = m_2(X) = 0$ for any non-singleton X, $m_1 \oplus m_2(X) = 0$. Then

$$1 - \sum \{m_1(Y).m_2(Z) \text{ s.t. } Y, Z \subseteq S, Y \cap Z = \emptyset\}$$
$$= 1 - \sum \{m_1(\{s\}).m_2(\{s'\}) \text{ s.t. } s, s' \in S, s \neq s'\}$$
$$= \sum \{m_1(\{s\}).m_2(\{s\}) \text{ s.t. } s \in S\}$$
$$= o(\varepsilon^{\min\{\kappa_1(s) + \kappa_2(s) \text{ s.t. } s \in S\}})$$
$$= o(\varepsilon^{\min(\kappa_1 + \kappa_2)})$$

Therefore, $(m_1 \oplus m_2)(\{s\}) = o(\varepsilon^{\kappa_1(s) + \kappa_2(s) - \min(\kappa_1 + \kappa_2)})$.

$\square$

NOTES

* A premliminary and shorter version of this paper in the Proceedings of the 16th European Conference on Artificial Intelligence (ECAI-04), pp. 368–372 (Laverny and Lang 2004).

[1] For instance, the crucial property $K(\varphi \wedge \psi) \longleftrightarrow (K\varphi \wedge K\psi)$ is satisfied in S5 and KD45, but not in a probabilistic doxastic logic: if $P_\alpha \varphi$ expresses that $\text{Prob}(\varphi) \geq \alpha$, or if it expressed that $\text{Prob}(\varphi) = \alpha$, then in both cases $P_\alpha \varphi \wedge P_\alpha \psi$ is not equivalent to $P_\alpha(\varphi \wedge \psi)$.

[2] Note, however, that there exist so-called non-prioritized versions of belief revision, in which acceptance of the observation is not taken for granted, and especially [Boutilier et al. 1988], to which our approach is compared at the end of this Section.

[3] This has to be contrasted with *transmutations* (Williams 1994) where one enforces the new belief state to satisfy a constraint of the form $\kappa(\varphi) = i$. Probability theory has also both kinds of rules: Jeffrey's (without implicit reinforcement) and Pearl's (see (Chan and Darwiche 2003) for a discussion).

[4] For instance, suppose that we ask repeatedly to the same agent (say, Hans), and that this agent is rational enough to give the same answer each time; in this case, we want to totally block the reinforcement effect; then we add five mutually exclusive new variables $a_1, \ldots, a_5$, which means that for all i and $j \neq i$, $\mathbf{K}\neg(a_i \wedge a_j)$ is part of the agent's background knowledge, as well as $\mathbf{B}_2(a_1 \rightarrow r)$, $\mathbf{B}_2(a_2 \rightarrow$

$\neg r$), $\mathbf{B}_1(a_3 \to r)$ and $\mathbf{B}_1(a_4 \to \neg r)$ – therefore, a_1 means that Hans has a strong belief that r holds, etc. – and the possible observations sent as feedback by the asking action are $\mathbf{Only}(\mathbf{K}a_i)$ for $i \in \{1, \ldots, 5\}$. This ensures that (a) the answer given is always the same, and (b) no reinforcement occurs: if for instance, the agent observes a_3 many times, then $\mathbf{B}_1r$ holds but $\mathbf{B}_2r$ does not.

[5] In the particular case where α is deterministic, its effects can be described by *successor state axioms* of the form $x_{t+1} \leftrightarrow \varphi_t$ for each $x \in PS$.

REFERENCES

Aucher, G.: 2004, 'A combined system for update logic and belief revision', in *7th Pacific Rim Int. Workshop on Multi-Agents (PRIMA2004)*.

Bacchus, F., J. Halpern, and H. Levesque: 1999, 'Reasoning about noisy sensors and effectors in the situation calculus', *Artificial Intelligence* **111**, 171–208.

Baral, C. and J. Lobo: 1997, 'Defeasible specifications in action theories', in *Proceedings of IJCAI'97*.

Benferhat, S., D. Dubois, and H. Prade: 2001, 'A computational model for belief change and fusing ordered belief bases', in M. A. Williams and H. Rott, (eds.), *Frontiers in Belief Revision*, Kluwer Academic Publishers, pp. 109–134.

Bonet, B. and H. Geffner: 2001, 'Planning and control in artificial intelligence. A unifying perspective', *Applied Intelligence* **3**(14), 237–252.

Boutilier, C.: 1998, 'A unified model of qualitative belief change: A dynamical systems perspective', *Artificial Intelligence Journal* **98**(1–2), 281–316.

Boutilier, C., R. Brafman, H. Hoos, and D. Poole: 1999, 'Reasoning with conditional *ceteris paribus statements*', in *Proceedings of the 15th Conf. on Uncertainty in Artificial Intelligence (UAI'99)*, pp. 71–80.

Boutilier, C., N. Friedman, and J. Halpern: 1998, 'Belief revision with unreliable observations', in *Proceedings of the Fifteenth National Conference on Artificial Intelligence (AAAI-98)*, pp. 127–134.

Chan, H. and A. Darwiche: 2003, 'On the revision of probabilistic beliefs using uncertain evidence', in *Proceedings of the 18th International Joint Conference on Artificial Intelligence (IJCAI-03)*.

Darwiche, A., and J. Pearl: 1997, 'On the logic of iterated belief revision', *Artificial Intelligence* **87**(1–2), 1–29.

Fariñas del Cerro, L., and A. Herzig: 1991, 'Modal logics for possibility theory', in *Proceedings of the First International Conference on the Fundamentals of AI Research (FAIR'91)*, Springer Verlag.

Dempster, A. P.: 1967, 'Upper and lower probabilities induced by a multivaluated mapping', in *Annals Mathematics Statistics* **38**, 325–339.

Dubois, D., J. Lang, and H. Prade: 1994, 'Possibilistic logic', in D. M. Gabbay, C. J. Hogger, and J. A. Robinson (eds.), *Handbook of logic in Artificial Intelligence and Logic Programming*, volume 3, Clarendon Press – Oxford, pp. 439–513.

Dubois, D., and H. Prade: 1997, 'A synthetic view of belief revision with uncertain inputs in the framework of possibility theory', *International Journal of Approximate Reasoning* **17**(2–3), 295–324.

Fagin, R., J. Halpen, Y. Moses, and M. Vardi: 1995, *Reasoning About Knowledge*, MIT Press.

Giunchiglia, E., J. Lee, N. McCain, V. Lifschitz, and H. Turner: 2003, 'Nonmonotonic causal theories', *Artificial Intelligence* **153**, 49–104.

Goldszmidt, M. and J. Pearl: 1992, 'Rank-based systems: A simple approach to belief revision, belief update, and reasoning about evidence and actions', in *Proceedings of KR'92*, pp. 661–672.

Grosskreutz, H. and G. Lakemeyer: 2000, 'Turning high-level plans into robot programs in uncertain domains', in *Proc. ECAI-2000*, pp. 548–552.

Herzig, A., J. Lang, D. Longin, and Th. Polacsek: 2000, 'A logic for planning under partial observability', in *AAAI-00*, pp. 768–773.

Herzig, A., J. Lang, and P. Marquis: 2003, 'Action representation and partially observable planning in epistemic logic', in *Proceedings of IJCAI03*, pp. 1067–1072.

Herzig, A., J. Lang, and T. Polacsek: 2001, 'A modal logic for epistemic tests', in *Proceedings of ECAI'2000*, pp. 553–557.

Kaelbling, L. P., M. L. Littman, and A. R. Cassandra: 1998, 'Planning and acting in partially observable stochastic domains', *Artificial Intelligence* **101**, 99–134.

Lang, J., P. Liberatore, and P. Marquis: 2003, 'Propositional independence : Formula-variable independence and forgetting. *Journal of Artificial Intelligence Research*, **18**, 391–443.

Lang, J., P. Marquis, and M.-A. Williams: 2001, 'Updating epistemic states', in Springer-Verlag (ed.), Lectures Notes in Artificial Intelligence 2256, *Proceedings of 14th Australian Joint Conference on Artificial Intelligence*, pp. 297–308.

Laverny, N. and J. Lang: 2004, 'From knowledge-based programs to graded BBPs, part I: on-line reasoning', in *Proceedings of ECAI-04*, pp. 368–372.

Laverny, N. and J. Lang: 2004, 'From knowledge-based programs to graded BBPs, part II: off-line reasoning', in *Proceedings of IJCAI-05*.

Levesque, H.: 1996, 'What is planning in the presence of sensing?', in *AAAI 96*, pp. 1139–1146.

Levesque, H. and G. Lakemeyer: 2000, *The Logic of Knowledge Bases*, MIT Press.

Lin, F.: 1996, 'Embracing causality in specifying the indeterminate effects of actions', in *Proc. of AAAI'96*.

Lin, F. and R. Reiter: 1994, 'Forget it!, in *Proceedings of the AAAI Fall Symposium on Relevance*, New Orleans, pp. 154–159.

Reiter, R.: 2001a, *Knowledge in Action: Logical Foundations for Specifying and Implementing Dynamical Systems*. MIT Press.

Reiter, R.: 2001b, 'On knowledge-based programming with sensing in the situation calculus', *ACM Transactions on Computational Logic* **2**, 433–457.

Scherl, R. B. and H. J. Levesque: 1993, 'The frame problem and knowledge-producing actions', in *AAAI-93*, pp. 698–695.

Shapiro, S., M. Pagnucco, Y. Lesperance, and H. Levesque: 2000, 'Iterated belief change in the situation calculus', in *Proceedings of KR2000*, pp. 527–537.

Spohn, W.: 1988, 'Ordinal conditional functions: a dynamic theory of epistemic states', in William L. Harper and Brian Skyrms (eds.), *Causation in Decision, Belief Change and Statistics*, volume 2, Kluwer Academic Pub., pp. 105–134.

Thielscher, M.: 2001, 'Planning with noisy actions (preliminary report)', in M. Brooks, D. Powers, and M. Stumptner (eds.), *Proceedings of the Australian Joint Conference on Artificial Intelligence*, LNAI, Adelaide, Australia, December 2001, Springer.

van der Hoek, W. and J.-J.Ch. Meyer: 1991, 'Graded modalities for epistemic logic', *Logique et Analyse* **133–134**, 251–270.

van Ditmarsch, H.: 2004, *Prolegomena to Dynamic Belief Revision*. Technical report, University of Otago, New Zealand.

van Linder, B., W. van der Hoek, and John-Jules Ch. Meyer: 1994, 'Tests as epistemic updates', in *Proceedings of ECAI 1994*, pp. 331–335.

Williams, M.-A: 1994, 'Transmutations of knowledge systems', in *Proceedings of KR'94*, pp. 619–629.

Noël Laverny
IRIT, Université Paul Sabatier
31062 Toulouse Cedex
France
E-mail: Noel.Laverny@freesbee.fr

Jérôme Lang
IRIT, Université Paul Sabatier
31062 Toulouse Cedex
France
E-mail: lang@irit.fr

MARTIN PETERSON and SVEN OVE HANSSON

ORDER-INDEPENDENT TRANSFORMATIVE DECISION RULES

ABSTRACT. A transformative decision rule alters the representation of a decision problem, either by changing the set of alternative acts or the set of states of the world taken into consideration, or by modifying the probability or value assignments. A set of transformative decision rules is *order-independent* in case the order in which the rules are applied is irrelevant. The main result of this paper is an axiomatic characterization of order-independent transformative decision rules, based on a single axiom. It is shown that the proposed axiomatization resolves a problem observed by Teddy Seidenfeld in a previous axiomatization by Peterson.

1. INTRODUCTION

In *The Foundations of Statistics*, Leonard Savage pointed out that rational decision making can be divided into two phases. Expressed in Savage's terminology, the main objective of the first phase is to decide "which world to use in a given context",[1] that is, to choose an appropriate representation of states, consequences, and acts, and thereby obtain a "formal description, or model, of what the person is uncertain about".[2] The purpose of the second phase of rational decision making is to establish "criteria for deciding among possible courses of action", and choose an act prescribed by such a criterion.[3]

Transformative decision rules can be used for analyzing what Savage (and several contemporary decision theorists[4]) characterize as the first, representational phase of rational decision making. A transformative decision rule alters the representation of a decision problem, either by changing the set of alternative acts or the set of states of the world taken into consideration, or by modifying the probability or value assignments. A paradigmatic example is the principle of insufficient reason, which prescribes that in case there is no reason to believe that one state of the world is more probable than another, the decision maker should transform the initial

Synthese (2005) 147: 323–342
Knowledge, Rationality & Action 269–288
DOI 10.1007/s11229-005-1351-0

representation of the decision problem into another in which every state is assigned equal probability.

A set of transformative decision rules *is order-independent* in case the order in which the rules are applied is irrelevant. Order-independence is an interesting property for (sets of) transformative decision rules, since it may be hard to determine in which non-arbitrary order the elements in a set of transformative decision rules should be applied. In this paper we give an axiomatic characterization of order-independent transformative decision rules. The axiomatization is based on a single axiom, referred to as weak monotonicity.

Section 2 gives a brief introduction to the concept of transformative decision rules. In Section 3 the problem of how to represent a decision problem in a formal representation is restated in a more precise way. Section 4 is devoted to a discussion of the weak monotonicity axiom, and in Section 5 three examples of order-independent transformative rules are proposed. In Section 6 the proposed axiomatization is compared with a previous axiomatization. Finally, in Section 7, a representation theorem for transformative decision rules is stated.

2. TRANSFORMATIVE DECISION RULES

As mentioned above, the task faced by the decision maker in the representational phase is to determine which π in a set of possible representations Π is the best representation of the decision problem under consideration. In order to render this basic idea more precise, it is instrumental to introduce the concept of a transformative decision rule.[5] A transformative decision rule is, intuitively put, a function that alters the representation of a given decision problem by adding, modifying, or deleting information to the initial representation. The precise mathematical structure of such a representation is irrelevant in the axiomatization of transformative decision rules to be presented here. However, in order to facilitate the presentation of realistic examples of (order-independent) transformative decision rules, we shall make the following structural assumptions: Let $A = \{a, a', \dots\}$ be a non-empty set of acts, and let $S = \{s, s', \dots\}$ be a non-empty set of states. $P = \{p : A \times S \to [0, 1]\}$ is a set of probability functions, and

$U = \{u : A \times S \to \mathrm{Re}\}$ is a set of utility functions. Then consider the following definition.

DEFINITION 1. A formal representation of a decision problem is an ordered quadruple $\pi = \langle A, S, P, U \rangle$.

The set Π is a set of alternative representations $\pi, \pi', \pi'' \ldots$ of one and the same decision problem. Note that in each representation π, P and U are sets of functions, rather than single functions. This set-up thus allows decision makers to consider several alternative probability and utility measures. In this way, decision makers can model uncertainty about utilities and second-order uncertainty about probabilities. The intuitive motivation for this model is that sometimes several alternative probability and utility functions are needed for describing a given situation, e.g. when the data used for constructing the utility and probability functions are themselves uncertain (for some examples, see Levi 1980; Gärdenfors and Sahlin 1982[6]).

A formal representation of a decision problem under *risk* is a quadruple $\pi = \langle A, S, P, U \rangle$ in which each of P and U has exactly one element, whereas a formal representation of a decision problem under *ignorance* is a quadruple $\pi = \langle A, S, P, U \rangle$ in which $P = \emptyset$ and U has exactly one element.

The concept of an 'outcome' or 'consequence' of an act is not explicitly employed in the formal representation. Instead, utilities are assigned to ordered pairs of acts and states. Also note that it is not assumed that the elements in A (or S) have to be jointly exhaustive and mutually exclusive. Whether such requirements ought to be levied or not is an issue that should be left open by a sufficiently general theory of representation.

A transformative decision rule is now defined as follows.

DEFINITION 2. Let Π be a set of formal decision problems. $\mathbf{t}$ is a transformative decision rule in Π if and only if $\mathbf{t}$ is a function such that for all $\pi \in \Pi$, it holds that $\mathbf{t}(\pi) \in \Pi$. $\mathbf{t}_{id}$ denotes the identity rule such that $\mathbf{t}_{id}(\pi) = \pi$ for all $\pi \in \Pi$.

In Section 1 a well-known example of a transformative decision rule was mentioned, namely the principle of insufficient reason. Other examples include: (i) the merger of states rule adopted by Luce and Raiffa,[7] prescribing that if two or more states yield identical outcomes under all acts in a decision problem under ignorance, then these states should be collapsed into one state,[8] (ii) Levi's condition of E-Admissibility, prescribing that if an alternative act is not *E-admissible*,[9] then it should be deleted from the set of alternative acts, and (iii) the *de minimis* rule, prescribing that if the probability of a state is extremely small, then it should not be included in the set of states considered by the decision maker.[10] Other well-known decision rules, such as Kahneman and Tversky's prospect rule, contain elements that have a clear transformative structure.[11]

A *composite rule* is a transformative rule that is made up of other transformative rules. (Note that we use a non-standard notation for composite functions).

DEFINITION 3. If t_i and t_j are transformative decision rules, then $(t_i \circ t_j)(\pi) = t_j(t_i(\pi))$ is a composite transformative rule.

For an example of a composite transformative rule, consider the rule prescribing that a decision maker facing a formal decision problem under ignorance should first apply the merger of states rule (**ms**), and thereafter transform the obtained representation into a decision problem under risk by applying the principle of insufficient reason (**ir**). This composite rule is described by the composite function $(ms \circ ir)(\pi) = ir(ms(\pi))$.

A set of transformative rules can be closed under rule composition in the following manner:

DEFINITION 4. Let Π be a set of formal decision problems and **T** a set of transformative decision rules in Π. The composite closure of **T** is the smallest set **T*** of decision rules such that

(i) $T \cup \{t_{id}\} \subseteq T^*$
(ii) If $t, u \in T^*$, then $t \circ u \in T^*$

Furthermore, a set **T** of transformative decision rules in Π is closed under composition if and only if $T = T^*$.

3. A RESTATEMENT OF THE PROBLEM

The question 'What guidelines should a rational decision maker follow in the representational phase of decision making?' can now be replaced by a more precise question, namely: what sequence of transformative decision rules $(\mathbf{t} \circ \mathbf{u} \circ \mathbf{v} \circ \cdots)$ should a rational decision maker apply to the initial formal decision problem π?

In order to answer the new question one needs to separate those (sequences of) transformative decision rules that may be applied to a formal representation, from those that may not be applied. Let $\langle \Pi, \succeq \rangle$ be a comparison structure for formal representations, in which Π is a set of formal representations, and $\succeq$ is a relation in Π corresponding to the English phrase 'at least a reasonable representation as'.[12] Thus, all elements in Π are different formal representations of one and the same decision problem, and $\succeq$ orders the elements in that set with regard to a list of deliberative values, such as realizability, completeness, relevance and simplicity. (These values are discussed at length in Peterson 2003b.) We assume that $\succeq$ is reflexive and transitive. Of course, the relations $\succ$ and $\sim$ can be constructed in terms of $\succeq$ in the usual way.

It follows from trivial combinatorial considerations that if $\mathbf{T}$ contains n different transformative rules, then there are $n!$ sequences in which each transformative rule is used exactly once. Of course, if rules may be used more than once, then the number of permissible sequences is unlimited. It is unreasonable to maintain that decision makers should be able to decide between all different sequences of transformative decision rules by systematically checking all possible combinations. Thus, a reasonable theory about transformative decision rules should either (1) contain a well-motivated instruction for the order in which the rules should be applied,[13] or (2) be order-independent in the sense that it does not matter in which order the different rules are applied.

In the remaining sections of the present article we investigate the latter approach.

4. AN AXIOM FOR ORDER-INDEPENDENT RULES

The following axiom ensures that a set of transformative decision rules is order-independent, as well as that it satisfies several other attractive properties.

4.1. *Weak Monotonicity*

For all $\mathbf{t}, \mathbf{u} \in \mathbf{T}^*$ and all $\pi \in \Pi$:

$$(\mathbf{u} \circ \mathbf{t})(\pi) \succeq \mathbf{t}(\pi) \succeq (\mathbf{t} \circ \mathbf{t})(\pi)$$

The left inequality, $(\mathbf{u} \circ \mathbf{t})(\pi) \succeq \mathbf{t}(\pi)$, states that a rule $\mathbf{u}$ should not, metaphorically expressed, throw a spanner in the work carried out by another rule $\mathbf{t}$. Hence, the representation obtained by first applying $\mathbf{u}$ and then $\mathbf{t}$ has to be at least as good as the representation obtained by only applying $\mathbf{t}$. For example, suppose that $\mathbf{t}$ is a rule that increases the simplicity of a representation by reducing redundant states (i.e. exactly parallel states that yield the same outcomes for all alternative acts). Then, $\mathbf{u}$ must be constructed such that the gain in simplicity to be cashed in by later applying $\mathbf{t}$ is not outweighed by a loss caused by $\mathbf{u}$; for instance, $\mathbf{u}$ might be a rule that increases simplicity by reducing the number of redundant (exactly parallel) acts.

The right-hand inequality, $\mathbf{t}(\pi) \succeq (\mathbf{t} \circ \mathbf{t})(\pi)$, says that nothing can be gained by immediately repeating a rule. (This property is further discussed below in relation to Theorem 1, part 2.)

The following theorem shows that weak monotonicity is sufficient to ensure several attractive properties of a system of transformative decisions rules.

THEOREM 1. Let $\mathbf{T}$ be a set of transformative rules in Π that is closed under composition and satisfies weak monotonicity. Then, for all $\mathbf{t}$, $\mathbf{u}$ in Π:

(1) $\mathbf{t}(\pi) \succeq \pi$,
(2) $\mathbf{t}(\pi) \sim (\mathbf{t} \circ \mathbf{t})(\pi)$,
(3) $(\mathbf{u} \circ \mathbf{t})(\pi) \sim (\mathbf{t} \circ \mathbf{u})(\pi)$,
(4) $(\mathbf{t} \circ \mathbf{u} \circ \mathbf{t})(\pi) \sim (\mathbf{u} \circ \mathbf{t})(\pi)$.

Part 1 states that the application of a transformative decision rule to a formal representation will yield a formal representation that is at least as reasonable as the one it was applied to. It might perhaps be objected that this result is too strong. Suppose, for instance, that $\pi \succ \mathbf{t}(\pi)$ and that $(\mathbf{t} \circ \mathbf{u})(\pi) \succ \pi$. As stated here, the property under consideration does not permit the decision maker to carry out the transformation from π to $(\mathbf{t} \circ \mathbf{u})(\pi)$ in two separate steps. However, in response to this argument, note that Part 1 does not prevent the decision maker from treating the composite rule $(\mathbf{t} \circ \mathbf{u})$ as a single rule fulfilling this condition (in which case $\mathbf{t} \circ \mathbf{u}$ but not $\mathbf{t}$ is an element of $\mathbf{T}$). Therefore, this is not a counter-example to the intuition underlying Part 1, and hence not to weak monotonicity.

Part 2 states that the aggregated value of a formal representation will remain constant no matter how many times (≥ 1) a transformative decision rule is iterated. A detailed defense of the intuition underlying this property (formulated in a slightly different way) was presented in Peterson (2003a). The basic argument runs as follows. In order for a transformative rule to be applicable, decision makers cannot be required to apply a rule more than a finite number of times. Obviously, this means that the rule has to be convergent in the sense that for every $\pi \in \Pi$ there is some number n such that for all $m \geq 1$ it holds that $(\mathbf{t}\circ)^{n+m}(\pi) \sim (\mathbf{t}\circ)^{n}(\pi)$, where $(\mathbf{t}\circ)^{n}$ denotes the rule $\mathbf{t}$ iterated n times. Otherwise the rule could be repeated indefinitely and yet its full capacity for improving the decision problem not be used. But in case a rule is convergent in the sense just defined, then it can be replaced by a rule that satisfies $(\mathbf{t}\circ\mathbf{t})(\pi) \sim \mathbf{t}(\pi)$ for all π, thus satisfying Part 2.

Part 3 establishes that transformative decision rules are order-independent: they can be applied in any order. This is an essential result. The notion of order-independence studied here, $(\mathbf{u}\circ\mathbf{t})(\pi) \sim (\mathbf{t}\circ\mathbf{u})(\pi)$, should not be mixed up with $(\mathbf{u}\circ\mathbf{t})(\pi) = (\mathbf{t}\circ\mathbf{u})(\pi)$. It is very rare that sets of transformative rules satisfy the latter, strong notion of order-independence (see Peterson 2004).

Finally, Part 4 makes it clear that nothing can be gained by applying $\mathbf{t}$ more than once, no matter what other transformative rules were applied between the two applications of $\mathbf{t}$.

A permutation of $\mathbf{T}$ is any composite rule that makes use of every element in $\mathbf{T}$ exactly once. (Thus, the permutations of $\mathbf{T} = \{\mathbf{t}, \mathbf{u}\}$ are $(\mathbf{t}\circ\mathbf{u})$ and $(\mathbf{u}\circ\mathbf{t})$.) Theorem 3 below shows that all permutations obtained from the *largest* subset of rules satisfying weak monotonicity are optimal. Hence, the decision maker may safely apply *all* transformative decision rules that satisfy these conditions. Lemma 2 is instrumental in the proof of Theorem 3.

LEMMA 2. Let $\mathbf{T}^{*}$ be a finite set of transformative rules for Π that is closed under composition and satisfies weak monotonicity. Then all permutations p_a and p_b obtainable from $\mathbf{T}$ are of equal value, i.e. $p_a(\pi) \sim p_b(\pi)$.

THEOREM 3. Let $\mathbf{T}$ be a set of transformative rules for Π that is closed under composition and satisfies weak monotonicity and let $A \subseteq B \subseteq T$. Then, for every $\pi \in \Pi$ and every permutation p_a obtainable from $\mathbf{A}$ and every permutation p_b obtainable from $\mathbf{B}$, it holds that $p_b(\pi) \succeq p_a(\pi)$.

To sum up, the theorems in this section show that if a set of transformative decision rules satisfies weak monotonicity, then the decision maker may apply all transformative rules in this set, in any desired order he or she wishes, and there is no requirement to apply any rule more than once. Furthermore, the transformative rules in that set will improve the initial representation as much as can possibly be achieved; there is no other way in which these rules can be applied that would return a formal representation that is strictly better.

5. THREE EXAMPLES

Order-independence is an attractive property of transformative rules, but in many cases it is too much to hope for. One can easily construct examples of transformative rules that are not order-independent.[14] In such cases, it is an open question whether the rules in question are normatively reasonable or not. Weak monotonicity, and hence order-independence, is perhaps not a necessary condition that all transformative rules have to fulfill.

Rather than entering a debate on the normative statues of order-independence, we feel that in the present paper it suffices to point out that order-independence is not an empty property: there are many reasonable examples of order-independent transformative rules. Below three examples are given. In the first and the second example the proposed rules are not only order-independent in the sense that $(\mathbf{u} \circ \mathbf{t})(\pi) \sim (\mathbf{t} \circ \mathbf{u})(\pi)$, they are also order-independent in the stronger sense that $(\mathbf{u} \circ \mathbf{t})(\pi) = (\mathbf{t} \circ \mathbf{u})(\pi)$. The latter notion of order-independence is not implied by the weak monotonicity axiom.[15] In the third example, only the weak form of order-independence holds.

All three examples that we give are trivial from a technical point of view. However, just because the examples are so simple they indicate that order-independence is of greater significance than one might think at first glance. For the first example, consider the formal representation depicted below. This is a decision problem under ignorance – no probabilities for the states s_1, s_2, and s_3 are known, so the set of probability functions P is empty. The letters $a - d$ denote utilities.

	s_1	s_2	s_3
a_1	a	b	b
a_2	c	d	d
a_2	c	d	d

According to Luce and Raiffa's merger of states rule (**ms**) mentioned in Section 2, it holds that:

Merger of states. *If π is a formal decision problem in which two or more states yield identical outcomes under all acts, then these states should be collapsed into a single state. (And if there are any known probabilities of the states they should be added.)*

By applying **ms** to the formal representation above one obtains a representation in which s_2 and s_3 are merged into as a single state. This is an improvement of the initial representation, since it leads to a gain in simplicity. For an example, suppose that s_2 is the state "prices increase by 10% and the coin lands heads up", and s_3 is the state "prices increase by 10% and the coin lands tails". Then, by merging s_2 and s_3 into a single state, one obtains the less complex state "prices increase by 10%".

In order to complete the example, we introduce the merger of acts rule (**ma**). This rule is parallel to the merger of states rule, except that it operates on acts. More precisely, the **ma** rule prescribes that alternative acts that yield identical outcomes, no matter which state occurs, should be collapsed into a single act.

Arguably, the **ms** rule and the **ma** rule return new formal representations that are at least as reasonable as the original representations; hence, both rules satisfy on the left-hand side of the weak monotonicity axiom. Furthermore, since all parallel states and acts are detected by applying the **ms** and the **ma** rules, both rules are also iterative in the sense required by on the right-hand side of weak monotonicity. The set constituted by the **ms** and the **ma** rules is, therefore, an example of order-independent transformative decision rules.

For the second example, remember that uncertainty about utilities and second-order uncertainty about probabilities is modelled by a set of utility functions (U) and a set of probability functions (P). Suppose that a decision makers wishes to transform a formal representation containing sets with many probability functions and utility functions into a representation containing only one utility function and one probability function. (The motivation might be that this makes it possible to calculate the expected utilities of the alternative acts.) Now consider two hypothetical transformative

[277]

rules, the **u** rule and **p** rule, respectively. The **u** rule is a rule that aggregates the elements of U into a single utility function u, e.g. by calculating the mean utilities, or in some other way. (For present purposes it does not matter how the elements of U are aggregated.) Furthermore, the **p** rule is a rule that aggregates the elements of P into a single probability function p. (As before, it does not matter exactly how this is done.) It follows trivially that **u** and **p** satisfy on the right-hand side of the order-independence axiom, and given that the aggregation functions are reasonable, both rules also satisfy on the left-hand side of the axiom. Hence, the **p** rule and the **u** rule constitute an example of order-independent transformative decision rules – or rather a set of examples, since both rules can be specified in several different ways.

The third example is more controversial that the previous ones. Consider the following formulation of the principle of insufficient reason (**ir**), mentioned in Section 2:

The principle of insufficient reason. *If π is a decision problem under ignorance, then it should be transformed into a decision problem under risk π' in which equal probabilities are assigned to all states.*

The **ir** rule as well as the **ms** rule satisfy the weak monotonicity axiom. Nothing is gained by applying one of the rules more than once, and none of the rules throw a spanner in the work carried out by the other. It follows that $(\textbf{ms} \circ \textbf{ir})(\pi) \sim (\textbf{ir} \circ \textbf{ms})(\pi)$. However, it does not hold that $(\textbf{ms} \circ \textbf{ir})(\pi) = (\textbf{ir} \circ \textbf{ms})(\pi)$ for all π, i.e. the two representations need not be identical. In order to construct an example of this, we stipulate that if the antecedent in the formulations of **ir** and **ms** are false, then no transformation is carried out, i.e. they return the same representation that was used as input.

The following story illustrates how the decision problem corresponding to the formal representation π, depicted below, can arise: You are a paparazzi photographer, and rumor has it that actress Julia Roberts will show up in either New York (NY), Geneva (G), or Zürich (Z). Nothing is known about the probabilities for these three states of the world. You have to decide if you should stay in Switzerland or catch a plane to America. If you stay (a_1) and Ms. Roberts shows up in New York (NY), you receive zero utiles; otherwise, you get your photos and receive 10. If you catch a plane to America (a_2) and Ms. Roberts shows up in New York (NY) you receive five utiles, and if she shows up in Switzerland you receive six (because you are able to call a friend that takes even better photos).

$[\pi]$

	NY	G	Z
a_1	0	10	10
a_2	5	6	6

$[\pi']$

	1/3	2/3
a_1	0	10
a_2	5	6

$[\pi'']$

	1/2	1/2
a_1	0	10
a_2	5	6

Representation π' is obtained from π by first applying the **ms** rule and then the **ir** rule. Representation π'' is obtained from π by applying the two rules in the reversed order. Which formal representation is best: π, π' or π''? (For the sake of the argument, we assume that no alternative representation is to be considered.) Because of Theorem 1, we know that $(\mathbf{ms} \circ \mathbf{ir})(\pi) \sim (\mathbf{ir} \circ \mathbf{ms})(\pi) \succeq \pi$. Hence, $\pi' \sim \pi'' \succeq \pi$. However, observe that $EU(a_1) > EU(a_2)$ in π', but $EU(a_2) > EU(a_1)$ in π''. Thus, the principle of maximizing utility recommends one act in π' (stay in Switzerland) but another in π'' (catch a plane to America). How should a rational decision maker act? Anyone who accepts the two transformative rules **ms** and **ir** and considers them to satisfy weak monotonicity will consider the two formal representations to be equally reasonable, because of Theorem 1.

To some degree, the present problem resembles the problem of underdetermination in science, famously discussed by e.g. Quine.[16] According to Quine's notion of underdetermination, there might be several different scientific theories that explain all accumulated evidence equally-well. In such cases there are at least two possible positions one could take. *Ecumenists* think that both (all) of the incompatible theories should be regarded as "locally" true, whereas *sectarians* argue that every rivaling theory ought to be considered false. The main problem with the latter standpoint is that it forces us to make judgments about the truth or falsity of scientific theories that are not based on empirical evidence (or other rational considerations, e.g. simplicity, scope, etc.), since the accumulated evidence and all other relevant features of both theories are equal. The problem faced by advocates of the ecumenical position is to explain what it means for incompatible theories to be "locally" true; in that case truth as "correspondence to external facts" seems impossible.

In decision theory the ecumenical position is less problematic. In fact, it seems perfectly reasonable to maintain that in case one and the same act is judged as rational in one formal representation but non-rational in another formal representation, then there are good reasons both to regard that act as rational and to regard it as irrational. This is not contradictory, given that acts are here treated as rational only *relative* to a certain formal representation of a decision problem.

6. STRONG MONOTONICITY

It is interesting to compare the axiomatization proposed in Section 4 with a previous axiomatization, in which Parts 1 and 2 of Theorem 1 were adopted as axioms together with the following monotonicity condition:[17]

6.1. *Strong Monotonicity*

If $\pi \succeq \pi'$, then $\mathbf{t}(\pi) \succeq \mathbf{t}(\pi')$.

Teddy Seidenfeld has pointed out that the axiomatization based on this strong monotonicity condition has the following implication:[18] Whenever $\mathbf{t}$ is a "better" rule for π than $\mathbf{u}$ (i.e. $\mathbf{t}(\pi) \succ \mathbf{u}(\pi)$), if $\mathbf{t}$ has been applied to a representation π, yielding the representation $\pi' = \mathbf{t}(\pi)$, then it is never the case that $\mathbf{u}$ can improve π'. Hence, no substantial interaction is allowed among transformative decision rules: it never happens that one obtains a better representation by first applying $\mathbf{u}$ and then $\mathbf{t}$, compared to what is obtained by applying $\mathbf{t}$ directly. In order to spell out this difficulty more in detail, suppose that $\mathbf{T} = \{\mathbf{t}, \mathbf{u}\}$ satisfies strong monotonicity and Parts 1 and 2 of Theorem 1, and also suppose that $\mathbf{t}(\pi) \succ \mathbf{u}(\pi) \succeq \pi$. Now, if $\mathbf{t}$ is applied to both $\mathbf{u}(\pi)$ and $\mathbf{t}(\pi)$ it follows from strong monotonicity that $(\mathbf{t} \circ \mathbf{t})(\pi) \succeq (\mathbf{u} \circ \mathbf{t})(\pi)$. But according to Part 2 of Theorem 1 it holds that $(\mathbf{t} \circ \mathbf{t})(\pi) \sim \mathbf{t}(\pi)$, so $\mathbf{t}(\pi) \succeq (\mathbf{u} \circ \mathbf{t})(\pi)$, which means that $\mathbf{t}$ and $\mathbf{u}$ have not interacted in a way that opens up for a representation that is any better than what could be reached by $\mathbf{t}$ alone. Hence, no interaction between $\mathbf{t}$ and $\mathbf{u}$ can occur.

Of course, the left-hand side of weak monotonicity draws on the same intuition as strong monotonicity. Hence, it might be objected that weak monotonicity and strong monotonicity are problematic for the same reason. However, in order to derive strong monotonicity from weak monotonicity we have to add the following axiom (or some other axiom that is at least as strong):

6.2. *Achievability*

If $\pi' \succeq \pi$, then there is a set of rules $\{\mathbf{t}_1, \ldots, \mathbf{t}_n\}$ such that $(\mathbf{t}_1 \circ \ldots \circ \mathbf{t}_n)(\pi) = \pi'$.

OBSERVATION 4. Weak monotonicity and achievability imply strong monotonicity.

In our view achievability is highly questionable. There is no reason to believe that there always is a set of rules $\{\mathbf{t}_1, \ldots, \mathbf{t}_n\}$ that can take us from one (bad) representation to another specified (better) representation. Suppose, for instance, that the better representation contains some information (e.g. more alternative acts) that was not contained in the bad representation; then it is not certain that there exists a set of rules that can take us to the better representation.

The following weaker version of achievability is not sufficient for deriving strong monotonicity from weak monotonicity.

6.3. *Weak Achievability*

If $\pi' \succeq \pi$, then there is a set of rules $\{\mathbf{t}_1, \cdots, \mathbf{t}_n\}$ such that $(\mathbf{t}_1 \circ \cdots \circ \mathbf{t}_n)(\pi) \succeq \pi'$.

OBSERVATION 5. Weak monotonicity and weak achievability do not imply strong monotonicity.

Even though weaker than achievability, weak achievability, is not self-evident. For example, it presupposes that there are no *cul-de-sacs*, that is, non-optimal representations that cannot be improved. However, from a normative point of view it seems reasonable to require that transformative decision rules should have a structure that does not allow the decision maker to end up in a *cul-de-sac*.

Observations 4 and 5 together indicate that the axiornatization based on the weak monotonicity axiom avoids the problem identified by Seidenfeld. Further evidence for this conclusion is given in the next section.

[281]

7. A REPRESENTATION THEOREM

In this section we present a representation theorem for transformative decision rules. This theorem shows that a decision maker obeying weak monotonicity for transformative decision rules can be described *as if* he or she maps formal representations into a one-dimensional space, while taking certain restrictions into account. This gives a better understanding of what is, and is not, implied by the weak monotonicity axiom. As will be explained below, the representation theorem ensures that the kind of problem pointed out by Seidenfeld cannot occur if weak monotonicity is assumed to be the sole axiom governing the application of transformative decision rules.

Let $a, b, \ldots$ be elements in a set M and consider the following definitions.

DEFINITION 5. A vector $\langle a, b \rangle$ is an upvector if and only if $|b| \geq |a|$, where $|\ |$ is a function that assigns a real number to each element in M.

DEFINITION 6. An upvector-label is a set L of upvectors such that if $\langle a, b \rangle \in L$ and $\langle a, b' \rangle \in L$, then $b = b'$.

The semantic unit we use is a set of upvector-labels, or SEUL for short. It can be easily verified that sets of transformative decision rules can be represented by SEULs. Let $\mathcal{V}$ be a function from Π to M such that $|\mathcal{V}(\pi)| \geq |\mathcal{V}(\pi')|$ if and only if $\pi \succeq \pi'$. $\mathcal{V}$ represents the projections of the elements of Π to a scale that conforms with the relation $\succeq$. Non-identical elements of Π can be equivalent in terms of $\succeq$, in other words we can have $|a| = |b|$ and $a \neq b$. (Otherwise it will not be possible to avoid implausible results; as one example $(\mathbf{t} \circ \mathbf{u})(\pi) = (\mathbf{u} \circ \mathbf{t})(\pi)$ would follow from weak monotonicity, contrary to our strivings in Section 4 to avoid such postulates.)

It follows from what has been said above that every representation $\pi \in \Pi$ and set of rules **T** can be described as a model consisting of a set M, a SEUL and a function $|\ |$, as defined above.

DEFINITION 7. A transformative rule **t** in Π is representable by an upvector-label in a SEUL if and only if, for every transformation from π to $\mathbf{t}(\pi)$ in Π, $\langle \mathcal{V}(\pi), \mathcal{V}(\mathbf{t}(\pi)) \rangle$ is an upvector.

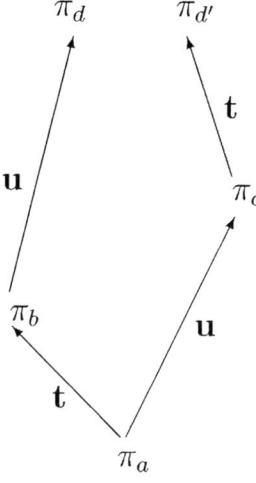

Figure 1.

The following observations shows that sets of transformative decision rules can be represented in a SEUL, and that order-independent transformative rules can be represented in a SEUL satisfying certain restrictions.

THEOREM 6. Let **T** be a set of transformative rules in Π that is closed under composition, such that $\mathbf{t}(\pi) \succeq \pi$ for all $\mathbf{t} \in \mathbf{T}$ and all $\pi \in$ Π. Then there exists a SEUL such that each $\mathbf{t} \in \mathbf{T}$ is represented by exactly one upvector-label, and vice versa.

THEOREM 7. Let **T** be a set of transformative rules in Π that is closed under composition and that satisfies weak monotonicity. Then there exists a SEUL such that each $\mathbf{t} \in \mathbf{T}$ is represented by exactly one upvector-label, and vice versa, with the following restrictions on the upvector-labels:

(1) There are no $|c| > |b| > |a|$ such that $\langle a, b \rangle$ and $\langle b, c \rangle$ are elements in the same upvector-label.
(2) If $|d| > |c| \geq |b| \geq |a|$, and $\langle a, d \rangle$ and $\langle b, c \rangle$ are elements in the same upvector-label, then there is no upvector-label that contains $\langle a, b \rangle$.

Theorem 7 can be graphically illustrated in an example (see Figure 1). In the model used in this example the only ways to reach one of the two optimal representations π_d and $\pi_{d'}$ is to either start with **t** and then apply **u**, or start with **u** and then apply **t**. (Of course, this model can be iterated and expanded in various ways.) The model is constructed by letting $\Pi = \{\pi_a, \pi_b, \pi_c, \pi_d, \pi_d' : \pi_d' \sim \pi_d \succ \pi_c \succ \pi_b \succ \pi_a\}$ and by assuming that $\mathbf{T} = \{\mathbf{t}, \mathbf{u}\}$, where : $\mathbf{t}(\pi_a) = \pi_b$, $\mathbf{t}(\pi_b) = \pi_b$, $\mathbf{t}(\pi_c) = \pi_d'$, $\mathbf{t}(\pi_d) = \pi_d$, $\mathbf{t}(\pi_d') = \pi_d'$, $\mathbf{u}(\pi_a) = \pi_c$, $\mathbf{u}(\pi_b) = \pi_d$, $\mathbf{u}(\pi_c) = \pi_c$, $\mathbf{u}(\pi_d) = \pi_d$, and $\mathbf{u}(\pi_d') = \pi_d'$. Since interaction between the rules is the only way to reach an optimal representation in this example, Theorem 7 shows that weak monotonicity is not too strong. Thus, Theorem 7 indirectly supports our claim that weak monotonicity satisfies reasonable normative requirements.

8. CONCLUSION

Transformative decision rules provide a formally satisfactory tool for analyzing the first, representational phase of rational decision making. Such a tool has been missing in decision theory. A reasonable theory about transformative decision rules should either (1) contain a well-motivated instruction for the order in which different transformative rules should be applied, or (2) be order-independent in the sense that it does not matter in which order the different rules are applied. In the present paper we have investigated the second approach, and shown that the weak monotonicity axiom is sufficient for order-independence. As we showed in Section 6, there are interesting models in which the weak monotonicity axiom is satisfied; in some models an optimal representation can only be reached by letting the available transformative rules interact.

PROOFS

THEOREM 1, Part 1:

According to Definition 4, $\mathbf{t}_{id} \in \mathbf{T}^*$. Substitute $\mathbf{t}_{id}$ for **t** in the leftmost part of the weak monotonicity axiom.

THEOREM 1, Part 2:

Substitute **t** for **u** in the leftmost part of the weak monotonicity axiom.

[284]

THEOREM 1, Part 3:

Let π be an arbitrary element in Π.

(1) $(\mathbf{t} \circ \mathbf{u} \circ \mathbf{t})(\pi) \succeq (\mathbf{u} \circ \mathbf{t})(\pi)$ Left-hand side of axiom
(2) $(\mathbf{t} \circ \mathbf{u} \circ \mathbf{t} \circ \mathbf{u})(\pi) \succeq (\mathbf{u} \circ \mathbf{t})(\pi)$ (1), Part 1 of present theorem
(3) $(\mathbf{t} \circ \mathbf{u})(\pi) \succeq (\mathbf{u} \circ \mathbf{t})(\pi)$ (2), Part 2 of present theorem
(4) $(\mathbf{u} \circ \mathbf{t} \circ \mathbf{u})(\pi) \succeq (\mathbf{t} \circ \mathbf{u})(\pi)$ Left-hand side of axiom
(5) $(\mathbf{u} \circ \mathbf{t} \circ \mathbf{u} \circ \mathbf{t})(\pi) \succeq (\mathbf{t} \circ \mathbf{u})(\pi)$ (4), Part 1 of present theorem
(6) $(\mathbf{u} \circ \mathbf{t})(\pi) \succeq (\mathbf{t} \circ \mathbf{u})(\pi)$ (5), Part 2 of present theorem
(7) $(\mathbf{u} \circ \mathbf{t})(\pi) \sim (\mathbf{t} \circ \mathbf{u})(\pi)$ (3), (6)

THEOREM 1, Part 4:

Let π be an arbitrary element in Π.

(1) $(\mathbf{u} \circ \mathbf{t} \circ \mathbf{u} \circ \mathbf{t})(\pi) \succeq (\mathbf{t} \circ \mathbf{u} \circ \mathbf{t})(\pi)$ Left-hand side of axiom
(2) $(\mathbf{u} \circ \mathbf{t})(\pi) \succeq (\mathbf{t} \circ \mathbf{u} \circ \mathbf{t})(\pi)$ Part 2 of present theorem
(3) $(\mathbf{t} \circ \mathbf{u} \circ \mathbf{t})(\pi) \succeq (\mathbf{u} \circ \mathbf{t})(\pi)$ Left-hand side of axiom
(4) $(\mathbf{t} \circ \mathbf{u} \circ \mathbf{t})(\pi) \sim (\mathbf{u} \circ \mathbf{t})(\pi)$ (2), (3)

LEMMA 2:

We prove this by induction: It follows from Theorem 1, Part 3 that the claim holds in case $\mathbf{T}'$ has two elements. (In case $\mathbf{T}'$ has only one element the claim is trivially true, because of reflexivity.) In order to prove the inductive step, suppose that the claim holds in case $\mathbf{T}'$ has $n\,(n \geq 2)$ elements. Let $\mathbf{t}$ be element $n + 1$, and let $\mathbf{F}$ be a sequence of v elements and let $\mathbf{G}$ be a sequence of w elements; $v + w = n$.

We need to show that $(\mathbf{t} \circ \mathbf{F} \circ \mathbf{G})(\pi) \sim (\mathbf{F} \circ \mathbf{t} \circ \mathbf{G})(\pi) \sim (\mathbf{F} \circ \mathbf{G} \circ \mathbf{t})(\pi) \sim (\mathbf{t} \circ \mathbf{G} \circ \mathbf{F})(\pi) \sim (\mathbf{G} \circ \mathbf{t} \circ \mathbf{F})(\pi) \sim (\mathbf{G} \circ \mathbf{F} \circ \mathbf{t})(\pi)$. First consider the case in which both $\mathbf{F}$ and $\mathbf{G}$ have a non-zero number of elements, i.e. $v, w \neq 0$. Note that the number of elements in $(\mathbf{t} \circ \mathbf{G})$ is $\leq n$. Hence, since the theorem was assumed to hold for up to n elements and $\mathbf{F}(\pi) \in \Pi$, it follows that $(\mathbf{F} \circ \mathbf{t} \circ \mathbf{G})(\pi) \sim (\mathbf{F} \circ \mathbf{G} \circ \mathbf{t})(\pi)$. We also need to show that $(\mathbf{t} \circ \mathbf{F} \circ \mathbf{G})(\pi) \sim (\mathbf{F} \circ \mathbf{G} \circ \mathbf{t})(\pi)$. In order to do this, we substitute $\mathbf{u}$ for $\mathbf{F} \circ \mathbf{G}$ in the proof of Part 3, Theorem 1. So far we have shown (since $\sim$ is transitive) that $(\mathbf{t} \circ \mathbf{F} \circ \mathbf{G})(\pi) \sim (\mathbf{F} \circ \mathbf{t} \circ \mathbf{G})(\pi) \sim (\mathbf{F} \circ \mathbf{G} \circ \mathbf{t})(\pi)$; by applying an analogous argument we find that $(\mathbf{t} \circ \mathbf{G} \circ \mathbf{F})(\pi) \sim (\mathbf{G} \circ \mathbf{t} \circ \mathbf{F})(\pi) \sim (\mathbf{G} \circ \mathbf{F} \circ \mathbf{t})(\pi)$. Finally, since the number of elements in $(\mathbf{F} \circ \mathbf{G})$ is $= n$, it follows from our assumption that $(\mathbf{t} \circ \mathbf{F} \circ \mathbf{G})(\pi) \sim (\mathbf{t} \circ \mathbf{G} \circ \mathbf{F})(\pi)$.

[285]

The second case, in which the number of elements in either **F** or **G** is zero, is trivial, since we have shown above that $(\mathbf{t} \circ \mathbf{F} \circ \mathbf{G})(\pi) \sim (\mathbf{F} \circ \mathbf{G} \circ \mathbf{t})(\pi)$.

THEOREM 3:

Let $\mathbf{C} = \mathbf{B} - \mathbf{A}$. From the right-hand side of the axiom it follows that for every $p_b(\pi)$ there is a permutation $p_c(\pi)$ such that $p_a \circ p_c(\pi) \sim p_b(\pi)$. Hence, because of Part 1 of Theorem 1, $p_b(\pi) \succeq p_a(\pi)$.

OBSERVATION 4:

Let us assume that $\pi' \succeq \pi$. We use achievability and define **u** as the composite rule $(u_1 \circ \cdots \circ u_n)(\pi) = \pi'$. Due to the left part of order-independence, i.e. $\mathbf{tu}(\pi) \succeq \mathbf{t}(\pi)$, it holds that $\mathbf{t}(\pi') = \mathbf{t}(\mathbf{u}(\pi)) \succeq \mathbf{t}(\pi)$. Hence, $\mathbf{t}(\pi') \succeq \mathbf{t}(\pi)$.

OBSERVATION 5:

This is proved by constructing a counter-example: in the model described in Section 7 (Figure 1) both the weak monotonicity axiom and weak achievability are satisfied, but strong monotonicity is not, because $\mathbf{u}(\pi) \succeq \mathbf{t}(\pi)$ but $(\mathbf{t} \circ \mathbf{u})(\pi) \succeq (\mathbf{u} \circ \mathbf{u})(\pi)$.

THEOREM 6:

In order to construct a SEUL such that each $\mathbf{t} \in \mathbf{T}$ is represented by exactly one upvector-label, let the upvector-label be constituted by the set of upvectors given by, for each transformation from π to $\mathbf{t}(\pi)$ in Π, $\langle \mathcal{V}(\pi), \mathcal{V}(\mathbf{t}(\pi)) \rangle$, as stated in Definition 7. That there is such an upvector-label follows from the assumption that $\mathbf{t}(\pi) \succeq \pi$ for all $\pi \in \Pi$ and Definition 2. The former assumption guarantees that there are upvectors. Definition 2, which states that transformative decision rules are functions, guarantees that $b = b'$ if $\langle a, b \rangle$ and $\langle a, b' \rangle \in L$. Hence, the upvector-labels exist.

THEOREM 7, Part 1:

Assume for *reductio* that $\langle a, b \rangle$ and $\langle b, c \rangle$ are elements in the upvector-label L. According to Theorem 6 there is exactly one $\mathbf{t} \in \mathbf{T}$ for each upvector-label; suppose that **u** is the transformative rule corresponding to L. Since $|c| > |b| \geq |a|$ in $\langle a, b \rangle$ and $\langle b, c \rangle$, it follows that for some $\pi \in \Pi$, $\mathbf{u} \circ \mathbf{u}(\pi) \succ \mathbf{u}(\pi) \succeq \pi$, which contradicts the right-hand side of the weak monotonicity axiom.

THEOREM 7, Part 2:

Assume for *reductio* that there is an upvector-label L that contains $\langle a, b \rangle$. Let L' be an upvector-label containing $\langle a, d \rangle$ and $\langle b, c \rangle$. From Theorem 6 it follows that there are some transformative decision rules corresponding to L and L'; suppose that **t** corresponds to L and suppose that **u** corresponds to L'. According to Theorem 1, Part 3, it holds that $\mathbf{t} \circ \mathbf{u}(\pi) \sim \mathbf{u} \circ \mathbf{t}(\pi)$. According to Theorem 6 there are some upvectors corresponding to these (composite) transformations, and the real numbers corresponding to these upvectors have to be equal (because of Theorem 1, Part 3). However, this implies a contradiction, since $|d|$ is strictly greater than $|c|$.

NOTES

[1] Savage 1954/72, p. 9.

[2] *Ibid.*, p. 7.

[3] *Ibid.*, p. 6.

[4] Cf. Clemen 1991, Joyce 1999, Resnik 1993.

[5] For further discussions of transformative decision rules (see Peterson 2003a, b, 2004).

[6] Gärdenfors and Sahlin only consider epistemic uncertainty about probabilities, not utilities. They model this uncertainty by introducing a quantitative measure of epistemic uncertainty.

[7] Luce and Raiffa 1957, p. 295 (Axiom 11).

[8] In a representation of decision problems, such as the present one, in which outcomes have been replaced by their utilities, this rule will have to be modified so that states with identical utility under all acts are merged.

[9] An act a is, roughly put, E-admissible just in case there is a "seriously permissible" probability function q in B (B is a set of probability functions) and a "seriously permissible" utility function u in G (G is a set of utility functions) such that the expected utility of a is optimal. For a precise definition (see Levi 1980, pp. 96).

[10] This rule is, of course, only applicable in case P contains exactly one element. For a discussions of the de minimis rule (see e.g. Whipple 1987; Peterson 2002).

[11] Cf. Kahneman and Tversky 1979.

[12] The concept of comparison structures is investigated in detail in Hansson 2001, Chapter 2.

[13] This is the approach taken in Levi 1980.

[14] For example, consider the de minimis rule (**dm**) mentioned in Section 2, which tells us to ignore sufficiently small probabilities, and the principle of maximizing expected utility (**eu**). When conceived of as a transformative rule, the **eu** rule (i) replaces the set of states with a single state that is assigned probability one, and (ii) replaces the original utility function that takes the utility of every act given the single state to be its original expected utility. If these two rules are accepted, it is only reasonable to first apply **dm** and then **eu**. If one first applies **eu**, this will

throw a spanner in the work carried out by **dm**, because then extremely improbable states are no longer ignored.

[15] For a discussion of the strong notion of order-independence (see Peterson 2004).

[16] See e.g. Quine 1992.

[17] See Peterson 2004. We assume that all conditions stated in this section hold for all $t, u \in T^*$ and all $\pi \in \Pi$, As before, the set Π is a set of alternative representations $\pi, \pi', \pi'' \ldots$ of one and the same decision problem.

[18] In conversation, May 16, 2003.

REFERENCES

Clemen, R. T.: 1991 *Making hard Decisions: An Introduction to Decision Analysis*, PWS-Kent Publishing Company.

Hansson, S. O.: 2001 *The Structure of Values and Norms*, Cambridge University Press, Cambridge.

Jeffrey, R.: 1983 *The Logic of Deciosion*, 2nd edn. (significant improvements from 1st), University of Chicago Press.

Joyce, J. M.: 1999 *The Foundations of Causal Decision Theory*, Cambridge University Press, Cambridge.

Kahneman, D. and A. Tversky: 1979 Prospect theory: an analysis of decisions under Risk, *Econometrica* **47**, 263–291.

Levi, I.: 1980 *The Enterprise of Knowledge*, MIT Press, Cambridge.

Luce, D. and H. Raiffa: 1957 *Games and Decisions: Introduction and Critical Survey*, Wiley, New York.

Peterson, M.: 2002 What is a de minimis risk?, *Risk Management* **4**(2), 47–55.

Peterson, M.: 2003a Transformative Decision Rules, *Erkenntnis* **58**, 71–85.

Peterson, M.: 2003b *Transformative Decision Rules: Foundations and Applications*, Theses in Philosophy from the Royal Institute ofTechnology No. 3.

Peterson, M.: 2004 Transformative Decision Rules, Permutability, and Non-Sequential Framing, *Synthese* **139**, 387–403.

Quine, W. V.: 1992 *Pursuit of Truth*, Harvard University Press.

Resnik, M.: 1993 *Choices. An Introduction to Decision Theory*, University of Minnesota Press.

Savage, L. J.: 1954 *The Foundations of Statistics*, Wiley 2nd edn 1972, Dover, New York.

Whipple, C. (ed.): 1987 *De Minimis Risk*, Plenum Press.

Department of Philosophy and the History of Technology
Royal Institute of Technology
E-mail: martinp@infra.kth.se

KATRIN SCHULZ

A PRAGMATIC SOLUTION FOR THE PARADOX OF FREE CHOICE PERMISSION

ABSTRACT. In this paper, a pragmatic approach to the phenomenon of free choice permission is proposed. Free choice permission is explained as due to taking the speaker (i) to obey certain Gricean maxims of conversation and (ii) to be competent on the deontic options, i.e. to know the valid obligations and permissions. The approach differs from other pragmatic approaches to free choice permission in giving a formally precise description of the class of inferences that can be derived based on these two assumptions. This formalization builds on work of Halpern and Moses (1984) on the concept of 'only knowing', generalized by Hoek et al., (1999, 2000), and Zimmermann's (2000) approach to competence.

1. INTRODUCTION

(1)   'You may go to the beach or go to the cinema'

I almost told my son Michael. But I thought better of it, and said:

(2)   'You may go to the beach.'

Boys shouldn't spend their afternoons in the stuffy dark of a cinema, especially not with such lovely weather as to-day's. Thus, what I did in fact permit was less than what I first intended to permit. We might even be inclined to say that the permission I contemplated, entailed, but was not entailed by, the permission I gave. (Kamp 1973, p. 57)

These are the starting lines of a paper of Kamp from 1973 with which he illustrated the well-known phenomenon of *free choice permission*: a sentence of the form 'You may A or B' seems to entail the sentences 'You may A' and 'You may B'.

According to the logical paradigm, a theory of interpretation should provide a formal description of the intuitive inferences a sentence of English comes with, thus, as we will say, it should lay down the *logic* of English.[1] As the extensive literature on the subject shows the inference of free choice permission poses a serious problem for this approach to interpretation. In fact, some students of the problem have argued that it is impossible to come up with a logic of English that treats free choice permission as valid.

Synthese (2005) 147: 343–377
Knowledge, Rationality & Action 289–323
DOI 10.1007/s11229-005-1353-y

Let us take a closer look at one of the central arguments brought forward to support this claim. One way to approach the logic of sentences
like (1) and (2) is to describe the meaning of the involved expressions as
'may' and 'or' by providing an axiomatization of the truth-maintaining
reasoning with sentences containing them. However, it seems impossible to find a reasonable set of axioms and derivation rules such that free
choice permission becomes a valid inference. As soon as one arrives at a
system that together with other necessary and uncontroversial assumptions takes free choice permission to be valid, a range of unintuitive
conclusions become derivable as well. For instance, the derivation rules
of modus ponens and necessitation, together with the classical tautologies and taking deontic 'may' and 'must' to be interdefinable[2] seem to
be very uncontroversical assumptions. But if the rule of free choice permission is added to this system it allows the following absurd argument
(see Zimmermann 2000).[3]

(3) a. Detectives may go by bus.
    b. Anyone who goes by bus goes by bus or boat.
    c. Thus, detectives may go by bus or boat.
    c. We conclude that detectives may go by boat.

The apparently unbridgeable misfit between what the logic of sentences like (1), (2), and those in (3) is supposed to look like and the
intuitive validity of free choice permission has led Wright (1969) to
speak of a *paradox* of free choice permission. But now one might
continue, if there is no convincing logic of English that captures the
validity of free choice permission, then the formal approach is not
an adequate strategy to describe the semantics of English. Conse-
quently, we should better dismiss the logical paradigm.

At least two assumptions involved in this line of argumentation have
been found deficient. First, one can question whether the 'necessary
and uncontroversial' assumptions about valid semantic inferences of
English involved in the argument (3) are actually that uncontroversial.
For instance, Zimmermann (2000) has argued that $A \rightarrow (A$ or $B)$ is
not valid for the semantics of English, thus, that English 'or' cannot be
translated as inclusive disjunction '$\vee$'. As a consequence, in the example
above the step from (3a) to (3c) is not admissible and the implausible
conclusion (3d) can no longer be derived.

A different kind of explanation for paradoxes similar to the paradox of free choice permission has been proposed by Grice (1957). He
addresses generally the observation that classical logic does not seem

to be able to describe the way we interpret English sentences. Grice admits that this is the case. However, he claims, this does not mean that it is not the appropriate logic to model the *semantics* of English. His point is that semantic meaning does not exhaust interpretation. There is also a contribution of contextual *use* to meaning. This information, the *pragmatic* meaning, then closes the gap between the classical logic of semantics and our intuitive understanding of English. Applied to the paradox of free choice permission this means that an axiomatization of the semantics of sentences like (1)–(3) as proposed by von Wright is on the right track. The fact that this logic is incompatible with free choice permission only suggests that this inference should better be analyzed as a pragmatic phenomenon. Grice's plan was then to provide a pragmatic theory that rescues the simple logical approach to language. This enterprise became known as the *Gricean Program*. Grice also outlined parts of such a pragmatic theory in his theory of conversational implicatures. According to this theory a speaker can derive additional information from taking the speaker to behave rationally and cooperatively in conversation. For Grice this means that the speaker will obey certain principles that govern such behavior: the *maxims of conversation*.

So far we have sketched two possible ways out of the paradox of free choice permission: first we can say that the notion of entailment on which the derivation of (3d) from (3a) is based is not the entailment of the semantics of English. Then, of course, we have to provide a better candidate that does not produce such infelicitous predictions. The second option is to follow the Gricean program: we keep the classical logical semantic analysis and propose free choice permission to be a pragmatic phenomenon. Then we are required to come up with a pragmatic theory that can account for the free choice inference. In this paper, we want to explore the second option. This choice has not been adopted based on an evaluation of free choice permission as pragmatic inference. While we will see that many characteristics of this inference speak for such an approach, observations pointing in the opposite direction can be found as well. The theoretical question driving the research was rather whether a satisfying pragmatic explanation for free choice permission *can* be given. There is a well-known and dreaded obstacle such an approach has to overcome. To show that a certain inference can be explained by Grice's theory of conversational implicatures, we first need a precise description of the conversational implicatures an utterance comes with. Grice himself did not provide such a thing. One of the main goals of pragmatics in the last decades has been to overcome this deficiency (e.g. Horn 1972; Gazdar 1979; Hirschberg 1985), but a completely satisfying proposal in this direction is still missing. One may ask for the reason of this lack of success. Perhaps Grice's program to rescue the logical approach to semantics only has shifted the problem to the realm of pragmatics. Now it is this part of interpretation that resists a formalization.

There are good reasons to believe that the mentioned attempts to improve on the clarity of Grice's theory did not exhaust their possibilities. When looking at the proposals made it emerges that a rather limited set of technical tools has been used. The main role is still played by classical deductive logic; the logic of Frege and Tarski. But also logic has had its revolutions since their times, among them the development of non-monotonic reasoning. Non-monotonicity has always been considered to be a central feature of conversational implicatures.[4] This suggests that techniques developed in non-monotonic logic may be of use to formalize the theory of Grice. In this paper we will try to use non-monotonic logic to formalize Grice's theory of conversational implicatures – at least to the extent that it allows us to give a pragmatic, Gricean explanation of the free choice permission.

Let us summarize the discussion so far. The aim of the present paper is to provide an explanation of the phenomenon of free choice permission. By 'explanation' we mean to come up with a formally precise and conceptually satisfying description of the semantic and pragmatic meaning of expressions like (1) and (2) such that we can explain why the second sentence follows from the first. In the framework of this paper we are not looking for *any* kind of explanation. The idea is to see how far we can get with a pragmatic explanation along the lines of the Gricean program. Thus, we want to maintain a simple approach to semantics that is based on classical logic. In particular, we will interpret utterance as in (1) and (2) as assertions, 'or' as inclusive disjunction, and 'may' as a unary modal operator. On the basis of such a semantics free choice permission will not come out as valid. Instead, this inference is to be explained as a conversational implicature. To overcome the lack of precision in the theory of Grice we will try to formalize parts of it using non-monotonic logic. Hopefully, this can be done in a way such that we can account for free choice permission.

The rest of the paper is structured as follows. In the following section we study in some more detail the phenomenon of free choice permission to get a clearer impression of what we have to explain. Afterwards a new Gricean approach to free choice permission is developed. Then we will discuss the proposal and compare it to other accounts of free choice permission. The paper will finish with conclusions and an outlook on future work.

2. FREE CHOICE INFERENCES

In this section we will have a closer look at the linguistic phenomenon we want to account for. The aim is to obtain a clear picture of the properties of free choice permission. We will also provide some linguistic motivation for the kind of approach we have adopted.

Part of the simple approach to the semantics of sentences as (1) adopted here is that we take them to be assertions. There have been doubts about such an analysis. Kamp (1973), for instance, defends a proposal that takes such sentences to be performatives, granting a permission. However, a closer look on the data reveals that we at least additionally need an approach to the free choice reading of (1) that treats the sentence as an assertion.

It seems to be quite clear that the problematic sentences *do* have a reportative reading and that also this reading allows to infer free choice permission. Assume one student asks another about the submission regularities concerning some abstract. The answer she gets is (4).

(4)   You may send it by post or by email.

This sentence also allows a free choice reading according to which both ways of submission are admitted. But in this context it is clear that it is not the speaker who is granting the permission. Thus, even if we could solve the paradox of free choice permission for the performative use, the problem would still exist for the assertive reading. A similar point is made by the observation that parallel inferences as free choice permission also exist for other constructions that cannot be analyzed as performatives (the examples stem from Kamp 1979).

(5) a. We may go to France or stay put next summer. (with the epistemic reading of '*may*')
    b. I can drop you at the next corner or drive you to the bus stop.

Similar to example (1), (5a) seems to entail 'We may go to France' and 'We may stay put next summer'. In the same way the use of (5b) allows the hearer to infer 'I can drop you at the next corner' and 'I can drive you to the bus stop'. Zimmermann has also argued that the inference of (6) that Peter may have taken the beer from the fridge and that Mary may have taken the beer from the fridge should be analyzed as belonging to the same family.

(6)   Peter or Marie took the beer from the fridge.

We will call all these inferences *free choice inferences*. Their similar structure suggests to treat them all as due to the same underlying mechanism. But then nothing of this mechanism should hinge on the possible performative use of (1).

The examples above also illustrate that free choice inferences can come with sentences of quite different forms. This makes it hard to find a semantic explanation of the phenomenon. Semantics would expect some part of the construction of (1) to trigger the free choice permission. But as (5a), (5b), and (6) show, an approach taking the sentence mood, the modal 'may', or modalities in general to be responsible for the inferences is doomed to fail.

Another item that immediately suggests itself as responsible for the free choice readings is the connector 'or'. Indeed, many semantic approaches to the problem take this starting point. They propose, for instance, that 'or' can function as conjunction, thus, that (1) semantically means, or can mean, (roughly) the same as 'You may go to the beach and you may go to the cinema'. One problem for such a proposal is that this conjunctive meaning of 'or' does not generalize to arbitrary linguistic contexts. For instance, the sentence (7) does not entail that Mr. X must take a boat and that he must take a taxi.

(7)   Mr. X must take a taxi or a boat.

It goes often unnoticed that also (7) comes with free choice inferences. The sentence has an interpretation from which one can conclude that Mr. X still may choose which disjunct of (7) he is going to fulfill, i.e. (7) allows us to infer that Mr. X may take a taxi and that he may take a boat. A similar reading also exists for epistemic 'must' (cf. Alonso-Ovalle 2004).

Another property of the free choice inferences that speaks in favor of a pragmatic approach is the fact that they are cancelable: they disappear in certain contexts.[5] For obvious reasons, context-dependence is difficult to handle for semantic approaches to the free choice inferences. But it is what you would expect when free choice inferences are pragmatic inferences, particularly conversational implicatures.

The first kind of context in which they disappear is the classical cancellation contexts: when they contradict semantic meaning or world knowledge. Consider, for instance (8).

(8)   Peter is in love or I'm a monkey's uncle.

From (8), in contrast to (6), one cannot infer that both sentences combined by 'or' are possibly true, and, thus, that the speaker might be a monkey's uncle. Intuitively, it is quite clear why this free choice inference is not admissible: because the (human) speaker cannot be (in the strict sense of the word) a monkey's uncle.

There is another class of situations where in particular deontic free choice inferences can be cancelled. These are contexts where it is known that the speaker is not competent on the topic of discourse. This can either be clear from the context or be explicitly said by the speaker, as in (9).

(9)   You may take an apple or a pear – but I don't know which.

This sentence does not convey that the addressee has the choice as to which fruit he picks. Instead, the sentence is interpreted as would be expected if 'or' means inclusive disjunction (plus the inference that the speaker takes both, taking an apple and taking a pear, to be possibly permitted; this is conveyed by the continuation 'but I don't know which'). This observation suggests that the competence of the speaker plays an important role in the derivation of free choice permission.[6]

As we have seen in this section, free choice permission is part of a wider class of free choice inferences that can come with quite different linguistic constructions. This form independence of the inferences plus their cancellability gives some linguistic support for the decision to try to come up with a pragmatic explanation for their existence. The goal of the next section is then to provide such a pragmatic approach that can account not only for free choice permission, but for free choice inferences and their properties in general.

3. THE APPROACH

3.1. *Introduction*

We come now to the main part of the paper. In the following, a pragmatic approach to the free choice inferences is developed. Given the intention of the paper to follow the Gricean program, we will adopt a simple and classical approach to the logic of semantic meaning, in particular, 'or' will be interpreted as inclusive disjunction and modal expressions are analyzed as unary modal operators. Because this semantics does not account for the free choice inferences, they have to be described as inferences of the pragmatic meaning of an utterance. We will try to describe them as conversational implicatures.

As pointed out in the introduction, if we want to explain certain inferences as conversational implicatures we first need to formalize the latter notion, i.e. to give a precise description of the conversational implicatures an utterance comes with. In order to do so we will use results from non-monotonic logic, particularly work from Halpern and Moses (1984) recently extended by Hoek et al. (1999, 2000).

3.2. *The semantics*

Before we can start looking for a pragmatic approach to the free choice inferences we first have to be entirely clear about what our classic approach to the semantics of English can do. Therefore, in this section a precise description of this semantics is given. We will introduce a formal language in which we can express sentences as (1) and (2), at least to that extent that we take to be relevant for the free choice inferences. Then, we will provide a model-theory for this language, and, thereby, a semantic theory for the sentences.

The Language. The semantics of the sentences giving rise to the free choice inferences is formulated in modal propositional logic. Our formal language $\mathcal{L}$ is generated from a finite set of propositional atoms $\mathcal{P} = \{\top, \bot, p, q, r, \ldots\}$, the logical connectives $\neg, \wedge, \vee$, and $\rightarrow$, and two unary modal operators $\{\Diamond, \triangle\}$. The diamond is used to formalize epistemic possibility (thus $\Diamond p$ stands for 'possibly p'). The intended reading of $\triangle p$ is roughly 'p is permitted'. We will use ∇ to shorten $\neg \triangle \neg$ and $\square$ abbreviates $\neg \Diamond \neg p$. $\square \phi$ is thus true if the speaker believes ϕ. This gives a very simplified picture of the modalities we can express in English. However, we hope that it will become clear that the approach to the free choice inferences we are going to propose applies as well to more complex modal systems.

We call $\mathcal{L}^0 \subseteq \mathcal{L}$ the language that contains the modal-free part of $\mathcal{L}$ i.e. the language defined by the BNF $\chi ::= p (p \in \mathcal{P}) | \chi \wedge \chi | \neg \chi$.[7] Furthermore, we introduce the following abbreviations for certain $\mathcal{L}$ sentence-schemes: $[D]$ for $\square \phi \rightarrow \Diamond \phi$, $[4]$ for $\square \phi \rightarrow \square \square \phi$, and $[5]$ for $\neg \square \phi \rightarrow \square \neg \square \phi$.

The Semantics. The model theory we assume for $\mathcal{L}$ is standard for modal propositional logic. A *frame for* $\mathcal{L}$ is a triple of a set of worlds W and two binary relations $R_\triangle$ and $R_\Diamond$ over W. A *model* for $\mathcal{L}$ is a tuple consisting of a frame for $\mathcal{L}$ and an interpretation function V for the non-logical vocabulary of $\mathcal{L}$: a function from $p \in \mathcal{P}$ to characteristic functions over W. Let $F = \langle W, R_\Diamond, R_\triangle \rangle$ be a frame for $\mathcal{L}$ and $M = \langle F, V \rangle$ a model. For $w \in W$, $R_\Diamond[w]$ denotes the set $\{v \in W | \langle w, v \rangle \in R_\Diamond\}$ and $R_\triangle[w]$ the set $\{v \in W | \langle w, v \rangle \in R_\triangle\}$. We call the tuple $s = \langle M, w \rangle$ for $w \in W$ a *state*. *Truth* of a sentence of $\mathcal{L}$ with respect to a state is defined along standard lines. We will give here only the definition of truth for a formula $\triangle \phi$: $\langle M, w \rangle \models \triangle \phi$ iff$_{def}$ there is a $v \in W$ such that $v \in R_\triangle[w]$ and $\langle M, v \rangle \models \phi$. A set of formulas Γ is *satisfiable* in a set S of states if there is some $s \in S$ where all elements of Γ are true. A set of formulas Γ *entails* a formula ϕ relative to a class of states S ($\Gamma \models_S \psi$) iff$_{def}$ for all $s \in S$: $s \models \Gamma$ implies $s \models \psi$. If $\Gamma = \{\phi\}$, we write $\phi \models_S \psi$. Because we intend the given model theory to describe the semantic meaning of $\mathcal{L}$-sentences, formulas entailed by $\models$ from a sentence q have to be understood as being entailed by the semantic meaning of ϕ.

Let $\mathcal{S}$ be the set of states that entail the sentence-schemes $[4]$, $[5]$, and $[D]$. It follows that $\mathcal{S}$ is the class of states $s = \langle M, w \rangle$ that have a locally (i.e. in w) transitive, euclidian and non-blind[8] accessibility relation $R_\diamond$.[9] In the following we will consider as domain of interpretation only subsets of S. Conceptually, this means that we assume that the speaker has positive and negative introspective power, and we exclude the absurd belief state.[10]

The Free Choice Inferences. Now we can formulate the different free choice inferences we came across in Section 2 in terms of the formal language $\mathcal{L}$. Let us write $\phi| \equiv_S \psi$ if ψ can be inferred from the utterance of ϕ in context S. Let p, q be $\mathcal{L}$-sentences that do not contain any modal operators, i.e. $p, q \in \mathcal{L}^0$. In order to model the free choice inferences, the following rules should be valid for $| \equiv_S$. ($\{A|B\}$ has to be read as 'A is the premise or B is the premise'.)

(D1) $\quad p \vee q| \equiv_S \diamond p \wedge \diamond q,$

(D2) $\quad \{\diamond(p \vee q)|\diamond p \vee \diamond q\}| \equiv_S \diamond p \wedge \diamond q,$

(D3) $\quad \{\square(p \vee q)|\square p \vee \square q\}| \equiv_S \diamond p \wedge \diamond q,$

(D4) $\quad \{\triangle(p \vee q)|\triangle p \vee \triangle q\}| \equiv_S \triangle p \wedge \triangle q,$

(D5) $\quad \{\nabla(p \vee q)|\nabla p \vee \nabla q\}| \equiv_S \triangle p \wedge \triangle q.$

As pointed out in the last section, however, free choice inferences are cancellable: certain additional information can suppress their derivation. That means that we do not want (D1)–(D5) to hold for all $S \subseteq S$. To take care of the observation that free choice inferences do not occur if inconsistent with other information in the context we should add to (D1)–(D3) '*iff $\diamond p \wedge \diamond q$ is satisfiable in S*'. Because of the special cancellation behavior of deontic free choice inferences we need for (D4) and (D5) the extended condition '*iff $\diamond p \wedge \diamond q$ is satisfiable in S and the speaker is not known to be incompetent in S*'.

We allow for the antecedent of the free choice inferences two different logical forms depending on the scope relation between '$\vee$' and the modal operators. The reason is that we do not see clear evidence that excludes one of the forms either from representing the underlying structure of a sentence like (10a) or from giving rise to the free choice inferences. Notice, for instance, that different authors have argued that sentences as (10b) where 'or' has explicitly wide scope over the modal expressions do have free choice readings as well.

(10) a. You may take an apple or a pear.
　　　b. You may take an apple or you may take a pear.

3.3. *Introducing the general ideas*

The central task of any approach to the free choice inferences is to find a notion of entailment that can take over the role of $|\equiv_S$ in (D1) to (D5). Of course, the first candidate that comes to mind is the semantic notion of entailment $\models$. However, the free choice inferences would not be a problem if $\models$ would do. Thus, and as we have observed already, the free choice conclusions of (D1) to (D5) are not valid on the semantic models of the respective premises. Following Grice's program, this means that we have to look for a pragmatic notion of entailment that does the job, i.e. we have to find a pragmatic interpretation function such that the conclusions of (D1) to (D5) are valid on the *pragmatic* models of the premises.

But which semantic models does the pragmatic interpretation function have to select to make the free choice inferences valid? Let us, for example, take the inference (D2). There are three types of states $s = \langle M, w \rangle$ where sentence $\Diamond(p \vee q)$ is true qua its semantic meaning. In a first class of states there are worlds accessible from w, where p is true but no worlds where q holds. This possibility is represented by s_1 in Figure 1. A second type of states has q-worlds accessible from w, but no p-worlds; for illustration see s_2. Finally, it may be the case that for both propositions p and q there are worlds in the belief state of the speaker in s where they are true. This type of states is exemplified by s_3 in Figure 1. Only on the last type of states is the conclusion of (D2) valid, i.e. $s_3 \models \Diamond p \wedge \Diamond q$. Thus, we need the pragmatic interpretation to be a function f that maps the class of semantic models of $\Diamond(p \vee q)$ on the set only containing states like s_3.

How can we characterize this function f? The central idea of the approach proposed here is that the state s_3 is special because while

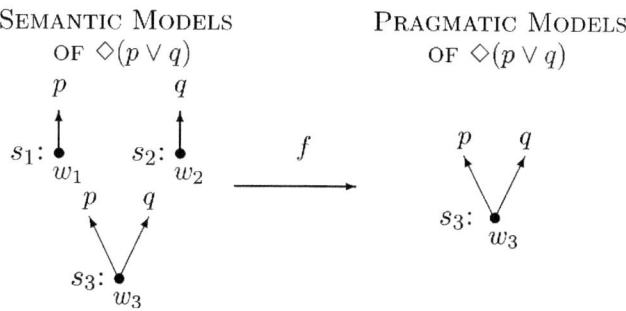

Figure 1.

the speaker believes her utterance to be true she believes less in s_3 than in every other semantic model where this is the case. In s_1, for instance, the speaker believes more than in s_3 because she holds the additional belief that p is true. In s_2 compared with s_3 the same holds for q. Thus, the pragmatic interpretation function f works as follows: besides $\Diamond(p \vee q)$ it takes some partial order $\preceq$ as argument that compares how much the speaker believes in different states, and then it selects those states (i) where $\Diamond(p \vee q)$ is true qua its semantic meaning, (ii) where the speaker believes her claim $\Diamond(p \vee q)$ to be true, and (iii) that are minimal with respect to the order $\preceq$. More precisely, the pragmatic interpretation $f_S^{\preceq}(\phi)$ of a sentence ϕ with respect to a set of states S and a partial order $\preceq$ is defined as the set $\{s \in S \mid s \models \phi \wedge \Box\phi \ \& \ \forall s' \in S : s' \models \phi \wedge \Box\phi \Rightarrow s \preceq s'\}$. Based on $f_S^{\preceq}(\phi)$ we can define the following notion of entailment: we say that sentence ϕ pragmatically entails sentence ψ with respect to S and $\preceq$, $\phi \mid \equiv_S^{\preceq} \psi$, if on all states in $f_S^{\preceq}(\phi)$, i.e. on all pragmatic models of the sentence, ψ is true.

DEFINITION 1 (The Inference Relation $\mid\equiv$).
Let $\preceq$ be a partial order on some class of states S. We define for sentences $\phi, \psi \in \mathcal{L} : \phi \mid\equiv_S^{\preceq} \psi \ iff_{\text{def}}$

$$\forall s \in S : \ [s \models \phi \wedge \Box\phi \ \& \ \forall s' \in S : s' \models \phi \wedge \Box\phi \Rightarrow s \preceq s'] \Rightarrow s \models \psi.$$

Let us reflect for a moment on the content of this definition. According to f the interpreter accepts only those models of the speaker's utterance as pragmatically well-formed where the speaker has no additional information that she withholds – by uttering $\Diamond(p \vee q)$ – from the interpreter. For instance, the interpreter does not take s_1 to be a proper pragmatic model of the sentence. Here, the speaker believes that p but nevertheless utters the weaker claim $\Diamond(p \vee q)$. The interpreter can be understood as taking the speaker to obey the following principle:

> *The contribution ϕ of a rational and cooperative speaker encodes all of the information the speaker has; she knows only ϕ.*

Readers familiar with Grice's theory of conversational implicatures will recognize the Gricean character of this assumption. It can be understood as a combination of his maxim of quality with the first sub-clause of the maxim of quantity. To base the free choice inferences on this assumption is to explain them as conversational implicatures. We will therefore call the above statement the *Gricean Principle* and refer to f as the pragmatic interpretation function *grice*.

3.4. *Working out the details*

So far everything has gone quite smoothly. We have localized a Gricean Principle that seems to be responsible for the free choice inferences. We were also able to propose a formalization of the notion of pragmatic entailment this principle gives rise to. But there is still something missing in Definition 1. We did not define the order $\preceq$, i.e. we have not said so far when in some state s' the speaker believes as least as much as in a state s. To find a satisfying definition will require some effort.

3.4.1. *The epistemic case*
Let us, for a moment, forget about the deontic modalities. When do we want to say that in state $s' = \langle M', w' \rangle$ the speaker believes as least as much as in state $s = \langle M, w \rangle$? The intuitive answer is that in s the speaker should be equally or less clear about how the actual world looks like as/than in s', thus, she should distinguish in s the same or a wider range of epistemic possibilities. Or, to be a little bit more precise, every state of affairs she considers possible in s' the speaker should also consider possible in s. Then, we have to say what it means that the speaker considers the same state of affairs possible in s and s'. Let us try the following: this is the case if there are $v \in R_\diamond[w]$ and $v' \in R_\diamond[w']$ that interpret the atomic propositions in the same way. Thus, we define the order comparing belief states of the speaker as follows.[11]

DEFINITION 2 (The basic order $\preceq^0$).
For $s = \langle M, w \rangle, s' = \langle M'w' \rangle \in S$ we define $s \preceq^0 s'$ iff:

$$\forall v' \in R'_\diamond[w'] \exists v \in R_\diamond[w] \quad (\forall p \in \mathcal{P} : V(p)(v) = V'(p)(v')).$$

With this definition at hand we can fill out the gap in Definition 1 and obtain the first concrete instance of a pragmatic entailment relation: $\phi| \equiv_S^{\preceq^0} \psi$, abbreviated $\phi| \equiv_S^0 \psi$ holds, if on the $\preceq^0$-minimal set of S where the speaker believes ϕ, ψ is valid. This finishes the formalization of the Gricean principle and brings us to the central question of the paper: can we account for the free choice inferences with this notion of entailment? That means, given that we only consider the epistemic modalities $\diamond$ and $\square$ in this subsection, are (D1)–(D3) valid for $\models^0_S$? This is not easily answered. To establish properties of minimal models is not straightforward. The problem is that we have no immediate access to these states, But it turns out that this is not necessary. The only thing we have to show is that any state where the inferences are not valid is not a minimal state.[12]

FACT 1. For any partial order $\preceq$, if $\forall_s \in S[s \models \phi \wedge \Box\phi \wedge \neg\psi \Rightarrow (\exists s' \in S : s' \models \phi \wedge \Box\phi \ \& \ s' \prec s)]$, then $\phi| \equiv^\preceq_S \psi$.

Fact 1 tells us that the only thing we have to do to establish, for instance (D2) (i.e. that for a set $\{p, q\} \subseteq \mathcal{L}^0$ satisfiable in S, $\Diamond(p \vee q)| \equiv^0_S \Diamond p \wedge \Diamond q$ is valid) is to show that for every state $s \in S$ that models $\Diamond(p \vee q) \wedge \Box\Diamond(p \vee q)$ but not the conclusion of (D2) we can find a state $s^* \in S$ where still $\Diamond(p \vee q) \wedge \Box\Diamond(p \vee q)$ is true and $s^* \prec^0 s$.

Let $s = \langle M, w \rangle \in \mathcal{S}$ be a state with the properties described above. Without loss of generality we assume $s \not\models \Diamond p$. How can we find the $s^* \in S$ we are looking for? This is quite simple: we take s^* to be a state in S that differs from s only in having an additional world $\tilde{v}$ in the belief state of the speaker $R_\Diamond[w^*]$ where p *does* hold.[13] It is easy to see that this state s^* has all the properties we need to prove the validity of (D2), i.e. (i) s^* still models $\Diamond(p \vee q) \wedge \Box\Diamond(p \vee q)$, (ii) s^* is $\preceq^0$-smaller than s: $s^* \not\succeq^0 s$, and (iii) s is not $\preceq^0$-smaller than s^*: $s \not\succeq^0 s^*$.

Ad (i): From $\langle M^*, \tilde{v} \rangle \models p$ it follows that $s^* \models \Diamond(p \vee q)$. Because $s^* \in S$ (in particular $s^* \models [5]$) we can conclude that $s^* \models \Box\Diamond(p \wedge q)$. Thus $s^* \models \Diamond(p \vee q) \wedge \Box\Diamond(p \vee q)$.

Ad (ii): The only difference between s^* and s is that s^* has one more $\Diamond$-accessible world: $\tilde{v}$. Thus, it will clearly be true that $\forall v \in R_\Diamond[w] \exists v^* \in R^*_\Diamond[w^*](\forall p \in \mathcal{P} : V(p)(v) = V(p)(v^*))$. We can conclude $s^* \preceq^0 s$.

Ad (iii): We know that there is a $v^* \in R^*[w^*]$ such that $\langle M^*, v^* \rangle \models p$ – this is $\tilde{v}$. Because $s \not\models \Diamond p$ there will be no $v \in R_\Diamond[w]$ such that $\langle M, v \rangle \models p$. Furthermore, because $p \in \mathcal{L}^0$ in no $v \in R_\Diamond[w]$ can the interpretation of the atomic propositions be the same as in $\tilde{v}$. But that means that $\forall v^* \in R^*_\Diamond[w^*] \exists v \in R_\Diamond[w] (\forall p \in \mathcal{P} : V(p)(v) = V(p)(v^*))$ cannot be true. Thus, $s \not\succeq^0 s^*$.

Using the same strategy we can also prove that for $p, q \in \mathcal{L}^0$ such that $\{p, q\}$ is satisfiable in S (D1): $p \vee q| \equiv^0_S \Diamond p \wedge \Diamond q$ and $\Box(p \vee q)| \equiv \Diamond p \wedge \Diamond q$ are valid. But what about the second antecedent of (D3)? Does $\Box p \vee \Box q | \equiv^0_S \Diamond p \wedge \Diamond q$ hold? Indeed, it does. Actually, we obtain $\Box p \vee \Box q | \equiv^0_S \bot$. The reason is that there is no $s \in S$ such that $s \models (\Box p \vee \Box q) \wedge \Box(\Box p \vee \Box q)$ and s is $\preceq^0$ smaller or equal to every other state in S with this property. Thus, we predict that the sentence has no pragmatic models, $grice^0_S(\Box p \vee \Box q)$ is empty.

To see that there can be no elements in $grice^0_S(\Box p \vee \Box q)$ notice that $\phi := (\Box p \vee \Box q) \wedge \Box(\Box p \vee \Box q)$ is, for instance, true in a state where the speaker believes that p and not q. Let $s_1 = \langle M_1, w_1 \rangle$ be a state where this is the case, i.e. $s_1 \models \Box(p \wedge \neg q)$. But the sentence is also true if the speaker believes that q and not p. Assume that this holds in $s_2 = \langle M_2, w_2 \rangle$, i.e. $s_2 \models \Box(\neg p \wedge q)$. It is not difficult to see that for s_1 and s_2 neither $s_1 \preceq^0 s_2$ nor $s_2 \preceq^0 s_1$ holds. If it were the case that $grice(\phi) \neq \emptyset$ (i.e. there would exists a state $s \in S$ that models ϕ and for all other states $s' \in S$ with this property: $s \preceq^0 s'$) then it would follow that $s \preceq^0 s_1$ and $s \preceq^0 s_2$. By the choice of s_1 ($s_1 \models \Box \neg q$) there are worlds in $R_\Diamond[w_1]$ where q does not hold. Because $q \in \mathcal{L}^0$, if $s \preceq s_1$, i.e. $\forall v_1 \in R_{1,\Diamond}[w_1] \, \exists v \in R_\Diamond[w](\forall p \in \mathcal{P} : V(p)(v) = V_1(p)(v_1))$, in $R_\Diamond[w]$ there have to be such worlds too. Thus $s \models \neg \Box q$. For the same reason, if $s \preceq^0 s_2$ there have to be worlds in $R_\Diamond[w]$ where p is false, and, hence, $s \models \neg \Box p$. But then $s \models \neg \Box p \wedge \neg \Box q$. This contradicts the condition $s \models \Box p \vee \Box q$. Thus $grice(\Box p \vee \Box q) = \emptyset$.

Conceptually, the fact that for logically independent $p, q \in \mathcal{L}^0$: $\Box p \vee \Box q | \equiv^0 \bot$ means that our theory predicts this sentence to be pragmatically not well-formed. But this seems to be – given Grice's theory and our formalization thereof – correct. If for a sentence ϕ satisfiable in S, $grice^0_S = \emptyset$, then there are incomparable $\preceq^0$-minimal states modeling $\phi \wedge \Box \phi$. This means that the speaker believes in minimal belief states for $\phi \wedge \Box \phi$ different things. Then, the speaker has to have in these minimal belief states beliefs she did not communicate. Thus, it is obvious for the interpreter that she did not obey the Gricean Principle. We follow Halpern and Moses (1984) in calling such sentences *dishonest*.

Dishonest sentences provide an interesting testing condition for the theory of Grice and the formalization thereof proposed here. Grice's theory predicts that dishonest sentences should be pragmatically out: they cannot be uttered by speakers that obey the Gricean Principle. Furthermore, because it is proposed here that the free choice inferences are conversational implicatures, another prediction that can be tested is that the dishonest sentence $\Box p \vee \Box q$

should not give rise to free choice inferences. And, indeed, sentences like (11a) and (11b) are reported to not allow a free choice reading. In addition, their use seems to be restricted to particular contexts.[14]

(11) a. ?Mr. X must be in Amsterdam or Mr. X must be in Frankfurt.
   b. ?I believe that A or I believe that B.

3.4.2. The deontic case

As we have seen in the last section we can formalize the Gricean Principle in a way such that we can account for the epistemic free choice inferences in context S. But it is easy to see that $|\equiv_S^0$ will not predict (D4) and (D5) to be valid as well. The reason is that the order $\preceq^0$ on which this notion of entailment is based and that is intended to compare the beliefs of the speaker does not compare what the speaker believes about the deontic accessibility relation. We said that we want to base the pragmatic interpretation on an order that calls a state $s \in S$ smaller than a state $s' \in S$ if in the first the speaker believes less/considers more possible than in the second. For the basic information order $\preceq^0$ (see Definition 2) the only thing that matters is that in the first state the speaker considers more interpretations of the propositional atoms possible than in the second. As a consequence, $\preceq^0$ compares only the speaker's belief about the interpretation of these atoms (and Boolean combinations thereof).[15] This suggest that to account for the deontic free choice inferences we should extend the order such that it respects also the speaker's beliefs about what holds on the deontic accessibility relation. Thus, we should rather say that in state $s = \langle M, w \rangle$ the speaker believes less (or equally much) than in state $s' = \langle M', w' \rangle$ if for every world the speaker considers possible in s there is some world the speaker considers possible in s' that not only agree on the interpretation of the propositional atoms but also on which interpretations are deontically possible. This is expressed in the definition of the following order.

DEFINITION 3. (The Objective Information Order $\preceq^n$)[16]
For $s = \langle M, w \rangle, s' = \langle M', w' \rangle \in S$ we define $s \preceq^n s'$ iff_{def}

$\forall v' \in R'_\diamond[w'] \exists v \in R_\diamond[w]:$

(i) $\forall p \in \mathcal{P} : V(p)(v) = V'(p)(v')$ &

(ii) $\forall u \in R_\triangle[v] \exists u' \in R'_\triangle[v'] (\forall p \in \mathcal{P} : V(p)(u) = V'(p)(u'))$ &

(iii) $\forall u' \in R'_\triangle[v'] \exists u \in R_\triangle[v] (\forall p \in \mathcal{P} : V(p)(u) = V'(p)(u')).$

By substituting $\preceq^n$ as order in Definition 1 we obtain a new notion of entailment $|\equiv$, shortly $|\equiv^n_S$. In the same way as in the last section one can show that the free choice inferences (D1)–(D3) are valid for $|\equiv^n_S$. The only difference between the orders $\preceq^0$ and $\preceq^n$ lays in the conditions (ii) and (iii) which concerns belief about the deontic options. Therefore, they make exactly the same predictions for sentences that do not contain $\triangle$ or ∇.

However, the deontic free choice inferences (D4) and (D5) do not hold for $|\equiv^n_S$. Given that (D2) and (D4) show a highly similar structure one may wonder why we can account with $|\equiv^n_S$ for one but not for the other. The reason is this. In $\mathcal{S}$ there is no connection between the actual deontic options and the speaker's beliefs about what is deontically accessible. Therefore, from minimizing the speaker's belief the interpreter will learn nothing about what is actually permitted and what not. But the deontic free choice inference $\triangle p \wedge \triangle q$ is about valid permissions. For the actual epistemic options and the speaker's beliefs about them such a connection is built into $\mathcal{S}$. We defined $\mathcal{S}$ as those states where the speaker has full introspective power. Thus, we assumed that the speaker knows about her beliefs and her uncertainty. This suggests that to make the deontic free choice inferences valid we would need something similar there too, i.e. the speaker has to know about the valid obligations and permissions. The speaker has to be *competent* on the deontic options.

This conclusion is also supported by an observations we made in Section 2. There, we have seen that the deontic free choice inferences are cancelled if it is known that the speaker is not competent on the deontic options. Thus, it seems that these inferences really depend on additional knowledge about the competence of the speaker.

3.4.3. *Competence*

The considerations at the end of the last section suggest that an additional assumption of the speaker's competence may be the missing link to obtain the deontic free choice inferences. For the formalization of this idea we will rely on Zimmermann (2000). He builds on a proposal of Groenendijk and Stokhof (1984) and defines competence by the following first-order model condition.[17]

DEFINITION 4 (Competence).
A speaker is competent in a state $\langle M, w \rangle \in \mathcal{S}$ with respect to a modality $\triangle$ *iff*$_{\text{def}}$

$$\forall v \in W^M \quad [v \in R^M_\diamond[w] \Rightarrow (R^M_\triangle[v] = R^M_\triangle[w])].$$

It is easy to prove that this condition is characterized in modal propositional logic by the two axioms $[C_1]: \nabla\phi \to \Box\nabla\phi$ and $[C_2]: \neg\nabla\phi \to \Box\neg\nabla\phi$, i.e. a speaker is competent in some state $s = \langle M, w \rangle$ if the underlying frame locally (hence, in w) satisfies $[C_1]$ and $[C_2]$. $[C_1]$ is a generalization of axiom [4] formalizing positive introspective power to the multi-modality case; it warrants that the speaker knows about all valid obligations. $[C_2]$, on the other hand, generalizes axiom [5] formalizing negative introspective power; it assures that the speaker also knows about the valid permissions.

Let us call $\mathcal{C}$ the set of states where additionally to the axioms [D], [4], and [5] also the competence axioms $[C_1]$ and $[C_2]$ are valid. Do we get the free choice inferences for $|\equiv^n_\mathcal{C}$? Unfortunately, this is not the case. The pragmatic interpretation we obtain this way is much too strong. It is predicted that every sentence $\phi \in \mathcal{L}$ satisfiable in $\mathcal{C}$ gives rise to an empty pragmatic interpretation, i.e. is dishonest. Or, in other words, given the way $|\equiv^n_\mathcal{C}$ interprets the Gricean Principle a speaker competent on $\triangle$ as formalized in $[C_1]$ and $[C_2]$ cannot utter any non-absurd sentence and be obeying this principle.

Let us have a closer look at why this is the case. Given the formalization of competence we have chosen, a competent speaker knows for every $\chi \in \mathcal{L}$ which of the sentences $\nabla\chi$ and $\neg\nabla\chi$ holds. Hence, in all states of $\mathcal{C}$ and for all sentences $\chi \in \mathcal{L}$ either $\Box\nabla\chi$ or $\Box\neg\nabla\chi$ is true. However, it is easy to see that for every $\chi \in \mathcal{L}^0$ a state where $\Box\nabla\chi$ holds is $\preceq^n$-incomparable with a state where $\Box\neg\nabla\chi$ holds. Thus, to prevent dishonesty, i.e. to warrant that the interpreter does not end up with different incomparable minimal states, for the sentence ϕ uttered by the speaker either $\phi \wedge \Box\phi \models_\mathcal{C} \Box\nabla\chi$ or $\phi \wedge \Box\phi \models_\mathcal{C} \Box\neg\nabla\chi$ has to hold. But the same argument applies for every $\chi \in \mathcal{L}^0$! Thus, for every sentence $\chi \in \mathcal{L}^0$ it has to be the case that ϕ entails semantically either that the speaker believes $\nabla\chi$ or that she believes $\neg\nabla\chi$. There can be no finite and satisfiable sentence that is that strong. Hence, every sentence $\phi \in \mathcal{L}$ satisfiable in $\mathcal{C}$ is dishonest.

3.4.4. *Solving the paradox of free choice permission*
One way to look at the problem we ended up with in the last section is that the formalization of the Gricean Principle given with $|\equiv^n$ is

too strong. By $|\equiv^n$ a speaker who wants to obey the principle has to give every bit of information about deontic accessible interpretations of the basic atoms that she has. Perhaps we can obtain a more natural notion of pragmatic entailment when we allow the speaker to withhold some of this information. The problem, then, becomes to find the right restriction that fits our intuitions.

To start with, we can ask ourselves which information about the deontic accessibility relation we can take to be not relevant for the order because it is accessible to the interpreter anyway. It turns out that if the speaker is competent on Δ, then which permissions the speaker believes to hold can be already concluded from taking her to convey all she knows about the valid obligations. If she is honest about this part of her beliefs, then if her utterance ϕ does not entail for some $\chi \in \mathcal{L}^0$ that she believes $\nabla \chi$ she cannot believe this obligation to be valid, i.e. $\neg\Box\nabla\chi$ holds. From her competence it follows that she has to believe that $\neg\chi$ is permitted. On the other hand, if for some $\chi \in \mathcal{L}$ it holds that the speaker believes χ to be permitted, then, by competence, $\neg\Box\nabla\neg\chi$ is true and because we assume her to believe in her utterance ϕ, ϕ cannot entail $\nabla\neg\chi$. Thus, a competent speaker believes some sentence $\chi \in \mathcal{L}^0$ to be permitted if and only if her utterance does not entail that χ is prohibited. This suggests that information about which permissions the speaker believes to be valid can be ignored by the order. It is enough to compare what a competent speaker believes to be a valid obligation.[18] We obtain such an order when we delete condition (ii) from the definition of $\preceq^n$.[19]

DEFINITION 5 (The Positive Information Order $\preceq^+$).[20]
For $s = \langle M, w \rangle, s' = \langle M', w' \rangle \in S$ we define $s \preceq^+ s'$ iff $_{\text{def}}$

$$\forall v' \in R'_\Diamond[w'] \exists v \in R_\Diamond[w]:$$
(i) $\forall p \in \mathcal{P}: V(p)(v) = V'(p)(v')$ &
(ii) $\forall u' \in R'_\Delta[v'] \exists u \in R_\Delta[v]\ (\forall p \in \mathcal{P}: V(p)(u) = V(p)(u')).$

By substituting $\preceq^+$ in Definition 1 we obtain a new notion of pragmatic entailment: $|\equiv_S^{\preceq^+}$, abbreviated $|\equiv_S^+$. It turns out that for $|\equiv_C^+$ not only the free choice inferences for the epistemic modality are valid, but (D4) and $\nabla(p \vee q)|\equiv_C^+ \Diamond\Delta p \wedge \Diamond\Delta q$ as well. Parallel to the epistemic case the sentence $\nabla p \nabla \vee q$ is predicted to be dishonest when uttered by a competent speaker that obeys the Gricean Principle.

Let us discuss the validity of (D4). The argumentation we employ has exactly the same structure as in Section 3.4.1. If for $p, q \in \mathcal{L}^0$ such that $\{p, q\}$ is satisfiable in $\mathcal{C}$ (D4): $\triangle(p \vee q)|\equiv_{\mathcal{C}}^+ \triangle p \wedge \triangle q$ were not valid then there would be a state $s \in \mathcal{C}$ minimal with respect to $\preceq^+$ such that $s \models \triangle(p \vee q) \wedge \square\triangle(p \vee q)$ but not $s \models \triangle p \wedge \triangle q$. Now, we show that this cannot be the case: every state $s \in \mathcal{C}$ that semantically entails $\triangle(p \vee q) \wedge \square\triangle(p \vee q)$ but where the consequence of (D4) is not true cannot be minimal with respect to $\preceq^+$.

Assume that for $s = \langle M, w \rangle \in \mathcal{C}$ we have $s \models \triangle(p \vee q) \wedge \square\triangle(p \vee q)$, but $s \not\models \triangle p \wedge \triangle q$. Without loss of generality $s \not\models \triangle p$. Let $s^* = \langle M^*, w^* \rangle \in \mathcal{C}$ be the state that is like s except that from w^* an additional world $\tilde{v}$ is $\triangle$-accessible where p is true.[21] Thus $s^* \models \triangle p$. We show that (i) $s^* \models \triangle(p \vee q) \wedge \square\triangle(p \wedge q)$, (ii) $s^* \preceq^+ s$, and (iii) $s \not\preceq^+ s^*$. Then s cannot be minimal because s^* is smaller.

Ad (i): We have seen already that $s^* \models \triangle p$. It follows $s^* \models \triangle(p \vee q)$. Because s^* is an element of $\mathcal{C}$ we can conclude from this (by $[C_2]$) that $s^* \models \square\triangle(p \vee q)$. This shows (i).

Ad (ii): We have to show that for all $v \in R_\diamond[w]$ we can find a $v^* \in R_\diamond^*[w^*]$ such that (i) $\forall p \in \mathcal{P}(V(p)(v) = V^*(p)(v^*))$ and (ii) $\forall u^* \in R_\triangle^*[v^*] \exists u \in R_\triangle[v] (\forall p \in \mathcal{P} : V(p)(u) = V(p)(u^*))$. (i) is simple, let us go directly to the interesting case: (ii). Because the difference between s^* and s is that s^* has one more $\triangle$-accessible world: $\tilde{v}$, we have $R_\triangle[w] \subset R_\triangle^*[w^*]$. From $s, s^* \in \mathcal{C}$ we conclude $\forall v \in R_\diamond[w] : R_\diamond[v] = R_\triangle[w]$ and $\forall v^* \in R_\diamond^*[w^*] : R_\diamond^*[v^*] = R_\triangle[w^*]$. Together, this gives: $\forall v \in R_\diamond[w] \forall v^* \in R_\diamond^*[w^*] : R_\diamond[v] \subset R_\triangle^*[v^*]$. Because by assumption s and s^* do not differ in the interpretation assigned in worlds of $R_\triangle[v]$ to elements of $\mathcal{P}$ this proves the claim.

Ad (iii): Finally, $s \not\preceq^+ s^*$. Because $s \not\models \triangle p$ we obtain by $[C_1]$ that $s \models \square\neg\triangle p$. Hence, for no $v \in R_\diamond[w]$ and no $v \in R_\triangle[v]$ we have $\langle M, u \rangle \models p$. But from $s^* \models \triangle p$ with $[C_2]$ it follows $s^* \models \square\triangle p$, and, thus, $\forall v^* \in R_\diamond^*[w^*] \exists u^* \in R_\triangle^*[v^*] : \langle M^*, u^* \rangle \models p$. Because $p \in \mathcal{L}^0$ condition (ii) of the definition of $\preceq^+$ is violated for $s \preceq^+ s^*$.

Thus, we see that adopting $|\equiv^+$ as a formalization of the Gricean Principle and applying it to the set of states $\mathcal{C}$ where the speaker is competent accounts for the free choice inferences.[22]

3.5. *The cancellation of free choice inferences*

In the last sections we have developed a pragmatic notion of entail-ment that with respect to the set S makes the epistemic free choice inferences (D1)–(D3) valid, and with respect to the more restricted context C additionally validates the deontic free choice inferences. Have we, thereby, achieved our initial goal to provide a Gricean account for the free choice inferences? No, there is still something to be done. As discussed in Section 3.2 the free choice inferences are non-monotonic inferences: they can be cancelled by additional information. It remains to be checked whether the approach devel-oped above predicts (D1)–(D5) to be valid exactly in those contexts where such canceling information is not given.

In Section 2 we have seen that there are two different types of information that may lead to a suspension of free choice infer-ences. Let us proceed by discussing both of them separately. Our first observation was that free choice inferences are cancelled in case they are inconsistent with information in the context or given by the speaker.[23] It is easy to see that this is also predicted by the sys-tem we propose. If one of the consequents of (D1)–(D5) is incon-sistent with information in some context S or the semantic meaning of the utterance made, then there will be no state where this conse-quent holds among those states in S where the utterance is true (by its semantic meaning). In particular, the states selected by our prag-matic interpretation function *grice* will not make such a consequent true. Thus, we see that the approach immediately accounts for this part of the non-monotonicity of the free choice inferences.

Now we come to the second observation. As we have seen in Section 2, the deontic free choice inferences can also be cancelled by information that the speaker is not fully competent on the topic of discourse. Therefore, we should derive these inferences only in contexts where such information has not been given. Whether the proposal made accounts for this observation is not clear yet. We predict the deontic free choice inferences to be valid in a context where the interpreter takes the speaker to be competent and to obey the Gricean Principle. Of course, information that the speaker is in some respects incompetent stands in conflict with taking the speaker to be competent (as described by $[C_1]$ and $[C_2]$). But we have not said anything so far about how the interpreter behaves in such a situation.

Let us sketch one position one could adopt. We can propose that taking the speaker to be competent is an assumption interpreters make – just as they assume the speaker to obey the Gricean Principle. Interpreters do not make this assumption if they are facing contradicting information.[24] This proposal predicts that if an interpreter who does not know the speaker to be competent encounters information contradicting the competence assumption, then she will not derive the deontic free choice inferences. If, however, no such conflicting information is given, the interpreter assumes the speaker to be competent on △ and the inferences become valid. So far the cancellation behavior of the deontic free choice inferences is captured correctly. It may, however, be the case that the interpreter knows that the speaker is competent in some respects and that this information does not contradict what she now learns about the incompetence of the speaker. In such a situation it does not seem to be plausible to take this independent information to be cancelled together with the competence assumption. If it is not dismissed then it depends on what exactly the interpreter knows about competence and incompetence of the speaker whether the deontic free choice inferences are derived. This approach needs to be evaluated by comparing its predictions with the interpretational behavior of native speakers. This has to be investigated in future work.[25]

3.6. *Conclusions*

In this section we have developed a formalization of the Gricean Principle that can (given standard assumptions about the introspective power of the speaker) account for the epistemic free choice inferences. However, this formalization on its own is not able to derive the deontic free choice inferences as well. They can be predicted if in the context it is additionally known that the speaker is competent on △. We adopted a strong notion of competence: the speaker is taken to know the valid obligations as well as as all permissions. With this system we can account for all free choice inferences.

Furthermore, we have seen that the proposal also models correctly the cancellation of free choice inferences when conflicting information is encountered. Whether it can also account for the suspension of the deontic free choice given information that the competence of the speaker is limited depends on how we understand the role of the competence assumption in interpreting utterances. We have sketched one possible position that promises to model the cancellation behavior correctly. Empirical investigations have to show whether this proposal is convincing.

4. DISCUSSION

In the last section we have seen that based on a classical logical approach to the semantics of English the free choice inferences can be described in a formally precise way as due to taking the speaker (i) to obey the Gricean Principle, and (ii) to be competent on the topic of discourse. Thus, the central goal with which we started the paper has been reached: we came up with an approach to the free choice inferences on the lines of the Gricean program. In the following section we will address some open questions concerning the introduced approach and relate the proposal to other approaches to the free choice inferences.

4.1. *An open problem*

Unfortunately, in the present form the approach predicts, along with the free choice inferences, many inferences that are not welcome. For instance, for arbitrary, in S logically independent $p, q, r \in \mathcal{L}^0$ it holds that $\triangle(p \vee q)| \equiv_S^+ \Diamond r \wedge \Diamond \neg r \wedge \Diamond \triangle r \wedge \Diamond \triangle \neg r$ and $\triangle(p \vee q)| \equiv_C^+ \triangle r \wedge \triangle \neg r$. Or, to use more natural examples, we obtain, for instance, that (12a) $| \equiv_C^+$-entails (12b) and (12c). These predictions are certainly wrong.

(12) a. You may take an apple or a pear.
     b. You may take a banana.
     c. Aunt Hefty may be making pie.

Where do these strange predictions come from? The pragmatic interpretation function $grice^+$ on which $| \equiv^+$ is based selects among the semantic models of a sentence those where the speaker believes the sentence to hold and has as few as possible other beliefs. This is what the Gricean Principle demands: a speaker does not withhold information – *any* information – she has from the hearer.[26] Therefore, it is not surprising that if a speaker utters a sentence like (12a) that does not exclude that aunt Hetty is making pie, then $| \equiv^+$ predicts that the speaker considers it as possible that she is: according to the Gricean Principle, if the speaker believed that aunt Hetty is not making apple pie, then she would have shared her belief with the audience. She did not do so when uttering (12a). Thus, she cannot hold this belief. The point is that when we interpret utterances, we certainly do not expect the speaker to convey *all* of her beliefs (that are not commonly known). The Gricean Principle underlying $|\equiv^+$ is too strong.

There is a way out of this problem already suggested in Grice's formulation of the first sub-clause of the maxim of quantity:[27] 'Make your contribution as informative as required (for the current purpose of exchange)' (Grice 1989, p. 26). What the Gricean Principle misses is some restriction to contextually required or *relevant* information. Thus, it should rather be formulated as follows.

> *The contribution ϕ of a rational and cooperative speaker encodes all of the relevant information the speaker has; she knows only ϕ.*

This suggests that to overcome the above mispredictions we have to formalize contextual relevance and build it into our pragmatic notion of entailment. Some ideas how this can be done can be found in van Rooy and Schulz (2004). In this paper the formalization of the Gricean Principle proposed here is used to give a pragmatic explanation for the phenomenon of *exhaustive interpretation*. 'Exhaustive interpretation' describes the often observed strengthening of the semantic meaning of answers to overt questions.[28] In the context of questions it is quite obvious which information is relevant: information that helps to answer the question. The authors propose a version of the interpretation function *grice* that respects such a notion of relevance. In future work it has to be seen whether this solution can be also applied to the modeling of the free choice inferences proposed here.

4.2. *Comparison*

4.2.1. *The approaches of Kamp and Zimmermann*
The proposal to the free choice inferences introduced in this paper is highly inspired by the work of Zimmermann (2000) and Kamp (1979) on this subject, particularly the outline of a pragmatic approach of the latter author. Zimmermann, as well as Kamp, bases the free choice inferences on two premisses. The first ingredient is that from a sentence giving rise to free choice inferences the interpreter learns something about the epistemic state of the speaker.

From a sentence 'You may take an apple or a pear' she learns, for instance, that the speaker takes both, 'You may take an apple' and 'You may take a pear' to be possibly true. Sometimes, this already accounts for the free choice observation, as for instance, for examples like (6): 'Mary or Peter took the beer from the fridge'. But for free choice permission this is not enough. Further information is necessary and both approaches take this to be due to the assumption that the speaker is competent on the deontic options.

The second part, the reliance on competence of the speaker, has been adopted here. But the way these two proposals accounted for the derivation of the first part, the epistemic information, has been found deficient. Zimmermann takes the semantics of 'or' to be responsible. Among other things this leads to unreasonable predictions when 'or' occurs embedded under other logical operators; Kamp derives the relevant assumptions on the belief state of the speaker via Grice's maxim of brevity. This approach is not general enough to extend to all contexts in which free choice inferences are observed.[29] Therefore, in the paper at hand the relevant epistemic inferences are derived in a different way: as conversational implicatures due to the first sub-clause of the maxim of quantity and the maxim of quality, summarized in the Gricean Principle.

4.2.2. Gazdar's approach to clausal implicatures

Already Gazdar (1979) analyzed the epistemic inferences that Peter may have taken the beer and Mary may have taken the beer from (6): *'Mary or Peter took the beer from the fridge'* as effects of the first sub-clause of Grice's maxim of quantity. Gazdar distinguishes two classes of implicatures due to this maxim. The first class, *scalar implicatures,* is not relevant for the discussion at hand. The inferences of (6) just mentioned fall in Gazdar's class of *clausal implicatures.* This rises the question how Gazdar's approach to these implicatures relates to the description of the inferences proposed here – and whether a combination with an assumption of competence of the speaker leads to the free choice inferences as well.

Gazdar (1979) describes the following procedure to calculate clausal implicatures. First, he defines the set of *potential* clausal implicatures (pcis) of a compound sentence ψ. The pcis of ψ are the sentences $\chi \in \{\Diamond\phi, \Diamond\neg\phi\}$ where ϕ is a subsentence of ψ such that ψ neither entails ϕ nor its negation $\neg\phi$.[30] But not all potential clausal implicatures are predicted by Gazdar to become part of the interpretation of an utterance. Gazdar proposes that first they have to pass a strict consistency check: add to the common ground the assumption that the speaker knows her utterance to be true. and a set of potential clausal implicatures that is satisfiable in this context. Only those pcis are predicted to be present that are satisfiable in all contexts that can be reached this way.

Given the similarity between both approaches it should not come as a surprise that the predictions made by Gazdar (1979) are strongly related to the ones we obtained in section 3. Gazdar is able to predict all epistemic free choice inferences (D1)–(D3). With a weaker notion of competence than used in Section 3 his approach is even able to derive the deontic free choice inferences (D4) and (D5) for competent speakers and, thus, to account for free choice permission.[31]

Let us run through the calculations for (D4). Gazdar can account for this inference only based on the antecedent giving the disjunction wide scope over the modality: $\triangle\phi \vee \triangle\psi$. For this sentence he predicts the following set of pcis: $\{\diamond\phi, \diamond\psi, \diamond\triangle\phi, \diamond\triangle\psi$ and the respective negations$\}$. If we assume the speaker to be competent, i.e take as context the set C, then we will not predict free choice permission. In C the pcis $\diamond\triangle p$ and $\diamond\neg\triangle p$, as well as $\diamond\triangle q$ and $\diamond\neg\triangle q$ contradict each other and, therefore, do not survive the consistency check. Those pcis that pass the test do not entail $\triangle p \wedge \triangle q$. However, free choice permission can be derived if we assume a weaker notion of competence: if we take as context the set of states C^+ where besides $[D]$, $[4]$, and $[5]$ only $[C_1]$ is valid but not $[C_2]$ then $\Box\triangle p$ passes the consistency check and entails $\triangle p$ – and the same is true for $\Box\triangle q$ and $\triangle q$.

As these considerations make clear, the ideas on which Gazdar's work and the account introduced in Section 3 are based are very similar. In the technical details, however, the approaches differ. For one thing, both proposals try to minimize the belief state of the speaker, however, they have different opinions about to which part of her beliefs this should be applied. The second discrepancy lays in the criteria the approaches apply to decide whether some belief state is a proper minimum. Below, both differences will be discussed in some detail.

Particularly the first difference is interesting for the discussion at hand. As we have seen in section 4.1, the approach introduced here takes too much of the belief state of the speaker to be relevant. Gazdar proposes a much more context-sensitive criterion to select relevant belief: relevant is what the speaker believes about the sentences that – in a very technical sense – the speaker is talking about:

the subsentences of the uttered sentence. We can try and build this idea into the approach developed here. Maybe this way we can overcome the problem of overgeneration.

As already mentioned in a footnote in section 3.4.4 the order $\preceq^+$ on which the notion of pragmatic entailment $|\equiv^+$ is based can be equivalently defined by comparing how many of a certain set of sentences the speaker believes.

FACT 3. Let $\mathcal{L}^+ \subseteq \mathcal{L}$ be language defined by the BNF $\chi_+ ::= p(p \in \mathcal{L}_{(0)})|\chi_+ \wedge \chi_+|\chi_+ \vee \chi_+|\nabla \ p(p \in \mathcal{L}^0)$. Then we have for s, $s' \in \mathcal{C}$:

$$s \preceq^+ s' \Leftrightarrow \forall \chi \in \mathcal{L}^+ : s \models \Box\chi \Rightarrow s' \models \Box\chi.$$

This representation of the order suggests a way how we can use Gazdar's idea in our approach: instead of $\mathcal{L}^+$ we take the subsentences of the uttered clause as the set of sentences defining the order. Thus, let $\mathcal{L}^+(\phi)$ be the set of sub-sentences of sentence ϕ. We define: $\forall s, s' \in \mathcal{S} : s \preceq^{g+} s'iff_{\text{def}} \forall \chi \in \mathcal{L}^+(\phi) : s \models \Box\chi \Rightarrow s' \models \Box\chi$. This order can then be used to define a respective notion of entailment $|\equiv_S^{g+}$. Applied to context $\mathcal{C}$ this relation still accounts for the free choice inferences – when in the sentence interpreted 'or' has wide scope over the modal expressions. Furthermore, $|\equiv_S^{g+}$ certainly predicts less false implicatures than does $|\equiv_S^+$. For instance, for arbitrary and logical independent $p, q, r \in \mathcal{L}^0$ we do not have $p \vee q|\equiv_S^{g+} \Diamond r \wedge \Diamond \neg r \wedge \triangle r \wedge \Diamond \neg \triangle r$ (the same is true for $|\equiv_{\mathcal{C}}^{g+}$). However, a restriction to subsentences does not completely solve the problem of overgeneration. $|\equiv_{\mathcal{C}}^{g+}$ will predict wrongly for $\triangle p \vee \triangle q$ the implicature $\Diamond p$.[32] Finally, there is also a conceptual problem with such an approach. $|\equiv_{\mathcal{C}}^{g+}$ is still intended to describe a class of conversational implicatures and to formalize Grice's theory thereof. But what kind of Gricean motivation can be given for such restrictions of the inferences to subsentences of the sentence uttered?

To explain the second difference between Gazdar's approach and the one introduced in Section 3 we should compare his approach with an even more Gazdarian variant of $|\equiv$. As the reader may have noticed, he considers not only the sub-sentences of an uttered sentence to be relevant but also their negations. Let us define $\mathcal{L}(\phi)$ as the closure of $\mathcal{L}^+(\phi)$ under negation. $|\equiv_S^g$ is obtained by substituting the order $\forall s, s' \in \mathcal{S} : s \preceq^g s' \ iff_{\text{def}} \forall \chi \in \mathcal{L}(\phi) : s \models \Box\chi \Rightarrow s' \models \Box\chi$ in Definition 1.

Intuitively, both Gazdar's description of clausal implicatures and $|\equiv^g$ do the same thing: making as many sentences $\Diamond\chi$ true for

$\chi \in \mathcal{L}(\phi)$ as they can. However, the predictions made are different and this difference is due to the consistency check pcis have to pass before they become actual clausal implicatures. As we have said above, Gazdar predicts those pcis not to be generated that together with the context, the statement that the speaker knows ϕ to hold, and some set of pcis satisfiable in the context lead to an inconsistency. What does $\models_S^g$ predict in such a case? If $\Diamond \chi$ for $\chi \in \mathcal{L}(\phi)$ and $\Diamond \Sigma = \{\Diamond \chi \mid \chi \in \Sigma\}$ for $\Sigma \subseteq \mathcal{L}(\phi)$ are not jointly satisfiable in the set of states $s \in S$ where $\Box \phi$ is valid, while $\Diamond \chi$ and $\Diamond \Sigma$ separately are satisfiable in this context, then this means that there are states $s_1 \models \Diamond \chi$ and $s_2 \models \bigwedge \Diamond \Sigma$, but that such states are incomparable which each other. For ϕ to be honest there has to be a state $s \in S, s \models \Box \phi$ such that $s \preceq^g s_1$ and $s \preceq^g s_2$. From this it follows that $s \models \Diamond \chi \wedge \bigwedge \Diamond \Sigma$. But this conjunction does not have any model. Thus ϕ has to be dishonest. The pragmatic interpretation breaks down, no implicatures are generated. Gazdar's predictions are less severe. According to him, sets of sentences on which the knowledge of the speaker cannot be minimized without resulting in inconsistencies are not minimized. They are taken out, so to say, of the set of relevant sentences. The Gricean interpreter modeled by Gazdar is more tolerant with the speaker than the interpreter modeled here.

This has consequences for the cancellation properties for free choice inferences that both approaches predict. While both proposals model the same behavior of free choice inferences in case they conflict with the context or the semantic meaning of the utterance that triggers them, they differ in their predictions in case pcis are inconsistent with each other (given a particular context). Gazdar's approach cancels only those implicatures that give rise to the inconsistency. According to the account presented here in this case the speaker disobeys the Gricean Principle. Therefore, no implicatures are derived that would rely on taking the speaker to obey the principle. Empirical investigations have to show which of these positions makes the better predictions.

5. CONCLUSIONS

Why can we conclude on hearing (1) 'You may go to the beach or go to the cinema' that the addressee may go to the beach and may go to the cinema? In this paper we have proposed that this is due to pragmatic reasons. Free choice permission is explained as a conversational implicature that can be derived if the speaker is taken (i) to obey the Gricean maxim of quality and the first sub-clause of the maxim of quantity,[33] and (ii) to be competent on the deontic options, i.e. to know the valid obligations and permissions.

The proposal made in this paper is not the first approach that tries to describe free choice permission as a conversational implicature.[34] What distinguishes it from others on the same line is that it provides a formally precise derivation of the free choice inferences. In particular, a formalization of the conversational implicatures that can be derived from the maxim of quality and the first sub-clause of the maxim of quantity is given. This part of the proposal essentially builds on work of Halpern and Moses (1984) on the concept of 'only knowing', generalized by Hoek et al. (1999, 2000).

A central feature of the presented account that distinguishes it from *semantic* approaches to the free choice inferences is that it maintains a simple and classical formalization of the semantics of English: modal expressions are interpreted as modal operators and 'or' as inclusive disjunction. This has the advantage that the approach is free of typical problems that many semantic approaches to the free choice inferences have to face. For instance, when embedded under other logical operators, 'or' behaves as if it means inclusive disjunction. Semantic aproaches often cannot account for this observation (cf. Zimmermann 2000; Geurts to appear; Alonso-Ovalle 2004). Furthermore, because with such an approach to semantics $\triangle(p \lor q)$ and $\triangle p \lor \triangle q$ are equivalent, the free choice inferences are predicted for both sentences, independent of whether 'or' has wide or narrow scope with respect to the modal expressions. This allows us to account for the observation that free choice inferences can come with sentences like (10b) 'You may take an apple or you may take a pear' as well. At the same time we are not forced to exclude a narrow scope analysis for 'You may take an apple or a pear' (cf. Zimmerann 2000; Geurts to appear).

To summarize, we can conclude that the central goal of the work presented here, to come up with a formally precise pragmatic account to free choice permission, has been achieved. But there are still many questions concerning the 'behavior of free choice inferences that remain unanswered by the present approach.

The most urgent question is, of course, how to get rid of the countless unwanted pragmatic inferences the account predicts. Closer considerations in section 4.1 have suggested that this problem is a consequence of the fact that the approach incorporates only parts of Grice's theory of conversational implicatures. In particular, contextual relevance does not play any role. Future work has to reveal whether an extension of the approach in this direction helps to get rid of the problem of overgeneration.

An important topic that has received only marginal attention here was the question in how much the behavior of the free choice inferences forces us to adopt a pragmatic approach towards them. We have already noted that this is not easily answered. Much depends on the concept of pragmatic inferences that is adopted, on the classification of the data, and other theoretical decisions. In section 2 we have seen a series of arguments that speak in favor of a pragmatic approach. But the evidence is not as clear as this might suggest. Some observations argue rather for a semantic treatment of free choice inferences. For instance, the pragmatic inferences a sentence ϕ comes with should be unaffected when in ϕ semantically equivalent expressions (having roughly the same complexity) are exchanged. A pragmatic approach to the free choice inferences would thus predict, one may argue, that with 'He may speak English or he may speak Spanish' 'He is permitted to speak English or he is permitted to speak Spanish' should also allow a free choice reading. This does not seem to be the case.[35] How serious a problem this is depends, of course, on the exact semantics assumed for 'permit' and 'may'. We cannot solve this issue here. The only point that we want to make is that the question whether the free choice inferences are semantic or pragmatic in character is essential for evaluating the pragmatic approach proposed here and, therefore, needs close attention in future work.

Another subject for future research is the additional and non-Gricean interpretation principle – assuming the speaker to be competent – that is part of the approach. It is not the first time that such a principle is taken to be relevant for interpretation. In the literature of conversational implicatures there is even a long tradition in describing certain implicatures as involving such a competence assumption.[36] On the other hand, competence as formalized here is a very strong concept. One may wonder how reasonable it is to ascribe (by default) such a property to speakers. Therefore, it is important, for instance, to investigate whether the competence principle also shows itself in other areas of interpretation.

ACKNOWLEDGEMENTS

This paper summarizes the findings of my master thesis submitted at the University of Amsterdam in winter 2003. I would like to thank my supervisor Frank Veltman, Paul Dekker and Robert van Rooij for comments and support during the preparation of the thesis and this paper. Furthermore, I am indebted to two anonymous referees whose comments have certainly helped to improve upon the paper. Finally, I thank Darrin Hindsill for checking the English. Of course, none of the people mentioned is responsible for any mistakes the article may still contain.

NOTES

[1] In this paper, we mean by the logic of a language a formally defined notion of entailment between the sentences of the language. The exact form of the definition is unspecified: it may be in terms of a proof system or a model-theoretic description.

[2] In the sense that 'you may A' means the same as 'it is not the case that you must not A'.

[3] The step from (3a) to (3c) is admissible because one can prove in such a system that from $A \rightarrow B$ it follows *may* $A \rightarrow may B$. (3d) is obtained from (3c) by an application of free choice permission.

[4] In the linguistic literature this property is not called non-monotonicity but known as the cancellability of conversational implicatures. This term has been also used by Grice himself.

[5] Thus, exactly speaking, when we say that a sentence gives rise to free choice inferences we mean that it does so in certain contexts.

[6] Notice that the epistemic free choice inferences cannot be cancelled in the same way. Adding 'but I don't know which' to a sentence like (6) is intuitively redundant and changes nothing (of relevance) about its interpretation.

[7] Of course, extra rules for $\rightarrow$ and $\vee$ can be suppressed because these logical operators can be defined in terms of $\wedge$ and $\neg$.

[8] A state $s = \langle M, w \rangle$ is non-blind in w with respect to $R_\diamond$ of M $iff_{\mathrm{def}} R_\diamond[w] \neq \emptyset$.

[9] For a proof see Blackburn et al. (2001).

[10] The reader may be surprised by the choice to ask only for the local validity of the schemes [D], [4], and [5]. One reason why we do not demand them to be valid in all points of a model is that in this paper we will never come in a situation where we will talk about belief embedded under other modalities. Furthermore, later on we will consider restrictions on frames that are only plausible when imposed locally.

[11] It is not difficult to prove that the following holds: $\forall \phi \in \mathcal{L}^0 : s_1 \preceq^0 s_2$ iff $s_1 \models \phi \Rightarrow s_2 \models \phi$. Thus the order $\preceq^0$ could have been defined as well by the condition that in s_2 the speaker believes as least as many $\mathcal{L}^0$-sentences as in s_1.

[12] Fact 1 holds because the order only compares the belief state for a finite modal depth and we have chosen a finite set of proposition letters. Therefore, we can assume that there are always minimal models.

[13] In Schulz (2004) a constructive description of s^* is given. s^* is 'obtained' from s by first adding a world to the model where p is true – this is possible if p is satisfiable in S – then making this world $\Diamond$-accessible from w, and, finally, close the accessibility relation $R_\Diamond$ under the axioms [4], [5], and [D] that characterize S such that the speaker again gains full introspective power. This closure is important because the state obtained by simply making an additional world $\Diamond$-accessible from w is not an element of S. (This also shows that in a strict sense s^* does not 'only' differ in what is $\Diamond$-accessible from w.)

[14] One context in which a sentence like (11b) intuitively can be used is when the speaker is known to withhold information and, hence, to be disobeying the Gricean Principle. This is exactly what is predicted by our approach. The following example has been provided by one of the referees.

(i)   I know perfectly well what I believe, but all I will say is this: I believe that A or I believe that B,

[15] Actually, this order also respects the speaker's beliefs about the $\mathcal{L}^0$-facts. This is due to the fact that in S the speaker has full introspective power.

[16] $\preceq^n$ compares only deontic information about basic facts. The order can easily be extended such that is respects all deontic information by using (restricted) bisimulation (see Schulz 2004). The reason why we do not give this more general definition here is that we do not need this complexity. We consider only sentences having in the scope of $\triangle$ and ∇ a modal free formula.

[17] The (intensional) predicate $\lambda w \lambda x. P(w)(x)$ in his definition is instantiated here by the characteristic function of $\triangle$-accessible worlds $\lambda w \lambda v. w R_\triangle v$.

[18] Of course, the same argument can be also used to show that the speaker does not have to convey all she believes about valid obligations, as long as she is honest about her beliefs concerning permissions. However, minimizing beliefs on permissions does not result in a convincing notion of pragmatic entailment. For instance, this one wrongly predicts that sentences like $\triangle(p \vee q)$ are dishonest. One would like to have some motivation for the choice of the order $\preceq^+$ besides the fact that it does the job, while some equally salient alternatives do not – particularly, given that we formalize a theory of rational behavior. But so far I am not aware of any conclusive arguments.

[19] Also for this order an equivalent definition using a set of sentences can be given (for a close discussion see Schulz 2004).

FACT 2. Let $\mathcal{L}^+ \subseteq \mathcal{L}$ be language defined by the BNF-form $\chi_+ ::= p(p \in \mathcal{L}^0) | \chi_+ \wedge \chi_+ | \chi_+ \vee \chi_+ | \nabla p(p \in \mathcal{L}^0)$. Then we have for s, $s' \in \mathcal{C}$:

$$s \preceq^+ s' \Leftrightarrow \forall \chi \in \mathcal{L}^+ : s \models \Box \chi \Rightarrow s' \models \Box \chi.$$

[20] Again, $\preceq^+$ only compares beliefs about formulas $\{\nabla \chi | \chi \in \mathcal{L}^0\}$, but an extension to sentences $\nabla \chi$ for $\chi \in \mathcal{L}$ is easily possible (see Schulz 2004). We use the simpler variant because the sentences we consider here are only of the former type.

[320]

[21] Again, Schulz (2004) provides a formally precise version of this proof, including a constructive description of s^*. s^* is obtained from s by first adding a world to the model where p is true – this is possible if p is satisfiable in C – then making this world $\triangle$-accessible from w, and, finally, close the resulting accessibility relations $R'_\diamond$ and $R'_\triangle$ under the axioms [4], [5], [D], [C_1], and [C_2] to obtain a state that belongs to C.

[22] There is another way to repair $|\equiv^n_C$ such that one can account for the deontic free choice inferences. Instead of weakening the order and thereby be less strict on what a speaker has to convey with her utterance, we can also take her to be less competent. It turns out that the competence axiom we have to drop is [C_2]: we weaken C to the set of states C^+ where [D], [4], [5], and [C_1] are valid. In this case, the speaker knows all valid obligations, but she may be not aware of certain permissions. While this accounts for the free choice inferences, other predictions made by $|\equiv^n_{C^+}$ are less convincing than what is predicted by $|\equiv^n_C$. For a more elaborate discussion the reader is referred to Schulz (2004).

Finally, it is interesting to note, that also the combination of $|\equiv^+$ with C^+, hence, the combination of weakening the order and weakening the notion of entailment allows us to derive the free choice inferences. Also this combination of a concept of competence with a formalization of the Gricean Principle does not work as well as $|\equiv^+_C$.

[23] This is probably the least disputed property characterizing conversational implicatures. Therefore, insofar as we claim to formalize conversational implicatures, all pragmatic inferences we predict should have this property.

[24] Given that the derivation of the free choice inferences appears to be the normal interpretation of sentences like (1) 'You may go to the beach or go to the cinema', this position is much more convincing than proposing that the interpreter *knows* the speaker to be competent when inferring free choice.

[25] There are other ways of how we can understand the role of competence in the derivation of the free choice inferences. In the scenario sketched above we took it to be an extra assumption that is cancelled *completely* if conflicting information is encountered. We might as well propose that in such a situation the interpreter tries to maintain as much of the competence assumption as she can. Such an approach has been adopted – for independent reasons – in van Rooy and Schulz (2004). In this case it depends on the kind of information about the incompetence of the speaker the interpreter has whether the deontic free choice inference are cancelled or not.

[26] The way we have defined the order $\preceq^+$ 'any information' means any information that can be expressed with the following sentences $x ::= p (p \in \mathcal{L}^0) | \chi \lor \chi | \chi \land \chi | \nabla p (p \in \mathcal{L}^0)$.

[27] Thus, our reformulation of this maxim in the Gricean Principle is not entirely faithful to Grice.

[28] For instance, in many contexts the answer 'John' to a question 'Who smokes?' is not only understood as conveying that John is among the smokers – what would be its semantic meaning – but it is additionally inferred that John is the only one who smokes.

[29] For a detailed discussion of these two approaches and their shortcomings see Schulz (2004).

[30] Gazdar adopts a slightly different interpretation of the modal operators as is proposed in section 3. He takes $S4$ to be the logic of the modal operator $\Diamond$. This is partly due to the fact that for Gazdar $\Box$ models knowledge and not belief. Gazdar's definition of pcis contains one further condition, but this one can be ignored for our purposes.

[31] Gazdar himself never discussed this application of his formalization of Grice's theory. In particular, it was not his intention to account for the free choice inferences this way.

[32] Although one (normally) infers from an utterance of 'You may A or B' that the speaker takes the asserted deontic options also to be epistemically possible, this inference should rather be analyzed as part of the appropriateness conditions (presuppositions) of permissions (and obligations).

[33] These two maxims where combined in the Gricean Principle.

[34] See e.g. Kamp (1979), Merin (1992), and van Rooy (2000).

[35] This type of argument against a pragmatic account of the free choice inferences has been brought forward at different places in the literature. The particular example used here can be found, for instance, in Forbes (2003), as pointed out by one of the referees.

[36] One of the oldest references may be Soames (1982).

REFERENCES

Alonso-Ovalle, L.: 2004, Equal rights for every disjunct! Quantification over alternatives or pointwise context chance?, handout, *Sinn und Bedeutung* **9**, *Nijmegen*.

Blackburn, P., et al.: 2001, *Modal Logic*, Cambridge University Press, Cambridge.

Forbes, G.: 2003, Meaning-postulates, inference, and the relation/notional ambiguity, *Facta Philosophica* **5**, 49–74.

Gazdar, G.: 1979, *Pragmatics*, Academic Press, London.

Geurts, B.:(to appear), Entertaining alternatives, *Natural Language Semantics*.

Grice, P.: 1989, *Studies in the Way of Words*, Harvard University Press, Cambridge.

Groenendijk, J. and M. Stokhof: 1984, *Studies in the Semantics of Questions and the Pragmatics of Answers*, Ph.D. thesis, University of Amsterdam.

Halpern, J.Y. and Y. Moses: 1984, Towards a theory of knowledge and ignorance, *Proceedings 1984 Non-monotonic reasoning workshop*, American Association for Artificial Intelligence, New Paltz, NY, pp. 165–193.

Hirschberg, J.: 1985, *A Theory of Scalar Implicature*, Ph.D. thesis, University of Pennsylvania.

Hoek, W. van der, et al.: 1999, Persistence and minimality in epistemic logic, *Annals of Mathematics and Artificial Intelligence* **27**, 25–47.

Hoek, W. van der, et al.: 2000, A General Approach to Multi-Agent Minimal Knowledge, in M. Ojeda-Aciego, I.P. Guzman, G. Brewka, and L.M. Pereira (eds.) *Proceedings JELLIA 2000*, LNAI 1919, Springer Verlag, Heidelberg pp. 254–268.

Horn, L.: 1972, *The Semantics of Logical Operators in English*, Ph.D. thesis, Yale University.

Horn, L.: 1989, *A Natural History of Negation*, University of Chicago Press, Chicago.

Kamp, H.: 1973, Free choice permission, in *Proceedings of the Aristotelian Society, N. S.* **74**, 57–74.

Kamp, H.: 1979, Semantics versus Pragmatics, in F. Guenther and S.J. Schmidt (eds.) *Formal Semantics and Pragmatics of Natural Languages*, Reidel, Dordrecht pp. 255–287.

Lewis, D.: 1970, A problem about permission, in Saarinen et al. (eds.) *Essays in Honor of Jaakko Hintikka,* Reidel, Dordrecht pp. 163–175.

Merin, A.: 1992, Permission Sentences stand in the Way of Boolean and other Lattice Theoretic Semantics, *Journal of Semantics* **9**, 95–152.

Rooy, R. van: 2000, Permission to Change, *Journal of Semantics* **17**, 119–145.

Rooy, R, van and K. Schulz: 2004, Exhaustive Interpretation of Complex Sentences, *Journal of Logic, Language, and Computation* **13**, 491–519.

Schulz, K.: 2004, *You May Read it Now or Later: A Case Study on the Paradox of Free Choice Permission*, master thesis, University of Amsterdam.

Soames, S.: 1982, How presuppositions are inherited: A solution to the projection problem, *Linguistic Inquiry* **13**, 483–545.

Wright, G. H. von: 1969, *An Essay on Deontic Logic and the Theory of Action*, Amsterdam.

Zimmermann, T.E.: 2000, Free choice disjunction and epistemic possibility, *Natural Language Semantics* **8**, 255–290.

University of Amsterdam
Dept. of philosophy (ILLC)
Nieuwe Doelenstraat 15
1012 CP Amsterdam
E-mail: K.Schulz@uva.nl

GIACOMO SILLARI

A LOGICAL FRAMEWORK FOR CONVENTION

1. INTRODUCTION

In this paper, I provide a logical framework for defining conventions, elaborating on the game-theoretic model proposed by David Lewis. The philosophical analysis of some of the key concepts in Lewis's model reveals that a modal logic formalization may be a natural one. The paper will develop on the analysis and critique of such concepts as those of common knowledge, indication, and the distinction between epistemic and practical rationality. In particular: (i) the analysis of Lewis's definition of common knowledge reveals that a suitable formalization can be obtained by adopting an approach analogous to that of awareness structures in modal logic; moreover (ii) the analysis of the notion of indication reveals that the agents may be required to make inductive inferences yielding probabilistic beliefs. I shall stress that such aspects, however, pertain to the sphere of epistemic rationality (i.e., they deal with the justification of the agents' beliefs) rather than to the sphere of practical rationality. Confounding the two spheres may lead to the wrong conclusion that, in order to make sense of, say, salience as a coordination device, one should incorporate psychological assumptions into an undivided notion of rationality. On the contrary, practical rationality stands as the usual notion of game-theoretic rationality, whereas epistemic rationality incorporates those aspects pointed out in (i) and (ii) above. This attempt to provide a formal framework for Lewis's theory of convention follows those of Vanderschraaf (1995, 1998) and Cubitt and Sugden (2003). In his work on Lewis, Vanderschraaf provides a characterization of convention as correlated equilibrium, adopting a formal framework close to the set-theoretical one proposed by Aumann (1976). Cubitt and Sugden point out that such a framework does not take into account certain elements that are however present in Lewis's original theory, and propose a different formal setup altogether. In this paper, I show how a formalization based on modal logic can incorporate those distinctive aspects introduced by David Lewis in *Convention*.

Synthese (2005) 147: 379–400
Knowledge, Rationality & Action 325–346
DOI 10.1007/s11229-005-1352-z

The paper is organized as follows: in the following section, I will provide an informal reconstruction of Lewis's account of convention. In Section 3, I will draw the distinction between epistemic and practical rationality and show that Lewis's concept of indication, and his analysis of how common knowledge and higher order expectations come about, pertain to epistemic rationality. In Section 4, I will argue and show that a modal logic formalization of belief, supplemented with awareness structures, can be a natural interpretation of the epistemic concepts involved in Lewis's analysis of convention.

2. LEWIS ON CONVENTION

Lewis's game-theoretic analysis of conventions starts with coordination games. In a pure coordination game, players' interests coincide. As Thomas Schelling[1] put it, games which represent social interactions can be placed along a continuum. One endpoint of such a scale contains games of pure conflict, that is to say, games in which the sum of the payoffs received by the players in every combination of strategies is null (the so called zero-sum games). At the other endpoint are games of pure coordination, in which players receive the same payoff in every strategy combination. Another far more frequent kind of coordination game is one in which players' interests do not exactly coincide, but they still prefer to coordinate with each other.

Lewis's intuition is that coordination problems (non-trivial, in the sense that they have more than one strict Nash equilibrium) underlie every convention, a convention being one particular recurrent equilibrium of such games. Lewis addresses the question of how a specific equilibrium can be reached. In general, in order to have a sufficient reason for choosing a particular action, an agent needs to have a belief (up to a sufficient degree) that the other agent will choose a certain action. Lewis argues that, in the case of coordination games, such sufficient degree of belief is reached by a *system of mutual expectations*. The focus of Lewis's study is on how conventions are sustained, rather than how they originate. In both cases, however, coordination is achieved by means of a system of mutual expectation. What differs is the means by which such systems are produced. There are several ways in which a system of mutual expectations can obtain. A natural one is, for example, that of agreeing to play a certain strategy profile. Another, common coordination device is salience. Facing a new coordination problem, agents recognize that an equilibrium has certain salient features, and each player expects them to be noticed by the other players, too. A particular kind of salience is precedent; in this case, the salient trait of the equilibrium is that it has served as a solution of a similar coordination problem in the past. Although Lewis does not state the point explicitly, it seems reasonable to conjecture that salience, in general, may serve as a coordination device for originating conventions, whereas precedent is the coordination device involved in their perpetuation.

Since systems of mutual expectations play such a fundamental role in Lewis's theory of convention, it is natural to investigate what mechanisms produce them. Lewis elucidates this point by providing a definition of "common knowledge." Although the term common knowledge[2] has had remarkable fortune in the literature of so many academic fields,[3] it is noteworthy that Lewis is interested in expectations that is to say, beliefs – rather than knowledge. Moreover, his definition does not even immediately deal with beliefs, but rather with reasons to believe, being in fact a definition of "common reason to believe" a certain proposition. However, possibly also due to the fact that the first (and seminal) mathematical formulation of the concept had a natural interpretation in terms of knowledge rather than belief, the expression "common knowledge" became prevalent. Lewis himself later acknowledges the incongruence: "That term [common knowledge] was unfortunate, since there is no assurance that it will be knowledge, or even that it will be true." (Lewis 1978, p. 44, n. 13) I take it that here Lewis is pointing to the fact that reasons to believe may fail to turn into actual beliefs (there is no assurance that it will be *knowledge*), or, even in the case that they do, the agent may entertain false beliefs about the world (there is no assurance that it will be *true*). To preserve conformity with Lewis's original terminology, I shall for now refer to common knowledge as well. This is Lewis's definition:

DEFINITION 2.1. A proposition *p* is common knowledge in the group *G* if a state of affairs *A* obtains such that

1. Everyone in *G* has reason to believe that *A* holds.
2. *A* indicates to everyone in *G* that everyone in *G* has reason to believe that *A* holds.
3. *A* indicates to everyone in *G* that *p*.

Textual evidence shows that the relation of indication should not be interpreted as material implication, but should allow for some kind of inductive inference: in general, if A indicates x to i, that means that, if i had reason to believe that A holds, i would thereby have reason to believe that x holds as well. Clauses (1)–(3), along with suitable assumptions about the agent's reasoning capabilities and inductive standards, originate an infinite series of epistemic propositions such that everyone in G has reason to believe that p, everyone has reason to believe that everyone has reason to believe that p, and so on. The state of affairs A, which allows the agents in the group G to have common knowledge of the proposition p, is said to be a *basis* for common knolwdge of p in G.

Lewis's definition of convention is then the following (cf. Lewis 1969, p. 58):

DEFINITION 2.2. A regularity R in the behavior of members of a population P when they are agents in a recurrent situation S is a convention if and only if it is true that, and it is common knowledge in P that, in any instance of S among members of P,

1. everyone conforms to R;
2. everyone expects everyone else to conform to R;
3. everyone prefers to conform to R on condition that the others do, since S is a coordination problem and uniform conformity to R is a coordination equilibrium in S.

Common knowledge, according to Lewis, should be included in the definition for two reasons: on one hand, on purely descriptive grounds, since it seems that common knowledge is a relevant characteristic of conventions;[4] on the other, because it prevents certain odd situations to count as conventions as they would in the absence of the common knowledge requirement. For example, if the agents (i) conform to R, (ii) expect everyone else to do so, (iii) prefer everyone else to do so, but at the same time they believe that no one conforms to R *because* they expect others to do the same, then conditions (1)–(3) would be satisfied, but they would not be common knowledge (nor would they, in fact, be first order knowledge). Notice that both such motives to incorporate common knowledge in the definition have nothing to do with the game-theoretic underpinnings of the definition itself.

3. PRACTICAL AND EPISTEMIC RATIONALITY

When trying to coordinate with someone else, one is presented with two different problems. The first is that of forming beliefs about what the other agent will do in the coordination game. The other is to choose an action, based on such beliefs. Hence, in his attempt to provide a rational reconstruction of conventions, Lewis confronts two different issues: one concerns the formation of (rational) beliefs, while the other concerns the choice of (rational) action. That is, certain processes of reasoning of the agents are directed at justifying the choice of a particular course of action, whereas others are directed toward the justification of agents' beliefs. Following Bicchieri (1993, p. 11), we should therefore consider both aspects of *practical* and *epistemic* rationality, the former being relative to the choice of the optimal action with respect to the agent's beliefs and preferences, the latter to the formation and justification of the agent's beliefs in light of available evidence.

It is extremely important to keep the two problems separate, or else the rational reconstruction becomes vulnerable to defeating counter-examples. For instance, substantial parts of Margaret Gilbert's critique[5] of Lewis's account of convention exploit the lack of a precise distinction of the two kinds of rationality. Consider for example her presumption that "it is natural to (take Lewis to be) assuming the usual game-theoretical approach (to rationality)". (cf. Gilbert 1989, p. 324). The characteristics of such an approach, as they are listed by Gilbert (cf. ib., pp. 321–322) are (i) that agents are perfect reasoners (they use all the relevant information in their possession and make no mistaken inferences), (ii) that they act as reason dictates, and (iii) that they act according to their preferences. Gilbert claims that if rationality is characterized as above, precedent and rationality together are not sufficient to model the behavioral regularities at the core of Lewis's idea of convention. According to her, common knowledge of both rationality and a successful precedent yield no reason to conform to such precedent in the future. If both agents know that there is a successful precedent, they both have a reason to act in accordance to precedent, given that the other will do so, which is the case only if each one knows that the other knows that the other will act in accordance to precedent, and so on ad infinitum. Thus, the infinite regress prevents the players to come up with a conclusive reason for action, unless, Gilbert claims, we incorporate a psychological (and hence foreign to rationality) element into the picture. Such psychological element would then be an a-rational tendency to follow precedent.

Lewis's analysis, to be sure, tends to keep the distinction between formation of belief and choice of action blurred. For example, cf. the following quote: "The more orders of expectation about action contribute to an agents decision, the more independent justifications the agents will have; and insofar as he is aware of those justifications, the more firmly his choice will be determined" (Lewis 1969, p. 33). This passage can be interepreted as consistent with Gilbert's criticism, which leads to the conclusion that the order of expectations needed to conclusively determine agents' choices is infinite. Lewis, however, does not at any point claim that mutual expectations of conformity to a successful precedent directly constitute a reason to act. He is, rather, saying that they give the players expectations about what other players will do (conform to the precedent), that is to say they provide a reason to believe (since expectation is a particular kind of belief) rather than a reason to act.[6]

In *Convention*, David Lewis does not explicitly characterize the features of epistemic as opposed to those of practical rationality. However, the distinction is crucial in order to avoid criticism *à la* Gilbert. Whereas the characteristics of practical rationality (optimality of action with respect to beliefs and preferences) can be seen as the usual game-theoretic notion of unqualified rationality, the characteristics of epistemic rationality should be more clearly spelled out. I believe that Lewis's text provides numerous insights about the nature of epistemic rationality – even beyond its relevance for explaining the phenomenon of convention – which deserve formal clarification. In particular, two notions elaborated in *Convention* concern the agents' epistemic rationality, that of *indication* and the distinction Lewis is keen to make between *reasons to believe* and *actual belief*. As recalled above, Lewis structures the relation of indication (i) by defining it in terms of reasons to believe and (ii) by implicitly differentiating it from material implication: a state of affairs A indicates proposition x to agent i if and only if, *if i* had reason to believe A, i would *thereby* have reason to believe x. The use of the expression *if ... thereby* denotes that Lewis is not thinking of material implication. Moreover, he states clearly that the relation of indication depends on the agents' inductive standards.

The feature of epistemic rationality I am going to focus on in this article is the one that Lewis introduces in his definition of common knowledge via the distinction between reasons to believe and actual beliefs. In the definition, Lewis gives a sufficient set of condition for an infinite number of epistemic clauses about reasons to believe to arise. Ideally, an agent would believe everything she has reason to, but in practice, obvious limitations occur. Lewis's analysis suggests that the expectations generated by the definition are not actual cognitive states, but merely potential ones. Such potential cognitive state, in my reading of Lewis, is a situation in which an agent may acquire a belief that is epistemically acceptable; that is to say, there exists for her a reason to believe that a certain proposition holds. An agent endowed with sufficient *rationality* acquires then an actual belief out of a potential one (cf. (Lewis 1969), p. 55: "Anyone who has reason to believe something will come to believe it, provided he has a sufficient degree of rationality."). Although Lewis does not qualify what kind of rationality an agent should possess in order to entertain actual beliefs out of her reasons to believe, here, of course he is not thinking of expected utility maximization. The aim the following section is to clarify to what, exactly, Lewis is referring when he assumes that the agents are endowed with this specific kind of rationality.

4. THE FORMAL FRAMEWORK

Is it possible to characterize the distinction sketched in the previous section by formalizing the elements pertaining to epistemic rationality? Cubitt and Sugden (2003) provide the syntax of such a formalization, and incorporate in a formal setup certain elements of the Lewisian analysis of common knowledge and convention overlooked by game theorists and economists in subsequent developements of such concepts. Cubitt and Sugden detect the complexity inherent in Lewis's account: "The concern is with those modes of human reasoning, whether deductive or inductive, that can properly be said to justify beliefs or actions". (Cubitt and Sugden 2003, p. 184.) And moreover: "Lewis' analysis is not, strictly speaking, about knowledge; it is about warranted belief. [...] A belief might be justified according to reasonable standards of inductive inference, yet not be true" (ib.). By and large, it is true that when Lewis's analysis, of common knowledge, has been first formulated in set-theoretical terms (Aumann 1976), its complexities have been levelled out, since the formal model proposed there does not allow for an object of common knowledge to be false. However, relaxing such assumption need not entail rejecting a set-theoretical formulation *tout court*, as Cubitt and Sugden do in their paper.[7] As I understand it, their rejection of a set-theoretical approach to modelling epistemic agents is based on the following grounds (cf. Cubitt and Sugden 2003, pp. 206 ff.):

(i) In partitional models, certain properties of the model are informally considered to be common knowledge among the agents. Since Lewis's analysis is concerned with the origin of common knowledge itself, such models cannot be appropriate.

(ii) In partitional models, we are modelling knowledge rather than belief. If agent i knows x, then x is the case.

(iii) Representing the indication relation as material implication (that is – in set-theoretical terms – as set inclusion) eliminates those aspects relative to Lewis's concerns about the agents' (inductive) reasoning.

In my view such objections are not conclusive grounds for dismissing epistemic models cast in set-theoretical (or modal logic) terms. As for (iii), I show in the following how indication can be incorporated in a modal logic setting. As for (ii), Cubitt and Sugden have a point that *partitional* models represent knowledge rather than belief, since the objects of agents' belief must be true in the actual world. However, such requirement, if too stringent, can be dropped. In particular, in the system described later in this section, formulas that are believed by the agents need not be true in the world in which they are believed; the agents can be mistaken.[8] As for (i), it is essential to stress the difference between the *informal* common knowledge of which the specifications of the set-theoretic model are object, and the idea of common knowledge formally defined and captured in the model. The properties of the model are common knowledge only in the informal sense that they are true in all possible states of the model.[9] Indeed, in the model there may not be an event expressing such properties, hence the fact that there is informal common knowledge of them need not detract from the relevance and validity of Lewis's formal definition of common knowledge.[10] To be sure, David Lewis's model is not cast in set-theoretical terms, and, indeed, his framework is, by and large, informal. Cubitt and Sugden's formal rendition of it (and especially of its characteristics missing in the usual formalizations) provides us with a very expressive model, although a purely syntactical one. It is the aim of this section to show that the richness of David Lewis's informal model, and of Cubitt and Sugden's rigorous syntax, may receive a natural and suitable semantical interpretation by means of Kripke (that is to say, set-theoretical) structures.

However, the formalization proposed by Cubitt and Sugden enjoys another feature that is not to be found in the usual set-theoretical models:

(iv) Following Lewis, they make a distinction between states of affairs and propositions, whereas partitional models deal with events only.

What is to be gained by distinguishing between state of affairs [roughly, as Cubitt and Sugden suggest 'states of the world' in the sense of Savage (1954)] and propositions? The gain in generality is only apparent, since it is trivial to translate the state of affairs A into the proposition 'A holds' and vice versa. Why, then, did Lewis introduce the distinction in the first place? It is my opinion that he did so in order to defend the idea that the indication relation is stronger than material conditional. Recall that saying "the state of affairs A indicates p to i" is tantamount to saying "if i has reason to believe that A holds, i thereby has reason to believe that p is the case." Since Lewis wants the indication relation to be stronger than material implication (at least, in his discussion of the definition of common knowledge he assumes that logical implication entails indication), that would entail that any vacuosly false formula indicates any proposition to any agent. By requiring that the indicating formula be a state of affairs, Lewis wants to avoid the paradoxes of material implication being carried over to the indication relation. An attenuation of the impact of the paradoxes can be obtained, without recurring to a sorted language, by requiring that material implication entails indication only in those cases in which the agent has reason to believe the antecedent of the implication (cf. axiom B2 below). Hence, since what I believe to be Lewis's concern motivating the introduction of state of affairs can be taken care axiomatically, I opt for simplicity and drop the distinction altogether.

In sum, the logical framework of *Convention*, including the properties of the indication relations, can be captured by modal axioms interpreted in a Kripke semantics. Agent i's reason to believe a proposition would thus be represented by means of the modal operator R_i. Notice that I do not intend to explicitly represent here the process of reasoning conducive to i's reason to believe a certain proposition p, nor do I intend to assert anything about that process. With the expression $R_i\varphi$ we only capture the fact that, somehow, i would be justified in believing φ. In this sense, an agent's access to a reason to believe a certain proposition can be treated as if it were an agent's propositional attitude towards that proposition, and can be given a natural modal interpretation.

Formally, let us define a language $\mathcal{L}_n^{\Rightarrow}$ whose *alphabet* is the typical alphabet of propositional calculus, augmented with n reason-to-believe operators $R_1, \ldots, R_n$, and indication operators $\Rightarrow_i, \ldots, \Rightarrow_n$.[11] We shall use the basic connectives $\wedge$ and $\neg$, and adopt the obvious abbreviations for the others. In particular, $\varphi \to \psi$ stands for $\neg(\varphi \wedge \neg\psi)$. The countably many *atomic propositions* of $\mathcal{L}_n^{\Rightarrow}$ are denoted by the metavariables p, q, r, etc. and they belong to the non-empty set Φ of atomic (primitive) propositions. The rules for the construction of *well-formed formulas* are the following:

(i) every atomic proposition p is a formula;
(ii) if φ is a formula, so is $\neg\varphi$;
(iii) if φ and ψ are formuals, so is $\varphi \wedge \psi$;
(iv) if φ is a formula, so is $R_i\varphi$;
(v) if φ and ψ are formuals, so is $R_i\varphi \Rightarrow_i R_i\psi$.

The formulas of the kind $R_i\varphi \Rightarrow_i R_i\psi$ render Lewis's indication relations, and are to be read "φ indicates ψ to agent i". The following system based on the language $\mathcal{L}_n^{\Rightarrow}$ captures the deductive core logic of Lewis's *Convention*:

B0	tautologies of propositional calculus,
B1	$(R_i\varphi \wedge R_i(\varphi \to \psi)) \to R_i\psi$,
B2	$(R_i\varphi \wedge (\varphi \to \psi)) \to (R_i\varphi \Rightarrow_i R_i\psi)$,
B3	$(R_i\varphi \wedge (R_i\varphi \Rightarrow_i R_i\psi)) \to R_i\psi$,
B4	$(R_i\varphi \Rightarrow_i R_i\gamma \wedge R_i\gamma \Rightarrow_i R_i\psi) \to (R_i\varphi \Rightarrow_i R_i\psi)$,
B5	from φ and $\varphi \to \psi$, infer ψ,
B6	from φ, infer $R_i\varphi$.

It is important to notice that axioms $B1$–$B6$ constitute a minimal system that needs to be further enriched. Although we are representing reasons to believe through modalities, we have not yet specified any property of the R_i operators. Moreover, the axioms proposed above capture the relation between the indication operator and the deductive capabilities of the agents *only*. Nothing is said specifically about the fact that the character of the indication operator is not strict of deductive.

[334]

Although modal logic has been used extensively to provide formal accounts of agents' propositional attitudes – knowledge, belief, desire, etc. – its use as an instrument of epistemological inquiry has undeservedly not received as much attention.[12] In this paper I consider modelling reasons to believe as modal operators. Consider the axioms above: $B1$ states that reasons-to-believe operators are normal, in that if an agent has reason to believe φ and that $\varphi \rightarrow \psi$, then the agent has reason to believe ψ as well. We find intuitively reasonable that an agent's reasons to believe be closed under *modus ponens*. $B2$ introduces the indication relation by linking it to material implication: it states that if φ materially implies ψ and agent i has reason to believe φ, then φ indicates ψ to i. The motivation behind this axiom is that we want to tie material implication and indication together, without letting the paradoxes of the former be carried on to the latter. By requiring that $R_i\varphi$ actually be the case among the premises,[13] we rule out those situations in which an agent would have a contradictory φ vacuosly indicate any proposition ψ to her. Axiom $B3$ requires that the indication relation be closed under *modus ponens*, whereas $B4$ requires that it be closed under substitution: those are deductive rules with which we want the agents in our system to be endowed, not only when they are dealing with the classical logical connectives, but also when they are considering indications and hence reasons to believe. $B5$ is *modus ponens* and, finally, $B6$ states that agents have reason to believe all logical truths. Again, it makes sense to require that an agent has reason to believe what logic dictates, although it is not the case that the agents in the model will come to *actually* believe all logical truths.

It is reasonable to add positive introspection to the axioms listed above:

B7 $R_i\varphi \rightarrow R_i R_i \varphi$,

since an agent that has reason to believe φ should have reason to believe that she has such reason as well. It also seems reasonable to require that the agents entertain consistent beliefs. This is captured by the axiom:

B8 $R_i\varphi \rightarrow \neg R_i \neg \varphi$.

Furthermore, to ease readability, we add the definitional axiom B9. It introduces the modal operator R_G, which stands for "every agent in the group G has reason to believe that ...":

B9 $\bigwedge_{i \in G} R_i x \leftrightarrow R_G x$.

Since we are using a standard modal logic, we can provide a natural semantics for our language in terms of Kripke structures. A Kripke structure is an $(n+2)$-tuple $M = \langle W, \mathcal{R}_1, \ldots, \mathcal{R}_n, \pi \rangle$ such that W is a set of possible worlds, $\mathcal{R}_1, \ldots, \mathcal{R}_n$ are n accessibility relations (one for each agent in the system) on $W \times W$, and π is a truth assignment $\pi := W \times \Phi \to \{\text{true, false}\}$ which assigns a truth value to each atom belonging to Φ for each possible world in W. The clauses defining the semantical relation of satisfaction will then be the usual, with $p \in \Phi$ and φ, ψ formulas of the language.

$(M, w) \models p$ iff $\pi(w, p) = \text{true}$

$(M, w) \models \neg\varphi$ iff $(M, w) \not\models \varphi$

$(M, w) \models \varphi \wedge \psi$ iff $(M, w) \models \varphi$ and $(M, w) \models \psi$

$(M, w) \models R_i\varphi$ iff, for all v such that $(w, v) \in R_i$, $(M, v) \models \varphi$

$(M, w) \models R_G\varphi$ iff $(M, w) \models R_i\varphi$ for all $i \in G$

$(M, w) \models R_i\varphi \Rightarrow_i R_i\psi$ iff (M, w)
$\models R_i\varphi$ and $(M, w) \not\models (\varphi \wedge \neg\psi)$[14]

How can we incorporate common reason to believe in the system? The standard way to do so in finitary systems is by adopting axioms that characterize common knowledge as a fixed point. In particular, introducing a new operator CR_G that stands for "members of G have common reason to believe that...", common reason to believe is defined by the axiom

B10 $CR_G\varphi \leftrightarrow R_G(\varphi \wedge CR_G\varphi)$,

while it is regulated by the rule

B11   If $\varphi \to R_G(\psi \wedge \varphi)$, then $\varphi \to CR_G\psi$.

It is convenient to express the semantic clause for common reason to believe by using the concept of reachability: we say that a world v is G-reachable in k steps from world w iff there is a path of length k from v to w such that the edges between adjacent worlds are labelled by the accessibility relations of members of G. It follows that

$(M, w) \models CR_G\varphi$ iff $(M, v) \models \varphi$ for all worlds v that are G-reachable from w in any number of steps.

For notational convenience, we define the operator $\Rightarrow_G$ as follows: for all $i, j \in G$, $(R_i\varphi \Rightarrow_G R_i\psi) \leftrightarrow (R_j\varphi \Rightarrow_j R_j\psi)$, that is to say, if φ indicates ψ to an agent $i \in G$, then it does so for any other agent

$j \in G$. Such an operator captures the idea that, although the indication relations differ among agents, in some cases inductive standards are shared by groups of agents, as, for example, in those cases in which common reason to believe comes about.

It is now easy to show that, in our system, Lewis's conditions for "common knowledge" give, in fact, rise to an infinite sequence of reasons to believe;

PROPOSITION 4.1. Let the following three conditions hold:

(a) $R_G \varphi$,
(b) $R_G \varphi \Rightarrow_G R_G R_G \varphi$,
(c) $R_G \varphi \Rightarrow_G R_G \psi$.

Then, the agents in G have common reason to believe that ψ.

Proof. We show by induction on the length of the path that if v is G-reachable from the actual world w, then $(M, v) \models \psi$. Let v be G-reachable from w in 1 step. By $B3$ at w all of $R_G \varphi, R_G R_G \varphi$, and $R_G \psi$ hold. Hence, at v all of $\varphi, R_G \varphi$, and, as desired, ψ hold. By induction hypothesis, if u is G-reachable from w in n steps, then all of $\varphi, R_G \varphi$ and ψ hold at u. However, from (b), (c) and the fact that $R_G \varphi$ holds at u, it follows that all of $R_G \varphi, R_G R_G \varphi$, and $R_G \psi$ hold at u, or that ψ hold at every world $u + 1$ which is reachable in one step from u. □

We have so far considered agents' reason to believe, rather than their actual beliefs. Lewis's analysis of "common knowledge", though centered on reasons to believe, serves the fundamental purpose of explaining how higher-order expectations (that is, actual beliefs) of the agents come about. Indeed, his rationale to introduce the distinction between reasons to believe and actual beliefs seems to be that of answering the question he poses at p. 52 of *Convention*: "And how is the process (of generating higher-order expectations) cut off – as it surely is – so that it produces only expectations of the first few orders?" The infinite chain of reasons to believe, to which the definition of common reason to believe gives rise, makes no harm descriptively, since it represents only potential epistemic states of the agents, rather than their actual reasoning or beliefs. Thus, Lewis introduces a tension between what a reasoner *should* believe (any proposition she has reason to) and what a reasoner *does* believe (a subset of the propositions she has reason to). An ideal agent, in this sense, would be unboundedly (epistemically) rational.

Such an agent would have no limitation in all three of her computational power, time, and storage capability. Therefore, she would believe every logical truth, and all consequences of the propositions she believed. What about her inductive capabilities? Whatever her inductive standards might be, she would believe any proposition yielded by such standards. In particular, if according to her inductive standards, a certain proposition were a basis for common knowledge, she would believe the whole sequence of beliefs of infinitely increasing order indicated by it. On the other hand, an actual agent, being boundedly rational, believes only a subset of the logical truths and of the consequences of the propositions she believes. Similarly, she believes only a subset of the propositions yielded by her inductive reasoning and, in particular, of the infinite series of propositions implied by a basis of common knowledge, she believes only those of the first few orders.

Only a portion of the potential (or implicit) knowledge an agent has is translated into actual beliefs. An agent might not focus on a proposition she has reason to believe, and therefore fail to entertain an actual belief about it. This may happen for psychological reasons, or because the proposition, though logically valid, is irrelevant. It may happen because the agent lacks the computational power to actually perform the reasoning necessary to deduce or induce it, or the time to perform the computation, etc. In the case of Lewis's definition of common knowledge, he requires that, for an agent to believe a proposition she has reason to, she possesses a "sufficient degree of rationality" (cf. Lewis 1969, pp. 55–56). Let us spell out the details of his idea. Suppose there is a basis for common knowledge of φ between agents i and j. Agent i has, then, reason to believe φ and, if i has a degree of rationality which is sufficient to realize first-order expectations, i actually believes φ. Also, i has reason to believe that j has reason to believe φ and, if i ascribes[15] to j a degree of rationality which is sufficient to realize first-order expectations, i has reason to believe that j actually believes φ. Provided that i has a degree of rationality which is sufficient to realize second-order expectations, i then actually believes that j actually believes φ. And so on, for all orders of belief. It seems that Lewis is suggesting that, if an agent is endowed with a first-order degree of (epistemic) rationality, she will come to believe everything she has reason to, if an agent is endowed with a second-order degree of rationality, she will come to believe everything that she has reason to believe that she has reason to, and so on. But this cannot be a satisfactory account of epistemic rationality: an agent may be sufficiently epistemically rational to actually entertain the first-order expectation yielded by a common knowledge basis, but it would be descriptively inadequate to claim that she actually translates in first-order beliefs *any* proposition she has reason to believe.

We can tweak Lewis's intuition about degrees of rationality capturing the relationship between reason to believe and actual belief by means of what is known in the literature as *awareness structures*. The idea of awareness structures is mainly used in the Artificial Intelligence community to represent the distinction between implicit and explicit knowledge[16] (or, as we put it above, between potential and actual belief, that is to say, between possessing a reason to believe and actually believing), while in the economics literature, models of unawareness seem useful in order to take into account unforeseen consequences[17]. An *awareness set* is associated to each agent and, intuitively, an agent is said to explicitly know a formula φ if she implicitly knows φ, *and* she is aware of φ (that is to say, if φ belongs to that agent's awareness set.) In our setting, the presence of a formula in the awareness set of a particular agent is witness of the fact that the agent is sufficiently rational to actually come to believe that formula, if she has reason to. According to the argument above, each formula that an ideal agent has reason to believe, up to any degree of epistemic nestedness, should be part of her awareness set and thus actually believed by the ideal agent herself. In practice, limitations dictated by physical constraints (but, possibly, also by costraints related to the agent's heuristics) entail that the set of formula actually believed by any agent is a proper subset of the set of formulas that the same agent, ideally, has reason to believe. By adding awareness structures to the model we gain the ability to formally take in account those limitations.

Formally, on the syntactic level we introduce n new modal operators A_i, one for each agent $i = 1, \ldots, n$ in the system, in such a way that the well formed formula $A_i\varphi$ has the intended meaning that agent i is aware of formula φ Furthermore, we introduce n modal operators B_i, one for each agent $i = 1, \ldots, n$ in the system, where the well-formed formula $B_i\varphi$ has the intended meaning that i actually believes φ. As for the semantics, we add to the Kripke structure defined above a set of formulas $\mathcal{A}_i(w)$ for each agent i and for each possible world w. The formulas belonging to $\mathcal{A}_i(w)$ represent those formulas that agent i is aware of at world w. We can then add the following semantical clauses:

$$(M, w) \models A_i \varphi \text{ iff } \varphi \in \mathcal{A}_i(w),$$
$$(M, w) \models B_i \varphi \text{ iff } (M, w) \models A_i \varphi \text{ and } (M, w) \models R_i \varphi.$$

As for the axioms regulating the behavior of the B_i operators, we add a definitory axiom:

B12 $B_i \varphi \leftrightarrow R_i \varphi \wedge A_i \varphi.$

B12 states that an agent actually believes a formula if and only if she has reason to believe it, and the formula is part of her awareness set.

Let us now return to Lewis's definition of common knowledge. Suppose that ψ is a basis for common reason between i and j to believe that φ. Suppose furthermore that the agents are rational up to a certain degree (say second-order rational) and that φ indicates to them that the both of them are first-order rational. By definition of common reason to believe, we then have:

(1R) $R_i \varphi,$
(2R) $R_j \varphi,$
(3R) $R_i R_j \varphi,$
(4R) $R_j R_i \varphi$

and so on. We assume that the agents are rational up to second-order rationality. By resorting to awareness structures, we can precisely spell out such rationality assumption. For the first order we have:

(1A) $A_i \varphi,$
(2A) $A_j \varphi,$

which, along with (1R) and (2R) yield

(1B) $B_i \varphi,$
(2B) $B_j \varphi.$

Moreover, ψ indicates to the agents that both of them are "first-order rational", The following proposition captures the indication of first-order rationality:

(a) $R_i \psi \Rightarrow_{\{i,j\}} R_i A_j \varphi,$
(b) $R_j \psi \Rightarrow_{\{i,j\}} R_j A_i \varphi.$

From (a), (b), (3R), (4R), and the fact that ψ is a basis for common reason to believe, it follows that

(3′) $R_i B_j \varphi$,
(4′) $R_j B_i \varphi$.

The assumption of second-order rationality is expressed by

(3A) $A_i B_j \varphi$,
(4A) $A_j B_i \varphi$,

which, with (3′) and (4′) yields

(3B) $B_i B_j \varphi$,
(4B) $B_j B_i \varphi$.

Since ψ does not provide an indication of epistemic rationality higher than first-order, no actual beliefs of an order higher than the second can be inferred.

The clarification of Lewis's assumptions about the epistemic rationality of agents above allows us to consider an example of succesful conventional coordination in formal terms. Recall how any instance of the coordination game on which a convention is based presents the agents with a problem of equilibrium selection, and how, according to the analysis developed in the previous sections, Lewis claims that such problem is solved for the agents by means of a system of mutual expectations, i.e., by means of what in general is called "common knowledge". I believe that, against the criticism of Gilbert, such an idea is not inconsistent with that of practically (game-theoretically) rational agents. In particular, what Gilbert calls "a-rational tendencies to follow precedent", are here seen as *epistemically rational* mechanisms to infer which action the other agent might choose. "Common knowledge", or more precisely, any φ that functions as a *basis* of common reason to believe that the other agent will choose a certain course of action, will make the agents aware (in the formal sense) of a solution for the equilibrium selection problem. Intuitively, if φ is a basis for common reason to believe in G that ψ, then we require that ψ belongs to the awareness set of each agent i in the group G. Say that φ represents the fact that there is a precedent according to which, in a situation S, all players conform to the regularity R: φ is then a basis for common reason to believe that ψ, where ψ represents the proposition that the agents conform to R. Then ψ is an element of the set A_i

for each $i \in G$ and, since there is common reason to believe that ψ holds, both $R_i\psi$ and $A_i\psi$, hold for all $i \in G$. According to axiom $B12$, every agent then actually believes that ψ is the case and such a fact, along with the fact that agents are (practically) rational, suffices to explain why players succeed in coordinating and perpetuating the convention. If we denote with φ_G^ψ the fact that φ is a basis for common reason to believe in G that ψ, then the feature of epistemic rationality with which we want to endow the agents is captured by the following requirement:

(∗) $\varphi_G^\psi \to A_i\psi,$ for all $i \in G.$

To see how this fits Lewis's definition of convention, consider a coordination problem S and a solution $\psi := \psi_1, \ldots, \psi_n$, where ψ_i stands for "agent i does her part in the coordination equilibrium ψ". Assume that agents know that ψ has worked in the past as a solution of S. They have reason to believe that ψ solved S in the past. If we denote "ψ has solved S in the past" with φ, we have that

(i) $R_G\varphi.$

Suppose such knowledge gives the agents reason to believe that they possess such knowledge (φ is public). We then have that

(ii) $R_G\varphi \Rightarrow_G R_G R_G\varphi.$

Finally, assume that the agents have reason to believe that the successful precedent has a bearing on the current situation:

(iii) $R_G\varphi \Rightarrow_G R_G\psi.$

Hence, φ is a basis for common reason to believe of ψ in G, and, because of (∗),

(iv) $A_i\psi$ for all $i \in G.$

From (i) and (iii), it follows that $R_G\psi$, hence

(v) $B_G\psi.$

In particular, for each $i \in G$, it is true that

(vi) $B_i\psi_1, \ldots, \psi_{i-1}, \psi_{i+1}, \ldots, \psi_n$

that is to say, each agent i (actually) believes that every other player will do her part in ψ. Assuming that agents are (practically) rational, (vi) implies that ψ_i obtains for all $i \in G$, or that the coordination equilibrium ψ will be played. Thus, we have that (1) everyone conforms to ψ, (2) everyone expects everyone else to conform to ψ (vi), and (3) everyone prefers everyone else to conform to ψ since (by our assumption) S is a coordination problem, and ψ is a coordination equilibrium for S. (1)–(3), moreover, are common knowledge in G, hence ψ is a tacit convention in G according to Lewis's definition.

5. CONCLUSION

What is to be gained by rendering Lewis's account of convention in a formal framework? On one hand, a rejoinder against possible criticisms for Lewis's theory of convention. On the other, and, more importantly, formalizing Lewis's framework allowes us to uncover relevant epistemological issues, as those related to both the indication relations and the distinction between reasons to believe and actual beliefs. The rational reconstruction of the social phenomenon of convention turns into a vantage point for investigating broader epistemological questions.

Focusing on the distinction between reasons to believe and actual beliefs, one could develop formal models of natural reasoning. Agents' heuristics could be formally captured and analyzed in terms of awareness structures. Awareness structures could prove useful for investigating agents' reasoning about other agents' (epistemic) rationality, yielding a richer approach to interactive epistemology. Focusing on the relations of indication, probabilities and inductive reasoning would enter the picture, providing a more precise account of convention and, again, possibly granting insights toward a philosophically relevant logical approach to epistemology. The basic framework displayed in this paper would of course profit from being enriched and complicated. Its most natural development would consist in incorporating dynamic aspects. One possible avenue by which this might be done would be adding temporal and dynamic modalities [cf. for instance van der Hoek and Wooldridge (2003), or Pauly and Wooldridge (2003)], while another possibilty might be by exploring in which way agents process information and revise their beliefs [cf. Bonanno (2005)].

ACKNOWLEDGMENTS

The author wishes to thank Cristina Bicchieri, Horacio Arló-Costa, Peter Vanderschraaf and Robert Sugden for comments, corrections and stimulating conversations. A first version of this paper was presented at the Sixth Conference on Logic and the Foundations of Game and Decision Theory (LOFT6), Leipzig, July 2004.

NOTES

[1] Cf. (Schelling 1960, p. 84): "If the zero-sum game is the limiting case of pure conflict, what is the other extreme? It must be the "pure collaboration" game in which players win or lose together, having identical preferencs regarding the outcome."

[2] *Common knowledge* is that epistemic state in which all agents know that p, all agents know that all agents know that p, all agents know that all agents know that all agents know that p, and so on ad infinitum. We shall see that, although Lewis characterizes this concept differently, his definition is equivalent to the iterative one just given.

[3] It is difficult to overestimate the influence that the introduction of such an idea has exerted in so many different fields, ranging from economics (Geanakoplos 1992) to computer science (Fagin et al. 1995; Meyer and van der Hoek 1995), from logic [besides the prepositional results from Fagin et al. (1995) and Meyer and Hoek (1995), cf. also issues of quantification in Wolter (1999) and Sturm et al. (2002), and the proof theoretical analysis of Alberucci and Jaeger (2005)] to linguistics (Clark 1996).

[4] Cf. (Lewis 1969), p, 59: "common knowledge of the relevant facts seems to be one (important feature common to our examples of conventions)".

[5] Gilbert has attacked Lewis's definition of convention in several articles. Her arguments are summed up in chapter 5 of Gilbert (1989).

[6] Cf. (Lewis 1969), p. 31: "Provided I go long enough [...] I eventually come out with a first order expectation about your action – which is what I need in order to know how I should act."

[7] Cf., among others (Geanakoplos 1989) (in which the possibility and the implications of representing epistemic states of the agents in set-theoretical models without partitions are explored); (Samet 1990) (in which it is shown that Aumann's "agreement theorem" holds in models weaker than partitional ones, provided that certain additional conditions are met); (Collins 1997) (which shows that Aumann's "agreement theorem" holds also when common belief rather than knowledge is assumed, provided that the agents do not entertain false beliefs about their own beliefs).

[8] Cf. also, with regards to this point (Vandreschraaf 1998), p. 362: "[...] Lewis's account applies to situations in which the agents' private information structures are not necessarily partitions".

[9] Cf. (Aumann 1999), p. 273, in which is argued that "common knowledge" of the agents' partitions holds only informally, and is by necessity included in the specifications of each possible world (and, in particular, of the true world).

[10] Even if there exists in the model an event X such that X expresses the properties of the model itself, and there is, thus, formal common knowledge of X, it would not be problematic that there may not exist a basis for common knowledge (in the formal sense) of X since, as Cubitt and Sugden themselves acknowledge in Cubitt and Sugden (2003), p. 190, there can be common knowledge of a proposition without there being a common knowledge basis for it.

[11] To avoid notational confusion, it is prudent to emphasize that the operators $\Rightarrow_i$ indexed by agents in the system stand for *indication* and not for material implication, which is represented by the symbol $\rightarrow$.

[12] The work of Vincent Hendricks is, under this respect an exception. (Cf. for instance Hendricks 2003).

[13] Since it seems to be the case that, in *Convention*, Lewis takes it a state of affair A to *indicate* to the agents that A holds, the requirement that the agents have reason to believe that A holds if the state of affairs A materially implies p is implicit in Lewis's account. By explicitly requiring it in axiom $B2$, we may dispense with the distinction and still avoid an indication relation vitiated by the paradoxes of material implication.

[14] Notice that the clause for indication is restricted to the deductive aspect of indication only, since in this article non-deductive capabilities of the agents are not taken into account.

[15] Lewis assumes here that the basis for common reason to believe also indicates the amount of rationality enjoyed by the agents.

[16] Cf. Fagin and Halpern (1988) and Halpern (2001).

[17] Cf. Modica and Rustichini (1994, 1999).

REFERENCES

Alberucci L. and Jaeger G.: 2005, About cut elimination for common knowledge logics. *Annals of Pure and Applied Logic*, **133**(1–3), 73–99.

Aumann R. J.: 1976, Agreeing to disagree. *Annals of Statistics* (4):1236–1239.

Aumann R. J.: 1999, Interactive epistemology i: Knowledge. *International Journal of Game Theory* (28), 263–300.

Bicchieri, C.: 1993, *Rationality and Coordination*. Cambridge University Press, Cambridge.

Bonanno, G.: 2005, A simple modal logic for belief revision. *Synthese*, this issue.

Clark, H. H.: 1996, *Using Language*. Cambridge University Press, Cambridge, MA.

Cubitt R. P. and Sugden R.: 2003, Common knowledge, salience and convention: a reconstruction of David Lewis' game theory. *Economics and Philosophy* (19), 175–210.

Fagin R. and Halpern J. Y.: 1988, Belief, awareness, and limited reasoning. *Artificial Intelligence* (34), 39–76.

Fagin R., Halpern J., Moses Y. and Vardi M.: 1995, *Reasoning about Knowledge*. MIT Press, Cambridge, MA.

Geanakoplos, J.: 1989, Game theory without partitions, and applications to specu-
lation and consensus. Technical report, Cowles Foundation Discussion Paper No.
914.

Geanakoplos, J.: 1992, Common knowledge. *Journal of Economic Perpsectives* (6),
53–82.

Gilbert, M.: 1989, *On Social Pacts*, Princeton University Press, Princeton.

Halpern, J. Y.: 2001, Alternative semantics for unawareness. *Games and Economic
Behavior* 37(2), 321–339.

Hendricks, V. F.: 2003, Active agents. *Journal of Logic, Language and Information*
(12), 469–495.

Lewis, D.: 1969, *Convention: A Philosophical Study*. Harvard University Press,
Cambridge, MA.

Lewis, D.: 1978, Truth in fiction. *American Philosophical Quarterly* 15, 37–46.

Modica, S. and Rustichini A.: 1994, Awareness and partitional information structures. *Theory and Decision* (37), 107–124.

Modica, S. and Rustichini A.: 1999, Unawareness and partitional information structures. *Game and Economic Behavior*, 27(2), 265–298.

Meyer, J. J-Ch. and van der Hoek W.: 1995, *Epistemic Logic for AI and Computer
Science*. Cambridge University Press, Cambridge, MA.

Pauly, M. and Wooldridge M. J. W.: 2003, Logic for mechanism design - a manifesto.
In *2003 Workshop on Game Theory and Decision Theory in Agent-based Systems
(GTDT-2003)*, Melbourne, Australia.

Samet, D.: 1990, Ignoring ignorance and agreeing to disagree. *Journal of Economic
Théory* (52), 190–207.

Savage, L.: 1954, *The Foundation of Statistics*. Wiley, New York, NY.

Schelling, T.: 1960, *The Strategy of Conflict*. Harvard University Press, Cambridge,
MA.

Sturm, H. Wolter F., and Zakharyashev M.: 2002, Common knowledge and quantification. *Economic Theory* 19, 157–186.

Vanderschraaf, P.: 1995, Convention as correlated equilibrium. *Erkenntnis* (42),
65–87.

Vandreschraaf, P.: 1998, Knowledge, equilibrium and convention. *Erkenntnis* (49),
337–369.

van der Hoek W. and Wooldridge M. J. W.: 2003, Cooperation, knowledge, and
time: Alternating-time temporal epistemic logic and its applications. *Studia Logica*
75(1), 125–157.

Wolter, F.: 1999, First order common knowlegde logics. *Studia Logica* 65(2),
249–271.

Index